Fundamentals of Physics

Part 5

7/e

Fundamentals of Physics

Part 5

David Halliday
University of Pittsburgh

Robert Resnick
Rensselaer Polytechnic Institute

Jearl Walker
Cleveland State University

John Wiley & Sons, Inc.

ACQUISITIONS EDITOR Stuart Johnson
SENIOR DEVELOPMENT EDITOR Ellen Ford
MARKETING MANAGER Bob Smith
SENIOR PRODUCTION EDITOR Elizabeth Swain
TEXT/COVER DESIGNER Madelyn Lesure
DUMMY DESIGNER Lee Goldstein
PHOTO EDITOR Hilary Newman
PHOTO RESEARCHER Elyse Rieder
ILLUSTRATION EDITOR Anna Melhorn
ILLUSTRATION STUDIO Radiant
MEDIA EDITOR Martin Batey
COVER PHOTO © Jeff Hunter/The Image Bank/Getty Images.

This book was set in 10/12 Times Ten by Progressive Information Technologies and printed and bound by Von Hoffmann Press. The cover was printed by Von Hoffmann Press.

This book is printed on acid free paper.

Library of Congress Cataloging-in-Publication Data

Halliday, David
Fundamentals of physics / David Halliday, Robert Resnick, Jearl Walker—7th ed.
p. cm.
Includes index.
ISBN 0-471-42965-1 (pt. 5 : pbk. : acid-free paper)
1. Physics. I. Resnick, Robert II. Walker, Jearl III. Title.

QC21.3.H35 2005
530—dc21 2003049663

Printed in the United States of America

10 9 8 7 6 5 4 3 2 1

Brief Contents

Volume 1

Part 1

Part 2

Volume 2

Part 3

Part 4

Part 5

Contents

Preface

The Goal

The principal goal of *Fundamentals of Physics*, seventh edition, is the one established by David Halliday and Robert Resnick in their first edition over 40 years ago: to provide instructors with a tool by which they can teach students how to effectively read scientific material, identify fundamental concepts, reason through scientific questions, and solve quantitative problems.

This process is not easy for either students or instructors. Indeed, the course associated with this book may be one of the most demanding of all the courses taken by a student. However, it can also be one of the most rewarding, because it reveals the world's fundamental clockwork from which all scientific and engineering applications spring.

Why a New Edition?

Some people question: "Because *Fundamentals of Physics* is so widely used and has been so well accepted, why write a new edition?" There are two primary reasons why this seventh edition was written:

➡ Many users of the sixth edition (both instructors and students) sent in comments and suggestions to improve the book. These improvements are now incorporated into the narrative and problems throughout the book. We (the current author Jearl Walker and the publisher John Wiley & Sons) regard the book as an ongoing project and encourage more input from users. You can send suggestions, corrections, and positive or negative comments to John Wiley & Sons (http:www.wiley.com/college/halliday) or Jearl Walker (mail address: Physics Department, Cleveland State University, Cleveland OH 44115 USA; fax number: (USA) 216 687 2424; or email address: physics@wiley.com). We may not be able to respond to all suggestions, but we keep and study each of them.

➡ In his full-time teaching, Jearl Walker gains first-hand experience with students as they learn from *Fundamentals of Physics*. After every lecture and exam, he carefully notes what concepts and procedures worked well with the students and (even more important for writing a book) which items confused the students, and improvements based on his notes are now incorporated throughout the book. His goal in his HRW courses is the same as the goal in the HRW book: To show students that they *can* reason from basic physics concepts all the way to valid conclusions.

Major Content Changes

➡ Several thousand of the end-of-chapter problems have been rewritten to streamline both the presentation and the answer.

➡ All of the chapter-opening puzzlers (the examples of applied physics designed to entice a reader into each chapter) are new and come straight from research journals in many different fields.

➡ The first section in every chapter introduces the subject of the chapter by asking and answering the question "What is physics?" as the answer pertains to the physics of the chapter.

➡ Chapters 9 and 10 in the sixth edition have been combined so that the current Chapter 9 contains the full discussion of momentum and the conservation of momentum for both explosions and collisions.

➡ The chapter on angular momentum (Chapter 11) now contains the physics of a gyroscope.

➡ The chapter on waves (Chapter 16) now contains a derivation of the wave equation.

➡ The chapter on AC circuits (Chapter 31) is now presented before the chapter on Maxwell's equations and magnetic materials (Chapter 32).

➡ Chapter 39 (in the extended version of the book) now contains a derivation of the Bohr model of the hydrogen atom.

Chapter Features

Opening puzzlers. A curious puzzling situation opens each chapter and is explained somewhere within the chapter, to entice a student to read the chapter. These features, which are a hallmark of *Fundamentals of Physics*, are based on current research as reported in scientific, engineering, medical, and legal journals.

What is physics? The narrative of every chapter now begins with this question, and with an answer that pertains to the subject of the chapter. (A plumber once asked Jearl Walker, "What do you do for a living?" Walker replied, "I teach physics." The plumber thought for several minutes and then asked, "What is physics?" The plumber's career was entirely based on physics, yet he did not even know what physics is. Many students in introductory physics do not know what physics is but assume that it is irrelevant to their chosen career.)

Checkpoints are stopping points that effectively ask the student, "Can you answer this question with some reasoning based on the narrative or sample problem that you just read?" If not, then the student should go back over that previous material before traveling deeper into the chapter. For example, see Checkpoint 2 on page 62 and Checkpoint 2 on page 280. ***Answers to all checkpoints are in the back of the book.***

Sample problems have been chosen to help a student organize the basic concepts of the narrative and to develop problem-solving skills. They are chosen to demonstrate how problems can be solved with reasoned solutions rather than quick and simplistic plugging of numbers into an equation with no regard for what the equation means.

Key Ideas in the sample problems focus a student on the basic concepts at the root of the solution to a problem. In effect, these key ideas say, "We start our solution by using this basic concept, a procedure that prepares us for solving many other problems. We don't start by grabbing an equation for a quick plug-and-chug, a procedure that prepares us for nothing."

Problem-solving tactics contain helpful instructions to guide the beginning physics student as to how to solve problems and avoid common errors.

Review & Summary is a brief outline of the chapter contents that contains the essential concepts but which is not a substitute for reading the chapter.

Questions are like the checkpoints and require reasoning and understanding rather than calculations. ***Answers to the odd-numbered questions are in the back of the book.***

Problems are grouped under section titles and are labeled according to difficulty. ***Answers to the odd-numbered problems are in the back of the book.***

Icons for additional help. When worked-out solutions are provided either in print or electronically for certain of the odd-numbered problems, the statements for those problems include a trailing icon to alert both student and instructor as to where the solutions are located. An icon guide is provided here and at the beginning of each set of problems:

SSM	Solution is in the Student Solutions Manual.
WWW	Solution is at http://www.wiley.com/college/halliday
ILW	Interactive LearningWare solution is at http://www.wiley.com/college/halliday

Additional problems. These problems are not ordered or sorted in any way so that a student must determine which parts of the chapter apply to any given problem.

On-line simulation problems. Nearly all the chapters have on-line problems involving simulations of the physics within the chapter. For those chapters, a notice with an icon appears at the very end of the homework problems.

Additional Features

Reasoning versus plug-and-chug. A primary goal of this book is to teach students to reason through challenging situations, from basic principles to a solution. Although some plug-and-chug homework problems remain in the book (on purpose), most homework problems emphasize reasoning.

Chapters of reasonable length. To avoid producing a book thick enough to stop a bullet (and thus also a student), we have made the chapters of reasonable length. We explain enough to get a student going but not so much that a student no longer must analyze and fuse ideas. After all, a student will need the skill of analyzing and fusing ideas long after this book is read and the course is completed.

Use of vector-capable calculators. When vector calculations in a sample problem can be performed directly on-screen with a vector-capable calculator, the solution of the sample problem indicates that fact but still carries through the traditional component analysis. When vector calculations cannot be performed directly on-screen, the solution explains why.

Graphs as puzzles. These are problems that give a graph and ask for a result that requires much more than just reading off a data point from the graph. Rather, the solution requires an understanding of the physical arrangement in a problem and the principles behind the associated equations. These problems are more like Sherlock Holmes puzzles because a student must decide what data are important. For examples, see problem 38 on page 79, problem 8 on page 108, and problem 21 on page 230.

Problems with applied physics, based on published research, appear in many places, either as the opening puzzler of a chapter, a sample problem, or a homework problem. For example, see the opening puzzler for Chapter 9 on page 201, Sample Problem 2-8 on page 28, and homework problem 30 on page 231. For an example of homework problems that build on a continuing story, see problems 4, 35, and 53 on pages 131, 134, and 135.

Problems with novel situations. Here is one of several hundred such problems: Problem 57 on page 112 relates a true story of how Air Canada flight 143 ran out of fuel at an altitude of 7.9 km because the crew and airport personnel did not consider the units for the fuel (an important lesson for students who tend to "blow off" units).

eGrade Plus (powered by EduGen)

eGrade Plus is a powerful, web-based course-management tool that allows instructors to provide their students with a dynamic, interactive learning environment.

eGrade Plus automates the process of assigning and grading homework, quizzes, and exams. It provides professors and students with numerous benefits including the ability to keep students on task and provide them with immediate feedback and scoring on their work without having to invest hours every week grading homework and quizzes. Every homework problem has a link to an on-line version of the text, giving the students immediate, structured information to help with problem solving. This context-sensitive help gives students quick access to a set of focused problem-solving aids.

For Instructors:

An Administration tool allows you to manage your class roster on-line.

A "Prepare and Present" tool contains a variety of the Wiley-provided resources to help make your preparation time more efficient. The specific resources include all of the *Instructor's Solutions Manual* in LaTex, MSWord, and pdf formats and all of the text illustrations in electronic format suitable both for classroom projection and for printing. You may easily adapt, customize, and add to this content to meet the needs of your course.

An "Assignment" tool allows you to create student homework and quizzes. You can use dynamic versions of end-of-chapter problems from *Fundamentals of Physics* or you can write your own dynamic questions. You may also assign readings, activities, and other work you want your students to complete. One of the most powerful features of eGrade Plus is that student assignments will be automatically graded and recorded in your gradebook. This will not only save you time but will provide your students with immediate feedback on their work, allowing them to determine right away how well they understand the course material.

A Gradebook tool will keep track of your students' progress and allow you to analyze individual and overall class results to determine their progress and level of understanding.

For Students

A "Study and Practice" area links directly to the multimedia version of *Fundamentals of Physics*, allowing students to review the text while they study and complete homework assignments. In addition to the complete text, students can also access the Student Solutions Manual, the Student Study Guide, interactive simulations, and the Interactive LearningWare Program.

An "Assignment" area keeps all the work you want your students to complete in one location, making it easy for them to stay "on task." Students will have access to a variety of interactive problem-solving tools, as well as other resources for building their confidence and understanding. In addition, the homework problems contain a link to the relevant section of the multimedia book, providing students with context-sensitive help that allows them to conquer problem-solving obstacles as they arise.

A Personal Gradebook for each student will allow students to view their results from past assignments at any time.

Versions of the Text

To accommodate the individual needs of instructors and students, the seventh edition of *Fundamentals of Physics* is available in a number of different versions.

The Regular Edition consists of Chapters 1 through 37 (ISBN 0-471-21643-7).

The Extended Edition contains six additional chapters on quantum physics and cosmology, Chapters 1–44 (ISBN 0-471-23231-9).

Both editions are available as single, hard-cover books, or in the following alternative versions:

- Volume 1 - Chapters 1–20 (Mechanics and Thermodynamics), hardcover, ISBN 0-471-42959-7
- Volume 2 - Chapters 21–44 (E&M, Optics, and Quantum Physics), hardcover, ISBN 0-471-42960-0
- Part 1 - Chapters 1–11, paperback, ISBN 0-471-42961-9
- Part 2 - Chapters 12–20, paperback, ISBN 0-471-42962-7
- Part 3 - Chapters 21–32, paperback, ISBN 0-471-42963-5
- Part 4 - Chapters 33–37, paperback, ISBN 0-471-42964-3
- Part 5 - Chapters 38–44, paperback, 0-471-42965-1

Instructor's Supplements

Instructor's Manual by J. Richard Christman, U.S. Coast Guard Academy. This manual contains lecture notes outlining the most important topics of each chapter; demonstration experiments; laboratory and computer projects; film and video sources; answers to all Questions, Exercises, Problems, and Checkpoints; and a correlation guide to the Questions, Exercises, and Problems in the previous edition. It also contains a complete list of all problems for which solutions are available to students (SSM, WWW, and ILW).

Instructor's Solutions Manual by Sen-Ben Liao, Massachusetts Institute of Technology. This manual provides worked-out solutions for all problems found at the end of each chapter.

Test Bank by J. Richard Christman, U.S. Coast Guard Academy. This manual includes more than 2200 multiple-choice questions. These items are also available in the Computerized Test Bank (see below).

Instructor's Resource CD. This CD contains:
- All of the *Instructor's Solutions Manual* in MSWord, and pdf files
- Computerized Test Bank, in both IBM and Macintosh versions, with full editing features to help you customize tests.
- All text illustrations, suitable for both classroom projection and printing.

Transparencies. More than 200 four-color illustrations from the text are provided in a form suitable for projection in the classroom.

On-line homework and quizzing. In addition to eGrade, *Fundamentals of Physics,* seventh edition also supports WebAssign and CAPA, which are other programs that give instructors the ability to deliver and grade homework and quizzes over the Internet.

WebCT and Blackboard. A variety of materials have been prepared for easy incorporation in either WebCT or Blackboard. WebCT and Blackboard are powerful and easy-to-use web-based course-management systems that allow instructors to set up complete on-line courses with chat rooms, bulletin boards, quizzing, student tracking, etc.

Student's Supplements

Student Web Site. This web site

http://www.wiley.com/college/halliday

was developed specifically for *Fundamentals of Physics*, seventh edition, and is designed to further assist students in the study of physics. The site includes solutions to selected end-of-chapter problems (which are identified with a **www** icon in the text); self-quizzes; simulation exercises; links to other helpful websites; and the Interactive LearningWare tutorials that are described below.

A Student's Companion by J. Richard Christman, U.S. Coast Guard Academy. This student study guide consists of an overview of the chapter's important concepts, hints for solving end-of-chapter questions and problems, and a practice quiz.

Student's Solutions Manual by J. Richard Christman, U.S. Coast Guard Academy and Edward Derringh, Wentworth Institute. This manual provides student with complete worked-out solutions to 30 percent of the problems found at the end of each chapter within the text. These problems are indicated with an **ssm** icon.

Interactive LearningWare. This software guides students through solutions to 200 of the end-of-chapter problems. These problems are indicated with an **ilw** icon. The solutions process is developed interactively, with appropriate feedback and access to error-specific help for the most common mistakes.

Multimedia Book. This web-based version of *Fundamentals of Physics*, seventh edition contains the complete, extended version of the text, *A Student's Companion*, the *Student's Solutions Manual*, the *Interactive LearningWare,* and numerous simulations, all connected with extensive hyperlinking.

Acknowledgments

A great many people have contributed to this book. J. Richard Christman, of the U.S. Coast Guard Academy, has once again created many fine supplements for us; his recommendations to this book have been invaluable. Sen-Ben Liao of MIT, James Whitenton of Southern Polytechnic State University, and Jerry Shi, of Pasadena City College, performed the Herculean task of working out solutions for every one of the homework problems in the book. We thank Renee M. Goertzen, now at the University of Maryland, for her checking many of the problems from a student perspective.

At John Wiley publishers, we have been fortunate to receive strong coordination and support from our editor Stuart Johnson, who constantly encourages our efforts and champions our work, Ellen Ford, who coordinated the developmental editing and multilayered preproduction process, Martin Batey, who coordinated the state-of-the-art media package, and Geraldine Osnato, who managed a super team to create an impressive supplements package.

We thank Lucille Buonocore and Pam Kennedy, our production managers, and Elizabeth Swain, our production editor, for pulling all the pieces together and guiding us through the complex production process. We also thank Maddy Lesure, for her design of both the text and the book cover; Helen Walden for her copyediting; Anna Melhorn for managing the illustration program; and Lilian Brady for her proofreading. Hilary Newman and Elyse Rieder were inspired in the search for unusual and interesting photographs.

We are very grateful to Irene Nunes who read every word in this book and made countless improvements to the clarity of the message in those words. Jearl Walker is also extremely grateful to Edward Millman who guided Walker through his first three editions of the book. (Walker dedicates this edition of the book to Ed Millman as a token of Walker's gratitude.)

Finally, our external reviewers have been outstanding and we acknowledge here our debt to each member of that team.

Maris A. Abolins
Michigan State University

Edward Adelson
Ohio State University

Barbara Andereck
Ohio Wesleyan University

Mark Arnett
Kirkwood Community College

Arun Bansil
Northeastern University

Albert Bartlett
University of Colorado

Michael E. Browne
University of Idaho

Timothy J. Burns
Leeward Community College

Joseph Buschi
Manhattan College

Philip A. Casabella
Rensselaer Polytechnic Institute

Randall Caton
Christopher Newport College

J. Richard Christman
U.S. Coast Guard Academy

Roger Clapp
University of South Florida

W. R. Conkie
Queen's University

Mike Crivello
San Diego State University

Peter Crooker
University of Hawaii at Manoa

William P. Crummett
Montana College of Mineral Science and Technology

Robert N. Davie, Jr.
St. Petersburg Junior College

Cheryl K. Dellai
Glendale Community College

Eric R. Dietz
California State University at Chico

N. John DiNardo
Drexel University

Eugene Dunnam
University of Florida

Robert Endorf
University of Cincinnati

F. Paul Esposito
University of Cincinnati

Jerry Finkelstein
San Jose State University

Alexander Firestone
Iowa State University

Alexander Gardner
Howard University

Andrew L. Gardner
Brigham Young University

John Gieniec
Central Missouri State University

Robert H. Good
California State University-Hayward

John B. Gruber
San Jose State University

Ann Hanks
American River College

Randy Harris
University of California-Davis

Samuel Harris
Purdue University

Harold B. Hart
Western Illinois University

Rebecca Hartzler
Edmonds Community College

Emily Haught
Georgia Institute of Technology

Laurent Hodges
Iowa State University

John Hubisz
North Carolina State University

Joey Huston
Michigan State University

Darrell Huwe
Ohio University

Shawn Jackson
University of Tulsa

Hector Jimenez
University of Puerto Rico

Sudhakar B. Joshi
York University

Claude Kacser
University of Maryland

Leonard M. Kahn
University of Rhode Island

Leonard Kleinman
University of Texas at Austin

Earl Koller
Stevens Institute of Technology

Arthur Z. Kovacs
Rochester Institute of Technology

Kenneth Krane
Oregon State University

Sol Krasner
University of Illinois at Chicago

Yuichi Kubota
Cornell University

Priscilla Laws
Dickinson College

Edbertho Leal
Polytechnic University of Puerto Rico

Peter Loly
University of Manitoba

Dale Long
Virginia Tech

Andreas Mandelis
University of Toronto

Robert R. Marchini
Memphis State University

David Markowitz
University of Connecticut

Paul Marquard
Caspar College

Howard C. McAllister
University of Hawaii at Manoa

W. Scott McCullough
Oklahoma State University

James H. McGuire
Tulane University

David M. McKinstry
Eastern Washington University

Joe P. Meyer
Georgia Institute of Technology

Roy Middleton
University of Pennsylvania

Irvin A. Miller
Drexel University

Eugene Mosca
United States Naval Academy

James Napolitano
Rensselaer Polytechnic Institute

Michael O'Shea
Kansas State University

Patrick Papin
San Diego State University

George Parker
North Carolina State University

Robert Pelcovits
Brown University

Des Penny
Southern Utah University

Oren P. Quist
South Dakota State University

Joe Redish
University of Maryland

Jonathan Reichart
SUNY—Buffalo

Timothy M. Ritter
University of North Carolina at Pembroke

Gerardo A. Rodriguez
Skidmore College

John Rosendahl
University of California at Irvine

Michael Schatz
Georgia Institute of Technology

Manuel Schwartz
University of Louisville

Darrell Seeley
Milwaukee School of Engineering

Bruce Arne Sherwood
Carnegie Mellon University

John Spangler
St. Norbert College

Ross L. Spencer
Brigham Young University

Harold Stokes
Brigham Young University

Michael G. Strauss
University of Oklahoma

Jay D. Strieb
Villanova University

Dan Styer
Oberlin College

Marshall Thomsen
Eastern Michigan University

Fred F. Tomblin
New Jersey Institute of Technology

David Toot
Alfred University

J. S. Turner
University of Texas at Austin

T. S. Venkataraman
Drexel University

Gianfranco Vidali
Syracuse University

Fred Wang
Prairie View A & M

Robert C. Webb
Texas A & M University

B. R. Weinberger
Trinity College

William M. Whelan
Ryerson Polytechnic University

George Williams
University of Utah

David Wolfe
University of New Mexico

William Zimmerman, Jr.
University of Minnesota

7/e

Fundamentals of Physics

Part 5

Photons and Matter Waves

This image shows the rugged terrain on the surface of a piece of copper metal, but the magnification is far greater than could ever be achieved using light in any kind of microscope. For example, the spire rising above the plateau on the left is only 23 nm high, and the ripples at the cliff edge in the middle are only 1.5 nm apart, far less than the wavelengths of visible light. Yet waves are needed to make this type of image.

What kind of waves can give such extreme magnification?

The answer is in this chapter.

38-1 What Is Physics?

One primary focus of physics is Einstein's theory of relativity, which took us into a world far beyond that of ordinary experience—the world of objects moving at speeds close to the speed of light. Among other surprises, Einstein's theory predicts that the rate at which a clock runs depends on how fast the clock is moving relative to the observer: the faster the motion, the slower the clock rate. This and other predictions of the theory have passed every experimental test devised thus far, and relativity theory has led us to a deeper and more satisfying view of the nature of space and time.

Now you are about to explore a second world that is outside ordinary experience—the subatomic world. You will encounter a new set of surprises that, though they may sometimes seem bizarre, have led physicists step by step to a deeper view of reality.

Quantum physics, as our new subject is called, answers such questions as: Why do the stars shine? Why do the elements exhibit the order that is so apparent in the periodic table? How do transistors and other microelectronic devices work? Why does copper conduct electricity but glass does not? Because quantum physics accounts for all of chemistry, including biochemistry, we need to understand it if we are to understand life itself.

Some of the predictions of quantum physics seem strange even to the physicists and philosophers who study its foundations. Still, experiment after experiment has proved the theory correct, and many have exposed even stranger aspects of the theory. The quantum world is an amusement park full of wonderful rides that are guaranteed to shake up the commonsense world view you have developed since childhood. We begin our exploration of that quantum park with the photon.

38-2 The Photon, the Quantum of Light

Quantum physics (which is also known as *quantum mechanics* and *quantum theory*) is largely the study of the microscopic world. In that world, many quantities are found only in certain minimum (*elementary*) amounts, or integer multiples of those elementary amounts; these quantities are then said to be *quantized.* The elementary amount that is associated with such a quantity is called the **quantum** of that quantity (*quanta* is the plural).

In a loose sense, U.S. currency is quantized because the coin of least value is the penny, or $0.01 coin, and the values of all other coins and bills are restricted to integer multiples of that least amount. In other words, the currency quantum is $0.01, and all greater amounts of currency are of the form n($0.01), where n is always a positive integer. For example, you cannot hand someone $0.755 = 75.5($0.01).

In 1905, Einstein proposed that electromagnetic radiation (or simply *light*) is quantized and exists in elementary amounts (quanta) that we now call **photons.** This proposal should seem strange to you because we have just spent several chapters discussing the classical idea that light is a sinusoidal wave, with a wavelength λ, a frequency f, and a speed c such that

$$f = \frac{c}{\lambda}. \tag{38-1}$$

Furthermore, in Chapter 33 we discussed the classical light wave as being an interdependent combination of electric and magnetic fields, each oscillating at frequency f. How can this wave of oscillating fields consist of an elementary amount of something—the light quantum? What *is* a photon?

The concept of a light quantum, or a photon, turns out to be far more subtle and mysterious than Einstein imagined. Indeed, it is still very poorly understood.

In this book, we shall discuss only some of the basic aspects of the photon concept, somewhat along the lines of Einstein's proposal.

According to that proposal, the quantum of a light wave of frequency f has the energy

$$E = hf \quad \text{(photon energy)}. \tag{38-2}$$

Here h is the **Planck constant,** the constant we first met in Eq. 32-23, and which has the value

$$h = 6.63 \times 10^{-34}\ \text{J}\cdot\text{s} = 4.14 \times 10^{-15}\ \text{eV}\cdot\text{s}. \tag{38-3}$$

The smallest amount of energy a light wave of frequency f can have is hf, the energy of a single photon. If the wave has more energy, its total energy must be an integer multiple of hf, just as the currency in our previous example must be an integer multiple of \$0.01. The light cannot have an energy of, say, $0.6hf$ or $75.5hf$.

Einstein further proposed that when light is absorbed or emitted by an object (matter), the absorption or emission event occurs in the atoms of the object. When light of frequency f is absorbed by an atom, the energy hf of one photon is transferred from the light to the atom. In this *absorption event,* the photon vanishes and the atom is said to absorb it. When light of frequency f is emitted by an atom, an amount of energy hf is transferred from the atom to the light. In this *emission event,* a photon suddenly appears and the atom is said to emit it. Thus, we can have *photon absorption* and *photon emission* by atoms in an object.

For an object consisting of many atoms, there can be many photon absorptions (such as with sunglasses) or photon emissions (such as with lamps). However, each absorption or emission event still involves the transfer of an amount of energy equal to that of a single photon of the light.

When we discussed the absorption or emission of light in previous chapters, our examples involved so much light that we had no need of quantum physics, and we got by with classical physics. However, in the late 20th century, technology became advanced enough that single-photon experiments could be conducted and put to practical use. Since then quantum physics has become part of standard engineering practice, especially in optical engineering.

CHECKPOINT 1 Rank the following radiations according to their associated photon energies, greatest first: (a) yellow light from a sodium vapor lamp, (b) a gamma ray emitted by a radioactive nucleus, (c) a radio wave emitted by the antenna of a commercial radio station, (d) a microwave beam emitted by airport traffic control radar.

Sample Problem 38-1

A sodium vapor lamp is placed at the center of a large sphere that absorbs all the light reaching it. The rate at which the lamp emits energy is 100 W; assume that the emission is entirely at a wavelength of 590 nm. At what rate are photons absorbed by the sphere?

Solution: We assume that all the light emitted by the lamp reaches (and thus is absorbed by) the sphere. Then the **Key Idea** is that the light is emitted and absorbed as photons. The rate R at which photons are absorbed by the sphere is equal to the rate R_{emit} at which photons are emitted by the lamp. That rate is

$$R_{\text{emit}} = \frac{\text{rate of energy emission}}{\text{energy per emitted photon}} = \frac{P_{\text{emit}}}{E}.$$

We then have, from Eq. 38-2 ($E = hf$),

$$R = R_{\text{emit}} = \frac{P_{\text{emit}}}{hf}.$$

Using Eq. 38-1 ($f = c/\lambda$) to substitute for f and then entering known data, we obtain

$$\begin{aligned} R &= \frac{P_{\text{emit}}\lambda}{hc} \\ &= \frac{(100\ \text{W})(590 \times 10^{-9}\ \text{m})}{(6.63 \times 10^{-34}\ \text{J}\cdot\text{s})(2.998 \times 10^{8}\ \text{m/s})} \\ &= 2.97 \times 10^{20}\ \text{photons/s}. \qquad \text{(Answer)} \end{aligned}$$

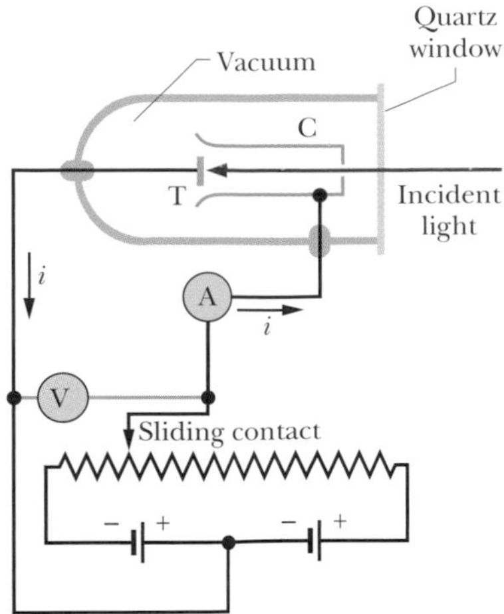

Fig. 38-1 An apparatus used to study the photoelectric effect. The incident light shines on target T, ejecting electrons, which are collected by collector cup C. The electrons move in the circuit in a direction opposite the conventional current arrows. The batteries and the variable resistor are used to produce and adjust the electric potential difference between T and C.

38-3 The Photoelectric Effect

If you direct a beam of light of short enough wavelength onto a clean metal surface, the light will cause electrons to leave that surface (the light will *eject* the electrons from the surface). This **photoelectric effect** is used in many devices, including TV cameras, camcorders, and night vision viewers. Einstein supported his photon concept by using it to explain this effect, which simply cannot be understood without quantum physics.

Let us analyze two basic photoelectric experiments, each using the apparatus of Fig. 38-1, in which light of frequency f is directed onto target T and ejects electrons from it. A potential difference V is maintained between target T and collector cup C to sweep up these electrons, said to be **photoelectrons.** This collection produces a **photoelectric current** i that is measured with meter A.

First Photoelectric Experiment

We adjust the potential difference V by moving the sliding contact in Fig. 38-1 so that collector C is slightly negative with respect to target T. This potential difference acts to slow down the ejected electrons. We then vary V until it reaches a certain value, called the **stopping potential** V_{stop}, at which point the reading of meter A has just dropped to zero. When $V = V_{stop}$, the most energetic ejected electrons are turned back just before reaching the collector. Then K_{max}, the kinetic energy of these most energetic electrons, is

$$K_{max} = eV_{stop}, \tag{38-4}$$

where e is the elementary charge.

Measurements show that for light of a given frequency, K_{max} *does not depend on the intensity of the light source.* Whether the source is dazzling bright or so feeble that you can scarcely detect it (or has some intermediate brightness), the maximum kinetic energy of the ejected electrons always has the same value.

This experimental result is a puzzle for classical physics. Classically, the incident light is a sinusoidally oscillating electromagnetic wave. An electron in the target should oscillate sinusoidally due to the oscillating electric force on it from the wave's electric field. If the amplitude of the electron's oscillation is great enough, the electron should break free of the target's surface—that is, be ejected from the target. Thus, if we increase the amplitude of the wave and its oscillating electric field, the electron should get a more energetic "kick" as it is being ejected. *However, that is not what happens.* For a given frequency, intense light beams and feeble light beams give exactly the same maximum kick to ejected electrons.

The actual result follows naturally if we think in terms of photons. Now the energy that can be transferred from the incident light to an electron in the target is that of a single photon. Increasing the light intensity increases the *number* of photons in the light, but the photon energy, given by Eq. 38-2 ($E = hf$), is unchanged because the frequency is unchanged. Thus, the energy transferred to the kinetic energy of an electron is also unchanged.

Second Photoelectric Experiment

Now we vary the frequency f of the incident light and measure the associated stopping potential V_{stop}. Figure 38-2 is a plot of V_{stop} versus f. Note that the photoelectric effect does not occur if the frequency is below a certain **cutoff frequency** f_0 or, equivalently, if the wavelength is greater than the corresponding **cutoff wavelength** $\lambda_0 = c/f_0$. This is so *no matter how intense the incident light is.*

This is another puzzle for classical physics. If you view light as an electromagnetic wave, you must expect that no matter how low the frequency, electrons can always be ejected by light if you supply them with enough energy—that is,

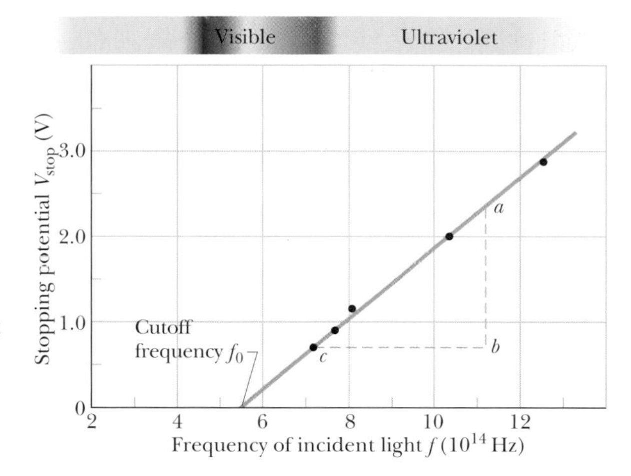

Fig. 38-2 The stopping potential V_{stop} as a function of the frequency f of the incident light for a sodium target T in the apparatus of Fig. 38-1. (Data reported by R. A. Millikan in 1916.)

if you use a light source that is bright enough. *That is not what happens.* For light below the cutoff frequency f_0, the photoelectric effect does not occur, no matter how bright the light source.

The existence of a cutoff frequency is, however, just what we should expect if the energy is transferred via photons. The electrons within the target are held there by electric forces. (If they weren't, they would drip out of the target due to the gravitational force on them.) To just escape from the target, an electron must pick up a certain minimum energy Φ, where Φ is a property of the target material called its **work function.** If the energy hf transferred to an electron by a photon exceeds the work function of the material (if $hf > \Phi$), the electron can escape the target. If the energy transferred does not exceed the work function (that is, if $hf < \Phi$), the electron cannot escape. This is what Fig. 38-2 shows.

The Photoelectric Equation

Einstein summed up the results of such photoelectric experiments in the equation

$$hf = K_{max} + \Phi \quad \text{(photoelectric equation).} \tag{38-5}$$

This is a statement of the conservation of energy for a single photon absorption by a target with work function Φ. Energy equal to the photon's energy hf is transferred to a single electron in the material of the target. If the electron is to escape from the target, it must pick up energy at least equal to Φ. Any additional energy $(hf - \Phi)$ that the electron acquires from the photon appears as kinetic energy K of the electron. In the most favorable circumstance, the electron can escape through the surface without losing any of this kinetic energy in the process; it then appears outside the target with the maximum possible kinetic energy K_{max}.

Let us rewrite Eq. 38-5 by substituting for K_{max} from Eq. 38-4 ($K_{max} = eV_{stop}$). After a little rearranging we get

$$V_{stop} = \left(\frac{h}{e}\right)f - \frac{\Phi}{e}. \tag{38-6}$$

The ratios h/e and Φ/e are constants, and so we would expect a plot of the measured stopping potential V_{stop} versus the frequency f of the light to be a straight line, as it is in Fig. 38-2. Further, the slope of that straight line should be h/e. As a check, we measure ab and bc in Fig. 38-2 and write

$$\frac{h}{e} = \frac{ab}{bc} = \frac{2.35\text{ V} - 0.72\text{ V}}{(11.2 \times 10^{14} - 7.2 \times 10^{14})\text{ Hz}}$$
$$= 4.1 \times 10^{-15}\text{ V}\cdot\text{s}.$$

Multiplying this result by the elementary charge e, we find

$$h = (4.1 \times 10^{-15}\ \text{V}\cdot\text{s})(1.6 \times 10^{-19}\ \text{C}) = 6.6 \times 10^{-34}\ \text{J}\cdot\text{s},$$

which agrees with values measured by many other methods.

An aside: An explanation of the photoelectric effect certainly requires quantum physics. For many years, Einstein's explanation was also a compelling argument for the existence of photons. However, in 1969 an alternative explanation for the effect was found that used quantum physics but did not need the concept of photons. Light *is* in fact quantized as photons, but Einstein's explanation of the photoelectric effect is not the best argument for that fact.

CHECKPOINT 2 The figure shows data like those of Fig. 38-2 for targets of cesium, potassium, sodium, and lithium. The plots are parallel. (a) Rank the targets according to their work functions, greatest first. (b) Rank the plots according to the value of h they yield, greatest first.

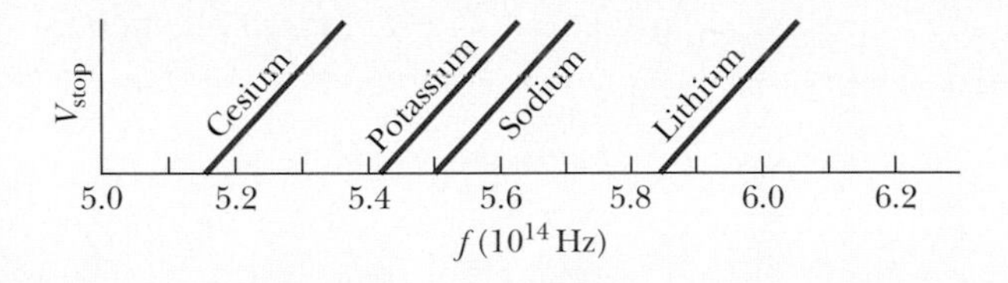

Sample Problem 38-2

A potassium foil is a distance $r = 3.5$ m from an isotropic light source that emits energy at the rate $P = 1.5$ W. The work function Φ of potassium is 2.2 eV. Suppose that the energy transported by the incident light were transferred to the target foil continuously and smoothly (that is, if classical physics prevailed instead of quantum physics). How long would it take for the foil to absorb enough energy to eject an electron? Assume that the foil totally absorbs all the energy reaching it and that the to-be-ejected electron collects energy from a circular patch of the foil whose radius is 5.0×10^{-11} m, about that of a typical atom.

Solution: The **Key Ideas** here are these:

1. The time interval Δt required for the patch to absorb energy ΔE depends on the rate P_{abs} at which the energy is absorbed:

$$\Delta t = \frac{\Delta E}{P_{abs}}.$$

2. If the electron is to be ejected from the foil, the smallest amount of energy ΔE it must gain from the light is equal to the work function Φ of potassium. Thus,

$$\Delta t = \frac{\Phi}{P_{abs}}.$$

3. Because the patch is totally absorbing, the rate of absorption P_{abs} is equal to the rate P_{arr} at which energy arrives at the patch; that is,

$$\Delta t = \frac{\Phi}{P_{arr}}.$$

4. With Eq. 33-23 [$I = (\text{power/area})_{avg}$], we can relate the energy arrival rate P_{arr} to the intensity I of the light at the patch and the area A of the patch:

$$P_{arr} = IA.$$

Then

$$\Delta t = \frac{\Phi}{IA}.$$

5. Because the light source is isotropic, the light intensity I at distance r from the source depends on the rate P_{emit} at which energy is emitted by the source, according to Eq. 33-27:

$$I = \frac{P_{emit}}{4\pi r^2}.$$

Thus, finally, we have

$$\Delta t = \frac{4\pi r^2 \Phi}{P_{emit} A}.$$

The detection area A is $\pi(5.0 \times 10^{-11}\ \text{m})^2 = 7.85 \times 10^{-21}$ m^2, and the work function Φ is 2.2 eV $= 3.5 \times 10^{-19}$ J. Substituting these and other data, we find that

$$\Delta t = \frac{4\pi(3.5\ \text{m})^2(3.5 \times 10^{-19}\ \text{J})}{(1.5\ \text{W})(7.85 \times 10^{-21}\ \text{m}^2)} = 4580\ \text{s} \approx 1.3\ \text{h}. \quad \text{(Answer)}$$

Thus, classical physics tells us that we would have to wait more than an hour after turning on the light source for a photoelectron to be ejected. The actual waiting time is less than 10^{-9} s. Apparently, then, an electron does *not* gradually absorb energy from the light arriving at the patch containing the electron. Rather, either the electron does not absorb any energy at all or it absorbs a quantum of energy instantaneously, by absorbing a photon from the light.

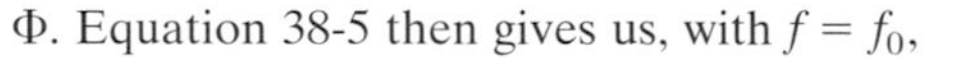

Sample Problem 38-3

Find the work function Φ of sodium from Fig. 38-2.

Solution: The **Key Idea** here is that we can find the work function Φ from the cutoff frequency f_0 (which we can measure on the plot). The reasoning is this: At the cutoff frequency, the kinetic energy K_{max} in Eq. 38-5 is zero. Thus, all the energy hf that is transferred from a photon to an electron goes into the electron's escape, which requires an energy of Φ. Equation 38-5 then gives us, with $f = f_0$,

$$hf_0 = 0 + \Phi = \Phi.$$

In Fig. 38-2, the cutoff frequency f_0 is the frequency at which the plotted line intercepts the horizontal frequency axis, about 5.5×10^{14} Hz. We then have

$$\Phi = hf_0 = (6.63 \times 10^{-34}\ \mathrm{J \cdot s})(5.5 \times 10^{14}\ \mathrm{Hz}) = 3.6 \times 10^{-19}\ \mathrm{J} = 2.3\ \mathrm{eV}. \quad \text{(Answer)}$$

38-4 Photons Have Momentum

In 1916, Einstein extended his concept of light quanta (photons) by proposing that a quantum of light has linear momentum. For a photon with energy hf, the magnitude of that momentum is

$$p = \frac{hf}{c} = \frac{h}{\lambda} \quad \text{(photon momentum)}, \tag{38-7}$$

where we have substituted for f from Eq. 38-1 ($f = c/\lambda$). Thus, when a photon interacts with matter, energy *and* momentum are transferred, *as if* there were a collision between the photon and matter in the classical sense (as in Chapter 9).

In 1923, Arthur Compton at Washington University in St. Louis carried out an experiment that supported the view that both momentum and energy are transferred via photons. He arranged for a beam of x rays of wavelength λ to be directed onto a target made of carbon, as shown in Fig. 38-3. An x ray is a form of electromagnetic radiation, at high frequency and thus small wavelength. Compton measured the wavelengths and intensities of the x rays that were scattered in various directions from his carbon target.

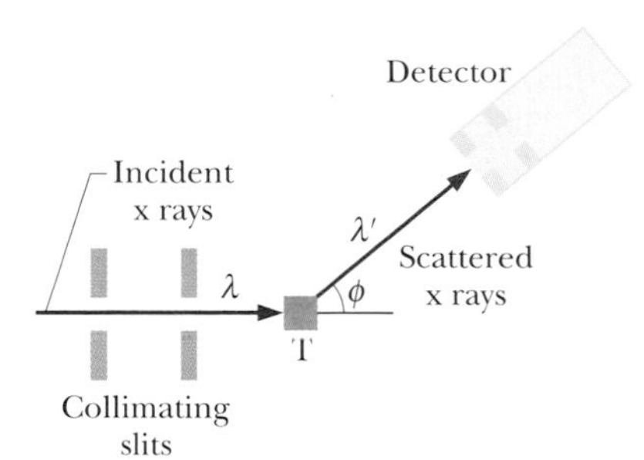

Fig. 38-3 Compton's apparatus. A beam of x rays of wavelength $\lambda = 71.1$ pm is directed onto a carbon target T. The x rays scattered from the target are observed at various angles ϕ to the direction of the incident beam. The detector measures both the intensity of the scattered x rays and their wavelength.

Figure 38-4 shows his results. Although there is only a single wavelength ($\lambda = 71.1$ pm) in the incident x-ray beam, we see that the scattered x rays contain a range of wavelengths with two prominent intensity peaks. One peak is centered

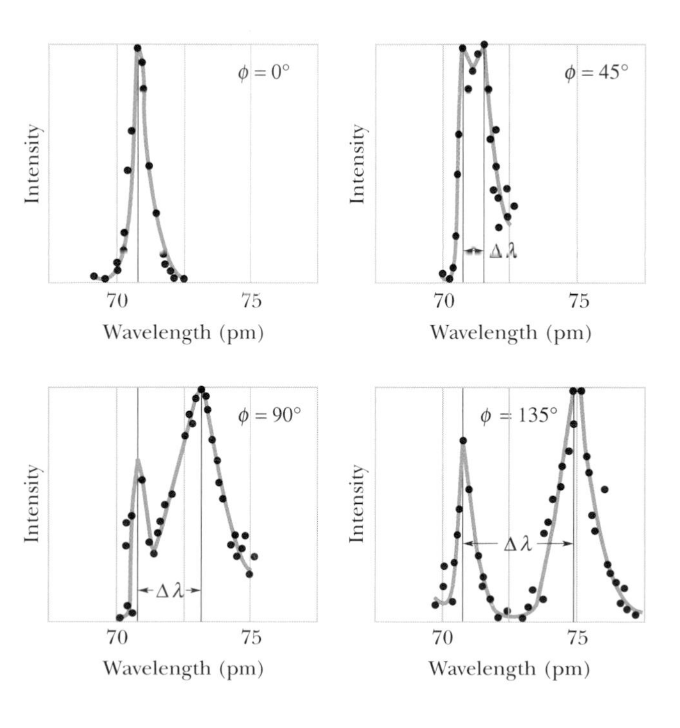

Fig. 38-4 Compton's results for four values of the scattering angle ϕ. Note that the Compton shift $\Delta\lambda$ increases as the scattering angle increases.

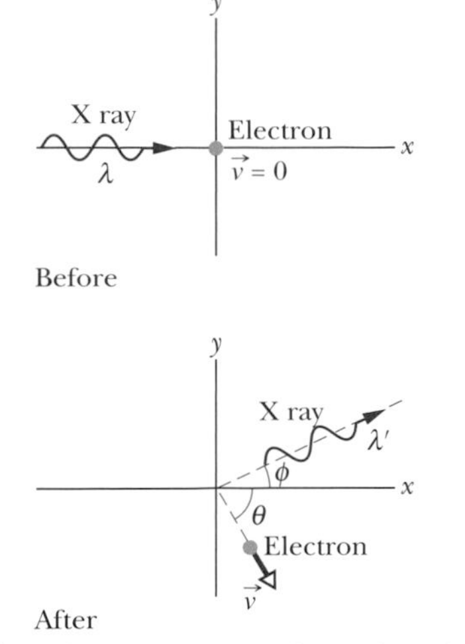

Fig. 38-5 An x ray of wavelength λ interacts with a stationary electron. The x ray is scattered at angle ϕ, with an increased wavelength λ'. The electron moves off with speed v at angle θ.

about the incident wavelength λ, the other about a wavelength λ' that is longer than λ by an amount $\Delta\lambda$, which is called the **Compton shift.** The value of the Compton shift varies with the angle at which the scattered x rays are detected.

Figure 38-4 is still another puzzle for classical physics. Classically, the incident x-ray beam is a sinusoidally oscillating electromagnetic wave. An electron in the carbon target should oscillate sinusoidally due to the oscillating electric force on it from the wave's electric field. Further, the electron should oscillate at the same frequency as the wave and should send out waves *at this same frequency,* as if it were a tiny transmitting antenna. Thus, the x rays scattered by the electron should have the same frequency, and the same wavelength, as the x rays in the incident beam—but they don't.

Compton interpreted the scattering of x rays from carbon in terms of energy and momentum transfers, via photons, between the incident x-ray beam and loosely bound electrons in the carbon target. Let us see, first conceptually and then quantitatively, how this quantum physics interpretation leads to an understanding of Compton's results.

Suppose a single photon (of energy $E = hf$) is associated with the interaction between the incident x-ray beam and a stationary electron. In general, the direction of travel of the x ray will change (the x ray is scattered), and the electron will recoil, which means that the electron has obtained some kinetic energy. Energy is conserved in this isolated interaction. Thus, the energy of the scattered photon ($E' = hf'$) must be less than that of the incident photon. The scattered x rays must then have a lower frequency f' and thus a longer wavelength λ' than the incident x rays, just as Compton's experimental results in Fig. 38-4 show.

For the quantitative part, we first apply the law of conservation of energy. Figure 38-5 suggests a "collision" between an x ray and an initially stationary free electron in the target. As a result of the collision, an x ray of wavelength λ' moves off at an angle ϕ and the electron moves off at an angle θ, as shown. Conservation of energy then gives us

$$hf = hf' + K,$$

in which hf is the energy of the incident x-ray photon, hf' is the energy of the scattered x-ray photon, and K is the kinetic energy of the recoiling electron. Because the electron may recoil with a speed comparable to that of light, we must use the relativistic expression of Eq. 37-52,

$$K = mc^2(\gamma - 1),$$

for the electron's kinetic energy. Here m is the electron's mass and γ is the Lorentz factor

$$\gamma = \frac{1}{\sqrt{1 - (v/c)^2}}.$$

Substituting for K in the conservation of energy equation yields

$$hf = hf' + mc^2(\gamma - 1).$$

Substituting c/λ for f and c/λ' for f' then leads to the new energy conservation equation

$$\frac{h}{\lambda} = \frac{h}{\lambda'} + mc(\gamma - 1). \tag{38-8}$$

Next we apply the law of conservation of momentum to the x-ray–electron collision of Fig. 38-5. From Eq. 38-7 ($p = h/\lambda$), the magnitude of the momentum of the incident photon is h/λ, and that of the scattered photon is h/λ'. From Eq. 37-41, the magnitude for the recoiling electron's momentum is $p = \gamma mv$. Because we have a two-dimensional situation, we write separate equations for the conservation of momentum along the x and y axes, obtaining

$$\frac{h}{\lambda} = \frac{h}{\lambda'} \cos\phi + \gamma mv \cos\theta \qquad (x \text{ axis}) \tag{38-9}$$

and

$$0 = \frac{h}{\lambda'} \sin\phi - \gamma mv \sin\theta \qquad (y \text{ axis}). \tag{38-10}$$

We want to find $\Delta\lambda$ ($= \lambda' - \lambda$), the Compton shift of the scattered x rays. Of the five collision variables (λ, λ', v, ϕ, and θ) that appear in Eqs. 38-8, 38-9, and 38-10, we choose to eliminate v and θ, which deal only with the recoiling electron. Carrying out the algebra (it is somewhat complicated) leads to an equation for the Compton shift as a function of the scattering angle ϕ:

$$\Delta\lambda = \frac{h}{mc}(1 - \cos\phi) \qquad \text{(Compton shift)}. \tag{38-11}$$

Equation 38-11 agrees exactly with Compton's experimental results.

The quantity h/mc in Eq. 38-11 is a constant called the **Compton wavelength.** Its value depends on the mass m of the particle from which the x rays scatter. Here that particle is a loosely bound electron, and thus we would substitute the mass of an electron for m to evaluate the *Compton wavelength for Compton scattering from an electron.*

A Loose End

The peak at the incident wavelength λ ($= 71.1$ pm) in Fig. 38-4 still needs to be explained. This peak arises not from interactions between x rays and the very loosely bound electrons in the target but from interactions between x rays and the electrons that are *tightly* bound to the carbon atoms making up the target. Effectively, each of these latter collisions occurs between an incident x ray and an entire carbon atom. If we substitute for m in Eq. 38-11 the mass of a carbon atom (which is about 22 000 times that of an electron), we see that $\Delta\lambda$ becomes about 22 000 times smaller than the Compton shift for an electron—too small to detect. Thus, the x rays scattered in these collisions have the same wavelength as the incident x rays.

Sample Problem 38-4

X rays of wavelength $\lambda = 22$ pm (photon energy $= 56$ keV) are scattered from a carbon target, and the scattered rays are detected at 85° to the incident beam.

(a) What is the Compton shift of the scattered rays?

Solution: The **Key Idea** here is that the Compton shift is the wavelength change of the x rays due to scattering from loosely bound electrons in a target. Further, that shift depends on the angle at which the scattered x rays are detected, according to Eq. 38-11. Substituting 85° for that angle and 9.11×10^{-31} kg for the electron mass (because the scattering is from electrons) in Eq. 38-11 gives us

$$\begin{aligned}\Delta\lambda &= \frac{h}{mc}(1 - \cos\phi) \\ &= \frac{(6.63 \times 10^{-34}\ \text{J}\cdot\text{s})(1 - \cos 85^\circ)}{(9.11 \times 10^{-31}\ \text{kg})(3.00 \times 10^{8}\ \text{m/s})} \\ &= 2.21 \times 10^{-12}\ \text{m} \approx 2.2\ \text{pm}. \qquad \text{(Answer)}\end{aligned}$$

(b) What percentage of the initial x-ray photon energy is transferred to an electron in such scattering?

Solution: The **Key Idea** here is to find the *fractional energy loss* (let us call it *frac*) for photons that scatter from the electrons:

$$frac = \frac{\text{energy loss}}{\text{initial energy}} = \frac{E - E'}{E}.$$

From Eq. 38-2 ($E = hf$), we can substitute for the initial energy E and the detected energy E' of the x rays in terms of frequencies. Then, from Eq. 38-1 ($f = c/\lambda$), we can substitute for those frequencies in terms of the wavelengths. We find

$$\begin{aligned}frac &= \frac{hf - hf'}{hf} = \frac{c/\lambda - c/\lambda'}{c/\lambda} = \frac{\lambda' - \lambda}{\lambda'} \\ &= \frac{\Delta\lambda}{\lambda + \Delta\lambda}.\end{aligned} \tag{38-12}$$

Substitution of data yields

$$frac = \frac{2.21 \text{ pm}}{22 \text{ pm} + 2.21 \text{ pm}} = 0.091 \quad \text{or} \quad 9.1\%. \quad \text{(Answer)}$$

Although the Compton shift $\Delta\lambda$ is independent of the wavelength λ of the incident x rays (see Eq. 38-11), the *fractional* photon energy loss of the x rays does depend on λ, increasing as the wavelength of the incident radiation decreases, as indicated by Eq. 38-12.

CHECKPOINT 3 Compare Compton scattering for x rays ($\lambda \approx 20$ pm) and visible light ($\lambda \approx 500$ nm) at a particular angle of scattering. Which has the greater (a) Compton shift, (b) fractional wavelength shift, (c) fractional energy loss, and (d) energy imparted to the electron?

38-5 Light as a Probability Wave

A fundamental mystery in physics is how light can be a wave (which spreads out over a region) in classical physics but be emitted and absorbed as photons (which originate and vanish at points) in quantum physics. The double-slit experiment of Section 35-4 lies at the heart of this mystery. Let us discuss three versions of that experiment.

The Standard Version

Figure 38-6 is a sketch of the original experiment carried out by Thomas Young in 1801 (see also Fig. 35-8). Light shines on screen B, which contains two narrow parallel slits. The light waves emerging from the two slits spread out by diffraction and overlap on screen C where, by interference, they form a pattern of alternating intensity maxima and minima. In Section 35-4 we took the existence of these interference fringes as compelling evidence for the wave nature of light.

Let us place a tiny photon detector D at one point in the plane of screen C. Let the detector be a photoelectric device that clicks when it absorbs a photon. We would find that the detector produces a series of clicks, randomly spaced in time, each click signaling the transfer of energy from the light wave to the screen via a photon absorption.

If we moved the detector very slowly up or down as indicated by the black arrow in Fig. 38-6, we would find that the click rate increases and decreases, passing through alternate maxima and minima that correspond exactly to the maxima and minima of the interference fringes.

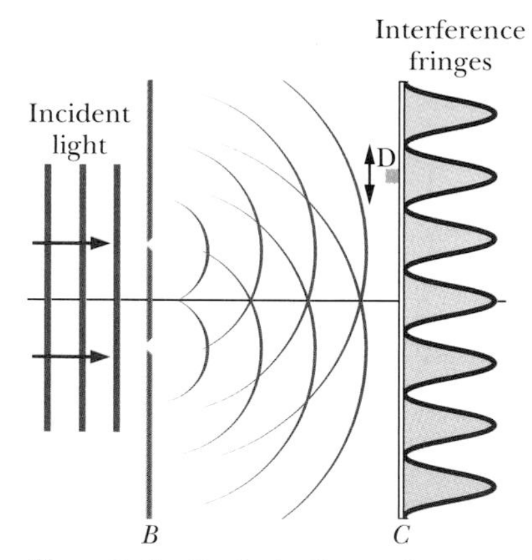

Fig. 38-6 Light is directed onto screen B, which contains two parallel slits. Light emerging from these slits spreads out by diffraction. The two diffracted waves overlap at screen C and form a pattern of interference fringes. A small photon detector D in the plane of screen C generates a sharp click for each photon that it absorbs.

The point of this thought experiment is as follows. We cannot predict when a photon will be detected at any particular point on screen C; photons are detected at individual points at random times. We can, however, predict that the relative *probability* that a single photon will be detected at a particular point in a specified time interval is proportional to the light intensity at that point.

We know from Eq. 33-26 ($I = E_{\text{rms}}^2/c\mu_0$) in Section 33-5 that the intensity I of a light wave at any point is proportional to the square of E_m, the amplitude of the oscillating electric field vector of the wave at that point. Thus,

> The probability (per unit time interval) that a photon will be detected in any small volume centered on a given point in a light wave is proportional to the square of the amplitude of the wave's electric field vector at that point.

We now have a probabilistic description of a light wave, hence another way to view light. It is not only an electromagnetic wave but also a **probability wave.** That is, to every point in a light wave we can attach a numerical probability (per unit time interval) that a photon can be detected in any small volume centered on that point.

The Single-Photon Version

A single-photon version of the double-slit experiment was first carried out by G. I. Taylor in 1909 and has been repeated many times since. It differs from the standard version in that the light source in the Taylor experiment is so extremely feeble that it emits only one photon at a time, at random intervals. Astonishingly, interference fringes still build up on screen *C* if the experiment runs long enough (several months for Taylor's early experiment).

What explanation can we offer for the result of this single-photon double-slit experiment? Before we can even consider the result, we are compelled to ask questions like these: If the photons move through the apparatus one at a time, through which of the two slits in screen *B* does a given photon pass? How does a given photon even "know" that there is another slit present so that interference is a possibility? Can a single photon somehow pass through both slits and interfere with itself?

Bear in mind that the only thing we can know about photons is when light interacts with matter—we have no way of detecting them without an interaction with matter, such as with a detector or a screen. Thus, in the experiment of Fig. 38-6, all we can know is that photons originate at the light source and vanish at the screen. Between source and screen, we cannot know what the photon is or does. However, because an interference pattern eventually builds up on the screen, we can speculate that each photon travels from source to screen *as a wave* that fills up the space between source and screen and then vanishes in a photon absorption at some point on the screen, with a transfer of energy and momentum to the screen at that point.

We *cannot* predict where this transfer will occur (where a photon will be detected) for any given photon originating at the source. However, we *can* predict the probability that a transfer will occur at any given point on the screen. Transfers will tend to occur (and thus photons will tend to be absorbed) in the regions of the bright fringes in the interference pattern that builds up on the screen. Transfers will tend *not* to occur (and thus photons will tend *not* to be absorbed) in the regions of the dark fringes in the built-up pattern. Thus, we can say that the wave traveling from the source to the screen is a *probability wave,* which produces a pattern of "probability fringes" on the screen.

The Single-Photon, Wide-Angle Version

In the past, physicists tried to explain the single-photon double-slit experiment in terms of small packets of classical light waves that are individually sent toward the slits. They would define these small packets as photons. However, modern experiments invalidate this explanation and definition. Figure 38-7 shows the arrangement of one of these experiments, reported in 1992 by Ming Lai and Jean-Claude Diels of the University of New Mexico. Source S contains molecules that emit photons at well separated times. Mirrors M_1 and M_2 are positioned to reflect light that the source emits along two distinct paths, 1 and 2, that are separated by an angle θ, which is close to 180°. This arrangement differs from the standard two-slit experiment, in which the angle between the paths of the light reaching two slits is very small.

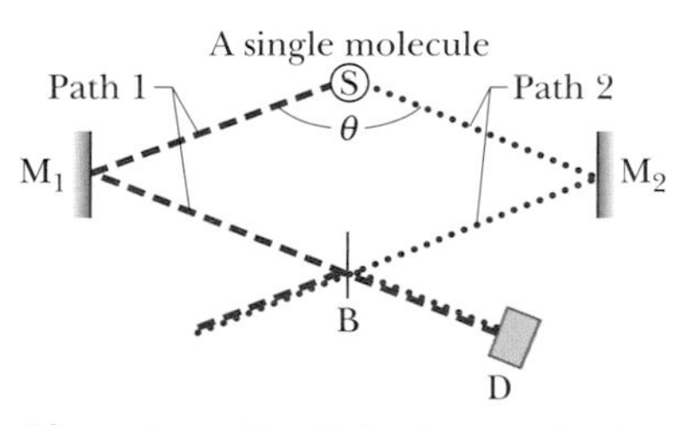

Fig. 38-7 The light from a single photon emission in source S travels over two widely separated paths and interferes with itself at detector D after being recombined by beam splitter B. (After Ming Lai and Jean-Claude Diels, *Journal of the Optical Society of America B*, **9**, 2290–2294, December 1992.)

After reflection from mirrors M_1 and M_2, the light waves traveling along paths 1 and 2 meet at beam splitter B. (A beam splitter is an optical device that transmits half the light incident upon it and reflects the other half.) On the right side of the beam splitter in Fig. 38-7, the light wave traveling along path 2 and reflected by B combines with the light wave traveling along path 1 and transmitted by B. These two waves then interfere with each other as they arrive at detector D (a *photomultiplier tube* that can detect individual photons).

The output of the detector is a randomly spaced series of electronic pulses, one for each detected photon. In the experiment, the beam splitter is moved slowly in a horizontal direction (in the reported experiment, a distance of only about 50 μm maximum), and the detector output is recorded on a chart recorder. Moving the beam splitter changes the lengths of paths 1 and 2, producing a phase shift between the light waves arriving at detector D. Interference maxima and minima appear in the detector's output signal.

This experiment is difficult to understand in traditional terms. For example, when a molecule in the source emits a single photon, does that photon travel along path 1 or path 2 in Fig. 38-7 (or along any other path)? Or can it move in both directions at once? To answer, we assume that when a molecule emits a photon, a probability wave radiates in all directions from it. The experiment samples this wave in two of those directions, chosen to be nearly opposite each other.

We see that we can interpret all three versions of the double-slit experiment if we assume that (1) light is generated in the source as photons, (2) light is absorbed in the detector as photons, and (3) light travels between source and detector as a probability wave.

38-6 Electrons and Matter Waves

In 1924, French physicist Louis de Broglie made the following appeal to symmetry: A beam of light is a wave, but it transfers energy and momentum to matter only at points, via photons. Why can't a beam of particles have the same properties? That is, why can't we think of a moving electron—or any other particle—as a **matter wave** that transfers energy and momentum to other matter at points?

In particular, de Broglie suggested that Eq. 38-7 ($p = h/\lambda$) might apply not only to photons but also to electrons. We used that equation in Section 38-4 to assign a momentum p to a photon of light with wavelength λ. We now use it, in the form

$$\lambda = \frac{h}{p} \quad \text{(de Broglie wavelength)}, \tag{38-13}$$

to assign a wavelength λ to a particle with momentum of magnitude p. The wavelength calculated from Eq. 38-13 is called the **de Broglie wavelength** of the moving particle. De Broglie's prediction of the existence of matter waves was first verified experimentally in 1927, by C. J. Davisson and L. H. Germer of the Bell Telephone Laboratories and by George P. Thomson of the University of Aberdeen in Scotland.

Figure 38-8 shows photographic proof of matter waves in a more recent experiment. In the experiment, an interference pattern was built up when electrons were sent, *one by one,* through a double-slit apparatus. The apparatus was like the ones we have previously used to demonstrate optical interference, except that the viewing screen was similar to a conventional television screen. When an electron hit the screen, it caused a flash of light whose position was recorded.

The first several electrons (top two photos) revealed nothing interesting and seemingly hit the screen at random points. However, after many thousands of electrons were sent through the apparatus, a pattern appeared on the screen, revealing fringes where many electrons had hit the screen and fringes where few had hit the screen. The pattern is exactly what we would expect for wave interference. Thus, *each* electron passed through the apparatus as a matter wave—the portion of the matter wave that traveled through one slit interfered with the portion that traveled through the other slit. That interference then determined the probability that the electron would materialize at a given point on the screen, hitting the screen there. Many electrons materialized in regions corresponding to bright fringes in optical interference, and few electrons materialized in regions corresponding to dark fringes.

Similar interference has been demonstrated with protons, neutrons, and various atoms. In 1994, it was demonstrated with iodine molecules I_2, which are not only 500 000 times more massive than electrons but far more complex. In 1999,

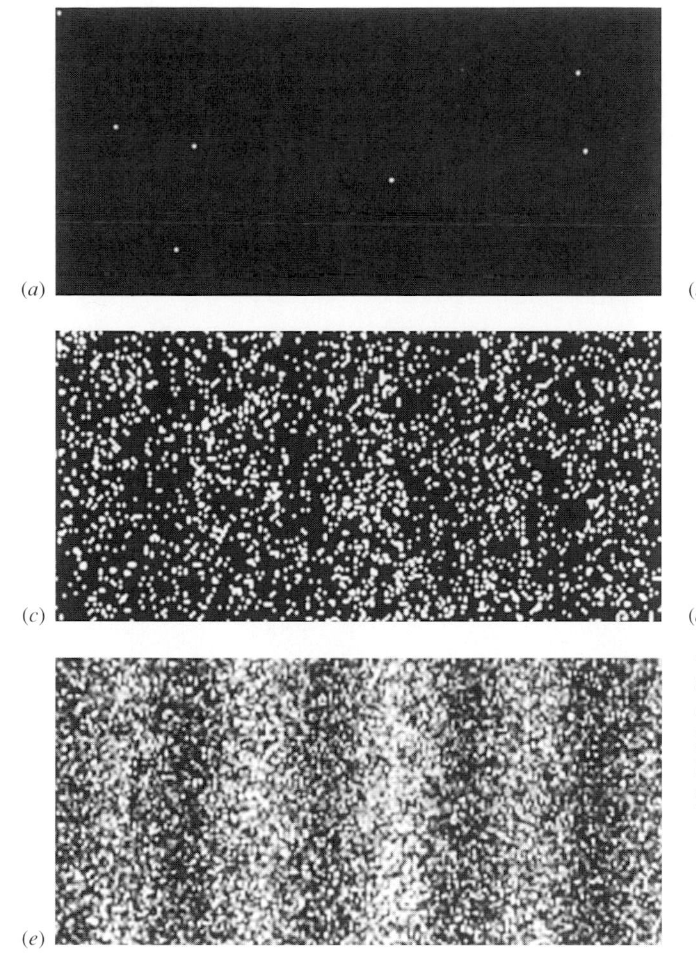

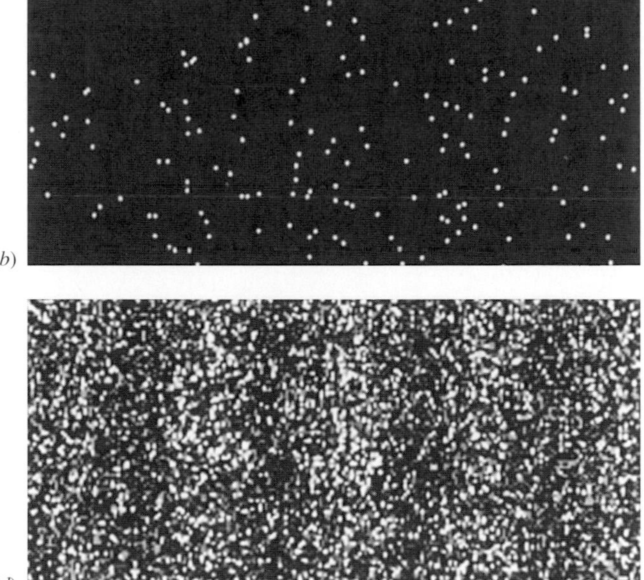

Fig. 38-8 Photographs showing the buildup of an interference pattern by a beam of electrons in a two-slit interference experiment like that of Fig. 38-6. Matter waves, like light waves, are *probability waves.* The approximate numbers of electrons involved are (*a*) 7, (*b*) 100, (*c*) 3000, (*d*) 20 000, and (*e*) 70 000.

it was demonstrated with the even more complex *fullerenes* (or *buckyballs*) C_{60} and C_{70}. (Fullerenes are molecules of carbon atoms that are arranged in a structure resembling a soccer ball, 60 carbon atoms in C_{60} and 70 carbon atoms in C_{70}.) Apparently, such small objects as electrons, protons, atoms, and molecules travel as matter waves. However, as we consider larger and more complex objects, there must come a point at which we are no longer justified in considering the wave nature of an object. At that point, we are back in our familiar nonquantum world, with the physics of earlier chapters of this book. In short, an electron is a matter wave and can undergo interference with itself, but a cat is not a matter wave and cannot undergo interference with itself (which must be a relief to cats).

The wave nature of particles and atoms is now taken for granted in many scientific and engineering fields. For example, electron diffraction and neutron diffraction are used to study the atomic structures of solids and liquids, and electron diffraction is used to study the atomic features of surfaces on solids.

Figure 38-9*a* shows an arrangement that can be used to demonstrate the scattering of either x rays or electrons by crystals. A beam of one or the other is directed onto a target consisting of a layer of tiny aluminum crystals. The x rays have a certain wavelength λ. The electrons are given enough energy so that their de Broglie wavelength is the same wavelength λ. The scatter of x rays or electrons by the crystals produces a circular interference pattern on a photographic film. Figure 38-9*b* shows the pattern for the scatter of x rays, and Fig. 38-9*c* shows the pattern for the scatter of electrons. The patterns are the same—both x rays and electrons are waves.

Fig. 38-9 (*a*) An experimental arrangement used to demonstrate, by diffraction techniques, the wave-like character of the incident beam. Photographs of the diffraction patterns when the incident beam is (*b*) an x-ray beam (light wave) and (*c*) an electron beam (matter wave). Note that the two patterns are geometrically identical to each other.

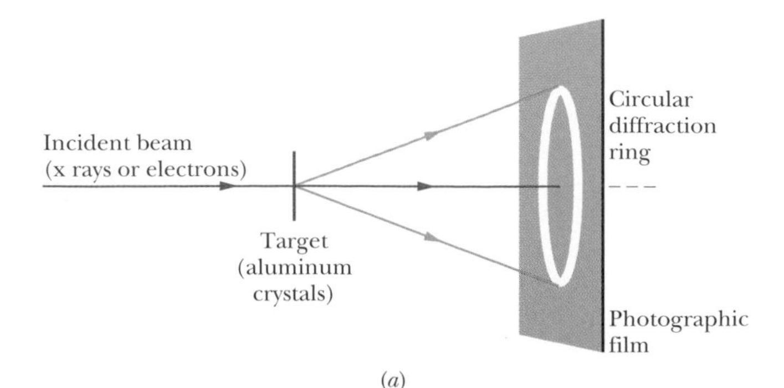

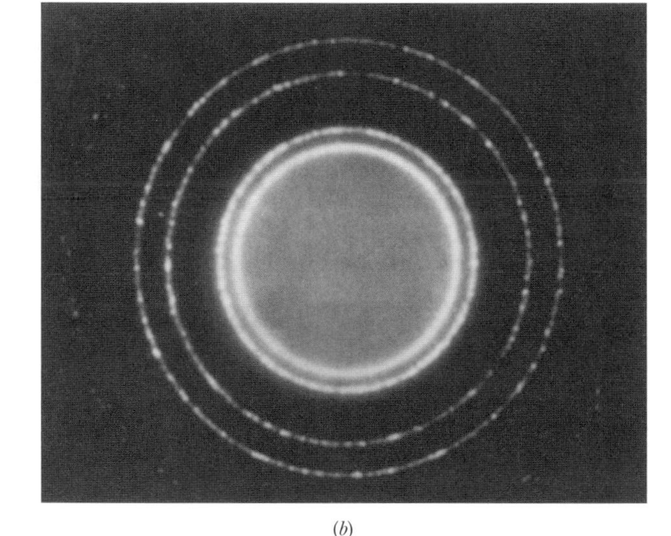

(*b*)

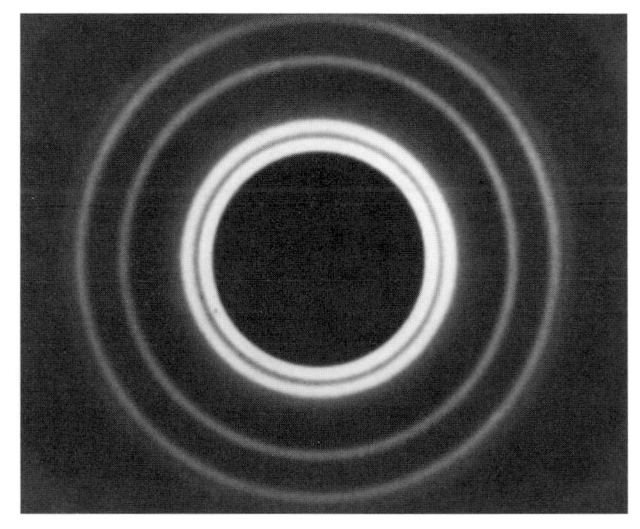

(*c*)

Waves and Particles

Figures 38-8 and 38-9 are convincing evidence of the *wave* nature of matter, but we have countless experiments that suggest its *particle* nature. Figure 38-10, for example, shows the tracks of particles (rather than waves) revealed in a bubble chamber. When a charged particle passes through the liquid hydrogen that fills such a chamber, the particle causes the liquid to vaporize along the particle's path. A series of bubbles thus marks the path, which is usually curved due to a magnetic field set up perpendicular to the plane of the chamber.

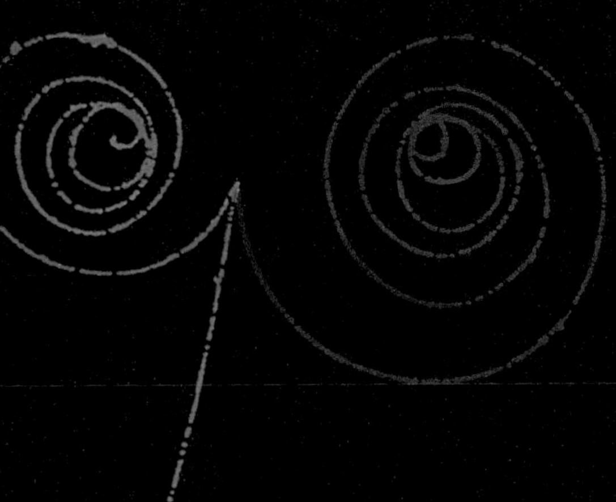

Fig. 38-10 A bubble-chamber image showing where two electrons (paths color coded green) and one positron (red) moved after a gamma ray entered the chamber.

In Fig. 38-10, a gamma ray left no track when it entered at the top because the ray is electrically neutral. However, it collided with one of the hydrogen atoms, kicking an electron out of that atom; the curved path taken by the electron to the bottom of the photograph has been color coded green. Simultaneous with the collision, the gamma ray transformed into an electron and a positron that then moved in tight spirals (color coded green for the electron and red for the positron) as they gradually lost energy in repeated collisions with hydrogen atoms. Surely these tracks are evidence of the particle nature of the electron and positron, but is there any evidence of waves in Fig. 38-10?

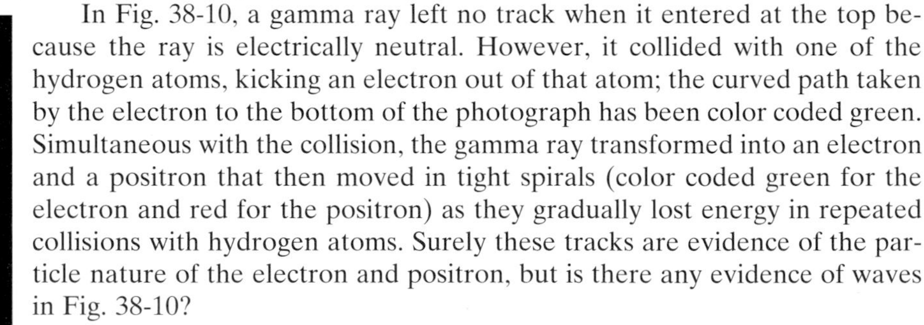

To simplify the situation, let us turn off the magnetic field so that the strings of bubbles will be straight. We can view each bubble as a detection point for the electron. Matter waves traveling between detection points such as I and F in Fig. 38-11 will explore all possible paths, a few of which are shown.

In general, for every path connecting I and F (except the straight-line path), there will be a neighboring path such that matter waves following the two paths cancel each other by interference. This is not true, however, for the straight-line path joining I and F; in this case, matter waves traversing all neighboring paths reinforce the wave following the direct path. You can think of the bubbles that

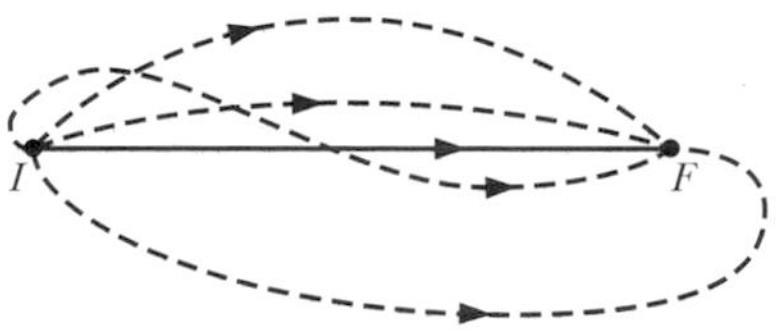

Fig. 38-11 A few of the many paths that connect two particle detection points I and F. Only matter waves that follow paths close to the straight line between these points interfere constructively. For all other paths, the waves following any pair of neighboring paths interfere destructively. Thus, a matter wave leaves a straight track.

form the track as a series of detection points at which the matter wave undergoes constructive interference.

Sample Problem 38-5

What is the de Broglie wavelength of an electron with a kinetic energy of 120 eV?

Solution: One **Key Idea** here is that we can find the electron's de Broglie wavelength λ from Eq. 38-13 ($\lambda = h/p$) if we first find the magnitude of its momentum p. A second **Key Idea** is that we find p from the given kinetic energy K of the electron. That kinetic energy is much less than the rest energy of an electron (0.511 MeV, from Table 37-3). Thus, we can get by with the classical approximations for momentum p ($= mv$) and kinetic energy K ($= \frac{1}{2}mv^2$).

Eliminating the speed v between these two equations yields

$$\begin{aligned} p &= \sqrt{2mK} \\ &= \sqrt{(2)(9.11 \times 10^{-31}\ \text{kg})(120\ \text{eV})(1.60 \times 10^{-19}\ \text{J/eV})} \\ &= 5.91 \times 10^{-24}\ \text{kg} \cdot \text{m/s}. \end{aligned}$$

From Eq. 38-13 then

$$\begin{aligned} \lambda &= \frac{h}{p} \\ &= \frac{6.63 \times 10^{-34}\ \text{J} \cdot \text{s}}{5.91 \times 10^{-24}\ \text{kg} \cdot \text{m/s}} \\ &= 1.12 \times 10^{-10}\ \text{m} = 112\ \text{pm}. \qquad \text{(Answer)} \end{aligned}$$

This wavelength associated with the electron is about the size of a typical atom. If we increase the electron's kinetic energy, the wavelength becomes even smaller.

CHECKPOINT 4 For an electron and a proton that have the same (a) kinetic energy, (b) momentum, or (c) speed, which particle has the shorter de Broglie wavelength?

38-7 Schrödinger's Equation

A simple traveling wave of any kind, be it a wave on a string, a sound wave, or a light wave, is described in terms of some quantity that varies in a wave-like fashion. For light waves, for example, this quantity is $\vec{E}(x, y, z, t)$, the electric field component of the wave. Its observed value at any point depends on the location of that point and on the time at which the observation is made.

What varying quantity should we use to describe a matter wave? We should expect this quantity, which we call the **wave function** $\Psi(x, y, z, t)$, to be more complicated than the corresponding quantity for a light wave because a matter wave, in addition to energy and momentum, transports mass and (often) electric charge. It turns out that Ψ, the uppercase Greek letter psi, usually represents a function that is complex in the mathematical sense; that is, we can always write its values in the form $a + ib$, in which a and b are real numbers and $i^2 = -1$.

In all the situations you will meet here, the space and time variables can be grouped separately and Ψ can be written in the form

$$\Psi(x, y, z, t) = \psi(x, y, z)\, e^{-i\omega t}, \tag{38-14}$$

where ω ($= 2\pi f$) is the angular frequency of the matter wave. Note that ψ, the lowercase Greek letter psi, represents only the space-dependent part of the complete, time-dependent wave function Ψ. We shall deal almost exclusively with ψ. Two questions arise: What is meant by the wave function? How do we find it?

What does the wave function mean? It has to do with the fact that a matter wave, like a light wave, is a probability wave. Suppose that a matter wave reaches a particle detector that is small; then the probability that a particle will be detected in a specified time interval is proportional to $|\psi|^2$, where $|\psi|$ is the absolute value of the wave function at the location of the detector. Although ψ is usually a complex quantity, $|\psi|^2$ is always both real and positive. It is, then, $|\psi|^2$, which we call the **probability density,** and not ψ, that has *physical* meaning. Speaking loosely, the meaning is this:

> The probability (per unit time) of detecting a particle in a small volume centered on a given point in a matter wave is proportional to the value of $|\psi|^2$ at that point.

Because ψ is usually a complex quantity, we find the square of its absolute value by multiplying ψ by ψ^*, the *complex conjugate* of ψ. (To find ψ^* we replace the imaginary number i in ψ with $-i$, wherever it occurs.)

How do we find the wave function? Sound waves and waves on strings are described by the equations of Newtonian mechanics. Light waves are described by Maxwell's equations. Matter waves are described by **Schrödinger's equation,** advanced in 1926 by Austrian physicist Erwin Schrödinger.

Many of the situations that we shall discuss involve a particle traveling in the x direction through a region in which forces acting on the particle cause it to have a potential energy $U(x)$. In this special case, Schrödinger's equation reduces to

$$\frac{d^2\psi}{dx^2} + \frac{8\pi^2 m}{h^2}[E - U(x)]\psi = 0 \quad \text{(Schrödinger's equation, one-dimensional motion),} \tag{38-15}$$

in which E is the total mechanical energy (potential energy plus kinetic energy) of the moving particle. (We do *not* consider mass energy in this nonrelativistic equation.) We cannot derive Schrödinger's equation from more basic principles; it *is* the basic principle.

If $U(x)$ in Eq. 38-15 is zero, that equation describes a **free particle**—that is, a moving particle on which no net force acts. The particle's total energy in this case is all kinetic, and thus E in Eq. 38-15 is $\frac{1}{2}mv^2$. That equation then becomes

$$\frac{d^2\psi}{dx^2} + \frac{8\pi^2 m}{h^2}\left(\frac{mv^2}{2}\right)\psi = 0,$$

which we can recast as

$$\frac{d^2\psi}{dx^2} + \left(2\pi\frac{p}{h}\right)^2\psi = 0.$$

To obtain this equation, we replaced mv with the momentum p and regrouped terms.

From Eq. 38-13 ($\lambda = h/p$) we recognize p/h in the equation above as $1/\lambda$, where λ is the de Broglie wavelength of the moving particle. We further recognize $2\pi/\lambda$ as the *angular wave number* k, which we defined in Eq. 16-5. With these substitutions, the equation above becomes

$$\frac{d^2\psi}{dx^2} + k^2\psi = 0 \quad \text{(Schrödinger's equation, free particle).} \tag{38-16}$$

The most general solution of Eq. 38-16 is

$$\psi(x) = Ae^{ikx} + Be^{-ikx}, \tag{38-17}$$

in which A and B are arbitrary constants. You can show that this equation is indeed a solution of Eq. 38-16 by substituting $\psi(x)$ and its second derivative into that equation and noting that an identity results.

If we combine Eqs. 38-14 and 38-17, we find, for the time-dependent wave function Ψ of a free particle traveling in the x direction,

$$\begin{aligned}\Psi(x, t) &= \psi(x)e^{-i\omega t} = (Ae^{ikx} + Be^{-ikx})e^{-i\omega t} \\ &= Ae^{i(kx-\omega t)} + Be^{-i(kx+\omega t)}.\end{aligned} \tag{38-18}$$

Finding the Probability Density $|\psi|^2$

In Section 16-5 we saw that *any function* F of the form $F(kx \pm \omega t)$ represents a traveling wave. This applies to exponential functions like those in Eq. 38-18 as well as to the sinusoidal functions we have used to describe waves on strings. In fact, these two representations of functions are related by

$$e^{i\theta} = \cos\theta + i\sin\theta \quad \text{and} \quad e^{-i\theta} = \cos\theta - i\sin\theta,$$

where θ is any angle.

The first term on the right in Eq. 38-18 thus represents a wave traveling in the direction of increasing x and the second term represents a wave traveling in the negative direction of x. However, we have assumed that the free particle we are considering travels only in the positive direction of x. To reduce the general solution (Eq. 38-18) to our case of interest, we choose the arbitrary constant B in Eqs. 38-18 and 38-17 to be zero. At the same time, we relabel the constant A as ψ_0. Equation 38-17 then becomes

$$\psi(x) = \psi_0\, e^{ikx}. \tag{38-19}$$

To calculate the probability density, we take the square of the absolute value of $\psi(x)$. We get

$$|\psi|^2 = |\psi_0\, e^{ikx}|^2 = (\psi_0^2)|e^{ikx}|^2.$$

Now, because

$$|e^{ikx}|^2 = (e^{ikx})(e^{ikx})^* = e^{ikx}\, e^{-ikx} = e^{ikx-ikx} = e^0 = 1,$$

we get

$$|\psi|^2 = (\psi_0^2)(1)^2 = \psi_0^2 \quad \text{(a constant)}.$$

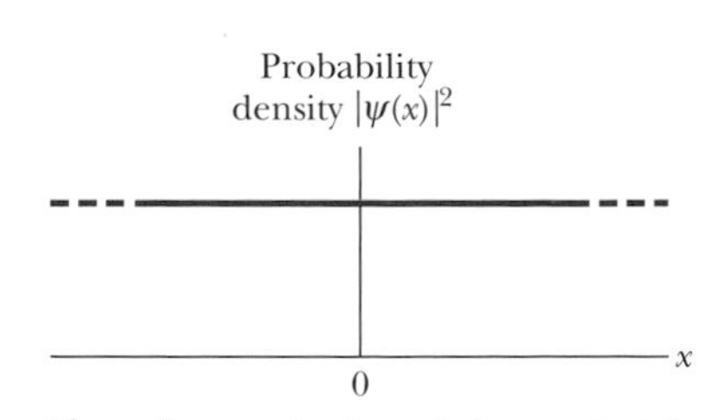

Fig. 38-12 A plot of the probability density $|\psi|^2$ for a free particle moving in the positive x direction. Since $|\psi|^2$ has the same constant value for all values of x, the particle has the same probability of detection at all points along its path.

Figure 38-12 is a plot of the probability density $|\psi|^2$ versus x for a free particle—a straight line parallel to the x axis from $-\infty$ to $+\infty$. We see that the probability density $|\psi|^2$ is the same for all values of x, which means that the particle has equal probabilities of being *anywhere* along the x axis. There is no distinguishing feature by which we can predict a most likely position for the particle. That is, all positions are equally likely.

We'll see what this means in the next section.

38-8 Heisenberg's Uncertainty Principle

Our inability to predict the position of a free particle, as indicated by Fig. 38-12, is our first example of **Heisenberg's uncertainty principle,** proposed in 1927 by German physicist Werner Heisenberg. It states that measured values cannot be assigned to the position $\vec{r}$ and the momentum $\vec{p}$ of a particle simultaneously with unlimited precision.

For the components of $\vec{r}$ and $\vec{p}$, Heisenberg's principle gives the following limits in terms of $\hbar = h/2\pi$ (called "h-bar"):

$$\begin{aligned} \Delta x \cdot \Delta p_x &\geq \hbar \\ \Delta y \cdot \Delta p_y &\geq \hbar \qquad \text{(Heisenberg's uncertainty principle).} \\ \Delta z \cdot \Delta p_z &\geq \hbar \end{aligned} \tag{38-20}$$

Here Δx and Δp_x represent the intrinsic uncertainties in the measurements of the x components of $\vec{r}$ and $\vec{p}$, with parallel meanings for the y and z terms. Even with the best measuring instruments that technology could ever provide, each product of a position uncertainty and a momentum uncertainty in Eq. 38-20 will be greater than $\hbar$; it can *never* be less.

The particle whose probability density is plotted in Fig. 38-12 is a free particle; that is, no force acts on it, and so its momentum $\vec{p}$ must be constant. We implied—without making a point of it—that we can determine $\vec{p}$ with absolute precision; in other words, we assumed that $\Delta p_x = \Delta p_y = \Delta p_z = 0$ in Eq. 38-20. That assumption then requires $\Delta x \to \infty$, $\Delta y \to \infty$, and $\Delta z \to \infty$. With such infinitely great uncertainties, the position of the particle is completely unspecified.

Do not think that the particle *really has* a sharply defined position that is, for some reason, hidden from us. If its momentum can be specified with absolute precision, the words "position of the particle" simply lose all meaning. The particle in Fig. 38-12 can be found *with equal probability* anywhere along the x axis.

Sample Problem 38-6

Assume that an electron is moving along an x axis and that you measure its speed to be 2.05×10^6 m/s, which can be known with a precision of 0.50%. What is the minimum uncertainty (as allowed by the uncertainty principle in quantum theory) with which you can simultaneously measure the position of the electron along the x axis?

Solution: The **Key Idea** here is that the minimum uncertainty allowed by quantum theory is given by Heisenberg's uncertainty principle in Eq. 38-20. We need only consider components along the x axis because we have motion only along that axis and want the uncertainty Δx in location along that axis. Since we want the minimum allowed uncertainty, we use the equality instead of the inequality in the x-axis part of Eq. 38-20, writing

$$\Delta x \cdot \Delta p_x = \hbar.$$

To evaluate the uncertainty Δp_x in the momentum, we must first evaluate the momentum component p_x. Because the electron's speed v_x is much less than the speed of light c, we can evaluate p_x with the classical expression for momentum instead of using a relativistic expression. We find

$$p_x = mv_x = (9.11 \times 10^{-31}\ \text{kg})(2.05 \times 10^6\ \text{m/s}) = 1.87 \times 10^{-24}\ \text{kg} \cdot \text{m/s}.$$

The uncertainty in the speed is given as 0.50% of the measured speed. Because p_x depends directly on speed, the uncertainty Δp_x in the momentum must be 0.50% of the momentum:

$$\Delta p_x = (0.0050)p_x = (0.0050)(1.87 \times 10^{-24}\ \text{kg} \cdot \text{m/s}) = 9.35 \times 10^{-27}\ \text{kg} \cdot \text{m/s}.$$

Then the uncertainty principle gives us

$$\Delta x = \frac{\hbar}{\Delta p_x} = \frac{(6.63 \times 10^{-34}\ \text{J} \cdot \text{s})/2\pi}{9.35 \times 10^{-27}\ \text{kg} \cdot \text{m/s}} = 1.13 \times 10^{-8}\ \text{m} \approx 11\ \text{nm}, \quad \text{(Answer)}$$

which is about 100 atomic diameters. Given your measurement of the electron's speed, it makes no sense to try to pin down the electron's position to any greater precision.

38-9 Barrier Tunneling

Suppose you slide a puck over frictionless ice toward an ice-covered hill (Fig. 38-13). As the puck climbs the hill, kinetic energy K is transformed into gravitational potential energy U. If the puck reaches the top, its potential energy is U_b. Thus, the puck can pass over the top only if its initial mechanical energy $E > U_b$. Otherwise, the puck eventually stops its climb up the left side of the hill and slides back to the left. For instance, if $U_b = 20$ J and $E = 10$ J, you cannot expect the puck to pass over the hill. We say that the hill acts as a **potential energy barrier** (or, for short, a **potential barrier**) and that, in this case, the barrier has a *height* of $U_b = 20$ J.

Figure 38-14 shows a potential barrier for a nonrelativistic electron traveling along an idealized wire of negligible thickness. The electron, with mechanical energy E, approaches a region (the barrier) in which the electric potential V_b is negative. Because it is negatively charged, the electron will have a positive po-

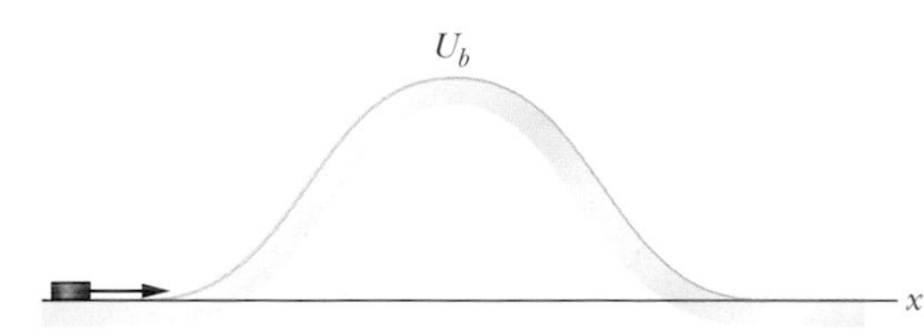

Fig. 38-13 A puck slides over frictionless ice toward a hill. The puck's gravitational potential at the top of the hill will be U_b.

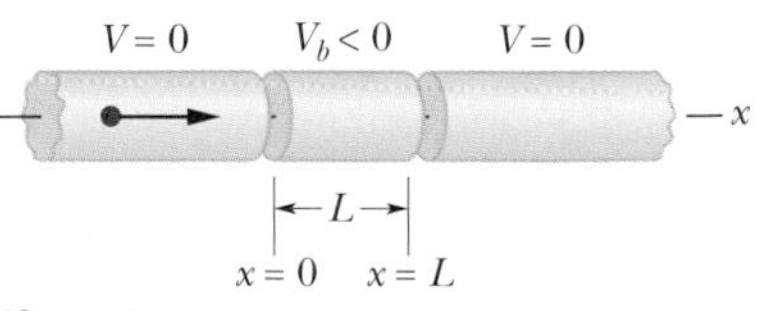

Fig. 38-14 The elements of an idealized thin wire in which an electron (the dot) approaches a negative electric potential V_b in the region $x = 0$ to $x = L$.

tential energy U_b (= qV_b) in that region (Fig. 38-15). If $E > U_b$, we expect the electron to pass through the barrier region and come out to the right of $x = L$ in Fig. 38-14. Nothing surprising there. If $E < U_b$, we expect the electron to be unable to pass through the barrier region. Instead, it should end up traveling leftward, much as the puck would slide back down the hill in Fig. 38-13 if the puck has $E < U_b$.

However, something astounding can happen to the electron when $E < U_b$. Because it is a matter wave, the electron has a finite probability of leaking (or, better, *tunneling*) through the barrier and materializing on the other side, moving rightward with energy E as though nothing (strange or otherwise) had happened in the region of $0 \leq x \leq L$.

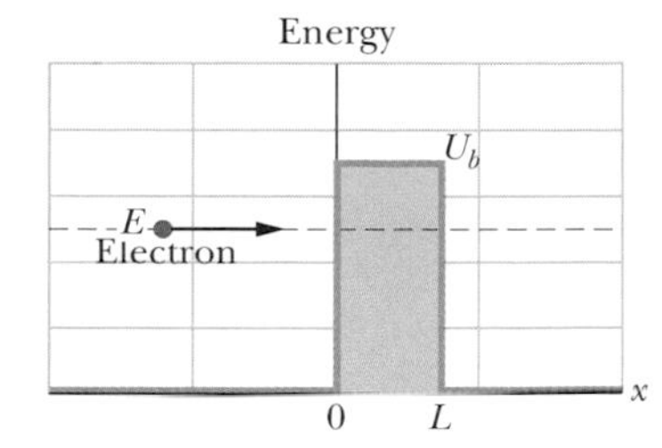

Fig. 38-15 An energy diagram containing two plots for the situation of Fig. 38-13: (1) The electron's mechanical energy E is plotted when the electron is at any coordinate $x < 0$. (2) The electron's electric potential energy U is plotted as a function of the electron's position x, *assuming* that the electron can reach any value of x. The nonzero part of the plot (the potential barrier) has height U_b and thickness L.

The wave function $\psi(x)$ describing the electron can be found by solving Schrödinger's equation (Eq. 38-15) separately for the three regions in Fig. 38-14: (1) to the left of the barrier, (2) within the barrier, and (3) to the right of the barrier. The arbitrary constants that appear in the solutions can then be chosen so that the values of $\psi(x)$ and its derivative with respect to x join smoothly (no jumps, no kinks) at $x = 0$ and at $x = L$. Squaring the absolute value of $\psi(x)$ then yields the probability density.

Figure 38-16 shows a plot of the result. The oscillating curve to the left of the barrier (for $x < 0$) is a combination of the incident matter wave and the reflected matter wave (which has a smaller amplitude than the incident wave). The oscillations occur because these two waves, traveling in opposite directions, interfere with each other, setting up a standing wave pattern.

Within the barrier (for $0 < x < L$) the probability density decreases exponentially with x. However, provided L is small, the probability density is not quite zero at $x = L$.

To the right of the barrier (for $x > L$), the probability density plot describes a transmitted (through the barrier) wave with low but constant amplitude. Thus, the electron can be detected in this region but with a relatively small probability. (Compare this part of the figure with Fig. 38-12 for a free particle.)

We can assign a **transmission coefficient** T to the incident matter wave and the barrier. This coefficient gives the probability with which an approaching electron will be transmitted through the barrier—that is, that tunneling will occur. As an example, if $T = 0.020$, then of every 1000 electrons fired at the barrier, 20 (on average) will tunnel through it and 980 will be reflected. The transmission coefficient T is approximately

$$T \approx e^{-2bL}, \tag{38-21}$$

in which

$$b = \sqrt{\frac{8\pi^2 m(U_b - E)}{h^2}}, \tag{38-22}$$

and e is the exponential function. Because of the exponential form of Eq. 38-21, the value of T is very sensitive to the three variables on which it depends: particle mass m, barrier thickness L, and energy difference $U_b - E$. (Because we do not include relativistic effects here, E does not include mass energy.)

Barrier tunneling finds many applications in technology, among them the tunnel diode, in which a flow of electrons produced by electrons by tunneling through a device can be rapidly turned on or off by controlling the barrier height. Because this can be done very quickly (within 5 ps), the device is suitable for applications demanding a high-speed response. The 1973 Nobel Prize in physics was shared by three "tunnelers," Leo Esaki (for tunneling in semiconductors), Ivar Giaever (for tunneling in superconductors), and Brian Josephson (for the Josephson junction, a rapid quantum switching device based on tunneling). The 1986 Nobel Prize was awarded to Gerd Binnig and Heinrich Rohrer to recognize their development of the scanning tunneling microscope.

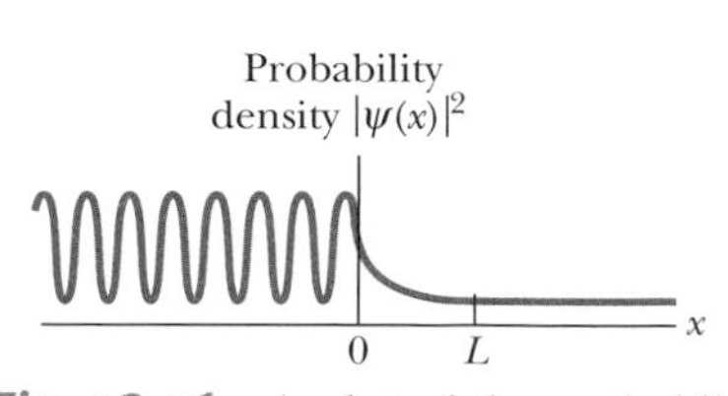

Fig. 38-16 A plot of the probability density $|\psi|^2$ of the electron matter wave for the situation of Fig. 38-15. The value of $|\psi|^2$ is nonzero to the right of the potential barrier.

CHECKPOINT 5 Is the wavelength of the transmitted wave in Fig. 38-16 larger than, smaller than, or the same as that of the incident wave?

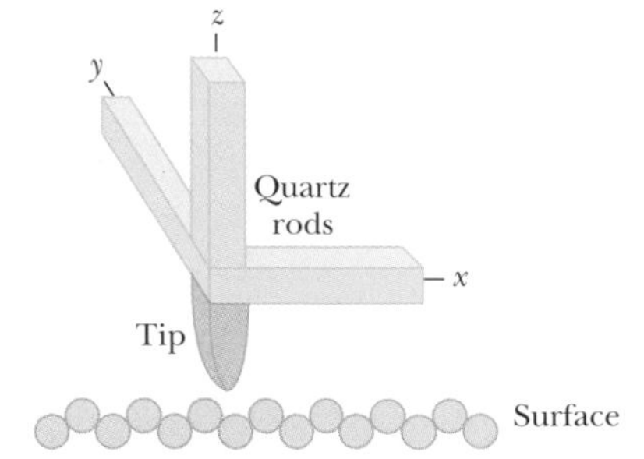

Fig. 38-17 The essence of a scanning tunneling microscope (STM). Three quartz rods are used to scan a sharply pointed conducting tip across the surface of interest and to maintain a constant separation between tip and surface. The tip thus moves up and down to match the contours of the surface, and a record of its movement is a map like that in the opening image of this chapter.

The Scanning Tunneling Microscope (STM)

The size of details that can be seen in an optical microscope is limited by the wavelength of the light the microscope uses (about 300 nm for ultraviolet light). The size of details that can be seen in the image that opens this chapter is far smaller and thus requires much smaller wavelengths. The waves used are electron matter waves, but they do not scatter from the surface being examined the way waves do in an optical microscope. Instead, the images we see are created by electrons tunneling through potential barriers at the tip of a *scanning tunneling microscope* (STM).

Images from an STM can be striking. For example, in this chapter's opening image of a copper surface, the ripples near each cliff edge are due to electrons that are free to roam the copper surface without being able to penetrate it. The electron waves reflecting from a cliff edge overlap the electron waves approaching the edge, to form a standing matter wave much like that drawn in Fig. 38-16.

Figure 38-17 shows the heart of the scanning tunneling microscope. A fine metallic tip, mounted at the intersection of three mutually perpendicular quartz rods, is placed close to the surface to be examined. A small potential difference, perhaps only 10 mV, is applied between tip and surface.

Crystalline quartz has an interesting property called *piezoelectricity:* When an electric potential difference is applied across a sample of crystalline quartz, the dimensions of the sample change slightly. This property is used to change the length of each of the three rods in Fig. 38-17, smoothly and by tiny amounts, so that the tip can be scanned back and forth over the surface (in the x and y directions) and also lowered or raised with respect to the surface (in the z direction).

The space between the surface and the tip forms a potential energy barrier, much like that plotted in Fig. 39-15. If the tip is close enough to the surface, electrons from the sample can tunnel through this barrier from the surface to the tip, forming a tunneling current.

In operation, an electronic feedback arrangement adjusts the vertical position of the tip to keep the tunneling current constant as the tip is scanned over the surface. This means that the tip–surface separation also remains constant during the scan. The output of the device is a video display of the varying vertical position of the tip, hence of the surface contour, as a function of the tip position in the xy plane.

Sample Problem 38-7

Suppose that the electron in Fig. 38-15, having a total energy E of 5.1 eV, approaches a barrier of height $U_b = 6.8$ eV and thickness $L = 750$ pm.

(a) What is the approximate probability that the electron will be transmitted through the barrier, to appear (and be detectable) on the other side of the barrier?

Solution: The **Key Idea** here is that the probability we seek is the transmission coefficient T as given by Eq. 38-21 ($T \approx e^{-2bL}$), where

$$b = \sqrt{\frac{8\pi^2 m(U_b - E)}{h^2}}.$$

The numerator of the fraction under the square-root sign is

$$(8\pi^2)(9.11 \times 10^{-31}\ \text{kg})(6.8\ \text{eV} - 5.1\ \text{eV}) \times (1.60 \times 10^{-19}\ \text{J/eV}) = 1.956 \times 10^{-47}\ \text{J}\cdot\text{kg}.$$

Thus, $$b = \sqrt{\frac{1.956 \times 10^{-47}\ \text{J}\cdot\text{kg}}{(6.63 \times 10^{-34}\ \text{J}\cdot\text{s})^2}} = 6.67 \times 10^9\ \text{m}^{-1}.$$

The (dimensionless) quantity $2bL$ is then

$$2bL = (2)(6.67 \times 10^9\ \text{m}^{-1})(750 \times 10^{-12}\ \text{m}) = 10.0$$

and, from Eq. 38-21, the transmission coefficient is

$$T \approx e^{-2bL} = e^{-10.0} = 45 \times 10^{-6}. \qquad \text{(Answer)}$$

Thus, of every million electrons that strike the barrier, about 45 will tunnel through it.

(b) What is the approximate probability that a proton with the same total energy of 5.1 eV will be transmitted through the barrier, to appear (and be detectable) on the other side of the barrier?

Solution: The **Key Idea** here is that the transmission coefficient T (and thus the probability of transmission) depends on the mass of the particle. Indeed, because mass m is one of the factors in the exponent of e in the equation for T, the probability of transmission is very sensitive to the mass of the particle. This time, the mass is that of a proton (1.67 ×

10^{-27} kg), which is significantly greater than that of the electron in (a). By substituting the proton's mass for the mass in (a) and then continuing as we did there, we find that $T \approx 10^{-186}$. Thus, although the probability that the proton will be transmitted is not exactly zero, it is barely more than zero. For even more massive particles with the same total energy of 5.1 eV, the probability of transmission is exponentially lower.

Review & Summary

Light Quanta—Photons An electromagnetic wave (light) is quantized, and its quanta are called *photons*. For a light wave of frequency f and wavelength λ, the energy E and momentum magnitude p of a photon are

$$E = hf \quad \text{(photon energy)} \tag{38-2}$$

and

$$p = \frac{hf}{c} = \frac{h}{\lambda} \quad \text{(photon momentum)}. \tag{38-7}$$

Photoelectric Effect When light of high enough frequency falls on a clean metal surface, electrons are emitted from the surface by photon–electron interactions within the metal. The governing relation is

$$hf = K_{max} + \Phi, \tag{38-5}$$

in which hf is the photon energy, K_{max} is the kinetic energy of the most energetic emitted electrons, and Φ is the **work function** of the target material—that is, the minimum energy an electron must have if it is to emerge from the surface of the target. If hf is less than Φ, electrons are not emitted.

Compton Shift When x rays are scattered by loosely bound electrons in a target, some of the scattered x rays have a longer wavelength than do the incident x rays. This **Compton shift** (in wavelength) is given by

$$\Delta\lambda = \frac{h}{mc}(1 - \cos\phi), \tag{38-11}$$

in which ϕ is the angle at which the x rays are scattered.

Light Waves and Photons When light interacts with matter, energy and momentum are transferred via photons. When light is in transit, however, we interpret the light wave as a **probability wave,** in which the probability (per unit time) that a photon can be detected is proportional to E_m^2, where E_m is the amplitude of the oscillating electric field of the light wave at the detector.

Matter Waves A moving particle such as an electron or a proton can be described as a **matter wave;** its wavelength (called the **de Broglie wavelength**) is given by $\lambda = h/p$, where p is the magnitude of the particle's momentum.

The Wave Function A matter wave is described by its **wave function** $\Psi(x, y, z, t)$, which can be separated into a space-dependent part $\psi(x, y, z)$ and a time-dependent part $e^{-i\omega t}$. For a particle of mass m moving in the x direction with constant total energy E through a region in which its potential energy is $U(x)$, $\psi(x)$ can be found by solving the simplified **Schrödinger equation:**

$$\frac{d^2\psi}{dx^2} + \frac{8\pi^2 m}{h^2}[E - U(x)]\psi = 0. \tag{38-15}$$

A matter wave, like a light wave, is a probability wave in the sense that if a particle detector is inserted into the wave, the probability that the detector will register a particle during any specified time interval is proportional to $|\psi|^2$, a quantity called the **probability density.**

For a free particle—that is, a particle for which $U(x) = 0$—moving in the x direction, $|\psi|^2$ has a constant value for all positions along the x axis.

Heisenberg's Uncertainty Principle The probabilistic nature of quantum physics places an important limitation on detecting a particle's position and momentum. That is, it is not possible to measure the position $\vec{r}$ and the momentum $\vec{p}$ of a particle simultaneously with unlimited precision. The uncertainties in the components of these quantities are given by

$$\begin{aligned} \Delta x \cdot \Delta p_x &\geq \hbar \\ \Delta y \cdot \Delta p_y &\geq \hbar \\ \Delta z \cdot \Delta p_z &\geq \hbar. \end{aligned} \tag{38-20}$$

Barrier Tunneling According to classical physics, an incident particle will be reflected from a potential energy barrier whose height is greater than the particle's kinetic energy. According to quantum physics, however, the particle has a finite probability of tunneling through such a barrier. The probability that a given particle of mass m and energy E will tunnel through a barrier of height U_b and thickness L is given by the transmission coefficient T:

$$T \approx e^{-2bL}, \tag{38-21}$$

where

$$b = \sqrt{\frac{8\pi^2 m(U_b - E)}{h^2}}. \tag{38-22}$$

Questions

1 According to the figure for Checkpoint 2, is the maximum kinetic energy of the ejected electrons greater for a target made of sodium or of potassium for a given frequency of incident light?

2 A metal plate is illuminated with light of a certain frequency. Which of the following determine whether or not electrons are ejected: (a) the intensity of the light, (b) how long the plate is exposed to the light, (c) the thermal conductivity of the plate, (d) the area of the plate, (e) the material of which the plate is made?

3 Of the electromagnetic waves generated in a microwave oven and in your dentist's x-ray machine, which has (a) the greater wavelength, (b) the greater frequency, and (c) the greater photon energy?

4 Of the following statements about the photoelectric effect, which are true and which are false? (a) The greater the frequency of the incident light is, the greater is the stopping potential. (b) The greater the intensity of the incident light is, the greater is the cutoff frequency. (c) The greater the work function of the target material is, the greater is the stopping potential. (d) The greater the work function of the target material is, the greater is the cutoff frequency. (e) The greater the light frequency is, the greater is the maximum kinetic energy of the ejected electrons. (f) The greater the energy of the photons is, the smaller is the stopping potential.

5 If you shine ultraviolet light on an isolated metal plate, the plate emits electrons for a while. Why does it eventually stop emitting electrons?

6 In the photoelectric effect (for a given target and a given frequency of the incident light), which of these quantities, if any, depend on the intensity of the incident light beam: (a) the maximum kinetic energy of the electrons, (b) the maximum photoelectric current, (c) the stopping potential, (d) the cutoff frequency?

7 In a Compton-shift experiment, light (in the x-ray range) is scattered in the forward direction, at $\phi = 0$ in Fig. 38-3. How much energy does the electron acquire?

8 According to Eq. 38-11, the Compton shift is the same for x rays and for visible light. Why is it that the Compton shift for x rays can be measured readily but that for visible light cannot?

9 Photon A is from an ultraviolet tanning lamp, and photon B is from a television transmitter. Which has the greater (a) wavelength, (b) energy, (c) frequency, and (d) momentum?

10 Photon A has twice the energy of photon B. (a) Is the momentum of A less than, equal to, or greater than that of B? (b) Is the wavelength of A less than, equal to, or greater than that of B?

11 The data shown in Fig. 38-4 were obtained by directing x rays onto a carbon target. In what essential way, if any, would these data differ if the target were sulfur?

12 (a) If you double the kinetic energy of a nonrelativistic particle, how does its de Broglie wavelength change? (b) What if you double the speed of the particle?

13 An electron and a proton have the same kinetic energy. Which has the greater de Broglie wavelength?

14 The following nonrelativistic particles all have the same kinetic energy. Rank them in order of their de Broglie wavelengths, greatest first: electron, alpha particle, neutron.

15 Figure 38-18 shows an electron moving (a) opposite an electric field, (b) in the same direction as an electric field, (c) in the same direction as a magnetic field, (d) perpendicular to a magnetic field. For each situation, is the de Broglie wavelength of the electron increasing, decreasing, or remaining the same?

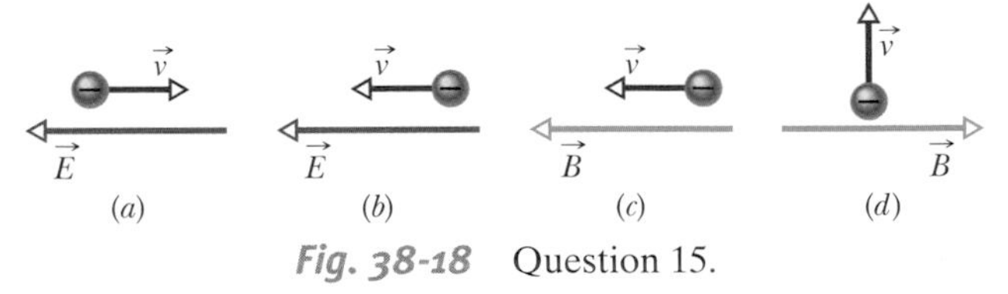

Fig. 38-18 Question 15.

16 At the left in Fig. 38-16, why are the minima in the values of $|\psi|^2$ greater than zero?

17 Which has the greater effect on the transmission coefficient T for electron tunneling through a potential energy barrier: (a) raising the barrier height U_b by 1% or (b) lowering the kinetic energy E of the incident electron by 1%?

18 A proton and a deuteron, each having a kinetic energy of 3 MeV, approach a potential energy barrier whose height U_b is 10 MeV. Which particle has the greater chance of tunneling through the barrier? (A deuteron is twice as massive as a proton.)

19 The table gives relative values for three situations for the barrier tunneling experiment of Figs. 38-14 and 38-15. Rank the situations according to the probability of the electron tunneling through the barrier, greatest first.

	Electron Energy	Barrier Height	Barrier Thickness
(a)	E	$5E$	L
(b)	E	$17E$	$L/2$
(c)	E	$2E$	$2L$

20 Suppose that the height of the potential energy barrier in Fig. 38-15 is infinite. (a) What value would you expect for the transmission coefficient of electrons approaching the barrier? (b) Does Eq. 38-21 predict your expected result?

Problems

SSM Solution is in the Student Solutions Manual.
WWW Solution is at http://www.wiley.com/college/halliday
• – ••• Number of dots indicates level of problem difficulty.

sec. 38-2 The Photon, the Quantum of Light

•1 Monochromatic light (that is, light of a single wavelength) is to be absorbed by a sheet of photographic film and thus recorded on the film. Photon absorption will occur if the photon energy equals or exceeds 0.6 eV, the smallest amount of energy needed to dissociate an AgBr molecule in the film. (a) What is the greatest wavelength of light that can be recorded by the film? (b) In what region of the electromagnetic spectrum is this wavelength located?

•2 The yellow-colored light from a highway sodium lamp is brightest at a wavelength of 589 nm. What is the photon energy for light at that wavelength?

•3 At what rate does the Sun emit photons? For simplicity, assume that the Sun's entire emission at the rate of 3.9×10^{26} W is at the single wavelength of 550 nm.

•4 A helium–neon laser emits red light at wavelength λ = 633 nm in a beam of diameter 3.5 mm and at an energy-emission rate of 5.0 mW. A detector in the beam's path totally absorbs the beam. At what rate per unit area does the detector absorb photons?

•5 The meter was once defined as 1 650 763.73 wavelengths of the orange light emitted by a source containing krypton-86 atoms. What is the photon energy of that light?

•6 How fast must an electron move to have a kinetic energy equal to the photon energy of sodium light at wavelength 590 nm?

••7 A special kind of lightbulb emits monochromatic light of wavelength 630 nm. Electrical energy is supplied to it at the rate of 60 W, and the bulb is 93% efficient at converting that energy to light energy. How many photons are emitted by the bulb during its lifetime of 730 h?

••8 Under ideal conditions, a visual sensation can occur in the human visual system if light of wavelength 550 nm is absorbed by the eye's retina at a rate as low as 100 photons per second. What is the corresponding rate at which energy is absorbed by the retina?

••9 An ultraviolet lamp emits light of wavelength 400 nm at the rate of 400 W. An infrared lamp emits light of wavelength 700 nm, also at the rate of 400 W. (a) Which lamp emits photons at the greater rate and (b) what is that greater rate? SSM WWW

••10 A satellite in Earth orbit maintains a panel of solar cells of area 2.60 m^2 perpendicular to the direction of the Sun's light rays. The intensity of the light at the panel is 1.39 kW/m^2. (a) At what rate does solar energy arrive at the panel? (b) At what rate are solar photons absorbed by the panel? Assume that the solar radiation is monochromatic, with a wavelength of 550 nm, and that all the solar radiation striking the panel is absorbed. (c) How long would it take for a "mole of photons" to be absorbed by the panel?

••11 A 100 W sodium lamp (λ = 589 nm) radiates energy uniformly in all directions. (a) At what rate are photons emitted by the lamp? (b) At what distance from the lamp will a totally absorbing screen absorb photons at the rate of 1.00 $photon/cm^2 \cdot s$? (c) What is the photon flux (photons per unit area per unit time) on a small screen 2.00 m from the lamp?

••12 The beam emerging from a 1.5 W argon laser (λ = 515 nm) has a diameter d of 3.0 mm. The beam is focused by a lens system with an effective focal length f_L of 2.5 mm. The focused beam strikes a totally absorbing screen, where it forms a circular diffraction pattern whose central disk has a radius R given by $1.22 f_L \lambda / d$. It can be shown that 84% of the incident energy ends up within this central disk. At what rate are photons absorbed by the screen in the central disk of the diffraction pattern?

sec. 38-3 The Photoelectric Effect

•13 Light strikes a sodium surface, causing photoelectric emission. The stopping potential for the ejected electrons is 5.0 V, and the work function of sodium is 2.2 eV. What is the wavelength of the incident light? SSM

•14 You wish to pick an element for a photocell that will operate via the photoelectric effect with visible light. Which of the following are suitable (work functions are in parentheses): tantalum (4.2 eV), tungsten (4.5 eV), aluminum (4.2 eV), barium (2.5 eV), lithium (2.3 eV)?

•15 The work function of tungsten is 4.50 eV. Calculate the speed of the fastest electrons ejected from a tungsten surface when light whose photon energy is 5.80 eV shines on the surface.

•16 Find the maximum kinetic energy of electrons ejected from a certain material if the material's work function is 2.3 eV and the frequency of the incident radiation is 3.0×10^{15} Hz.

••17 (a) If the work function for a certain metal is 1.8 eV, what is the stopping potential for electrons ejected from the metal when light of wavelength 400 nm shines on the metal? (b) What is the maximum speed of the ejected electrons?

••18 An orbiting satellite can become charged by the photoelectric effect when sunlight ejects electrons from its outer surface. Satellites must be designed to minimize such charging. Suppose a satellite is coated with platinum, a metal with a very large work function (Φ = 5.32 eV). Find the longest wavelength of incident sunlight that can eject an electron from the platinum.

••19 Light of wavelength 200 nm shines on an aluminum surface; 4.20 eV is required to eject an electron. What is the kinetic energy of (a) the fastest and (b) the slowest ejected electrons? (c) What is the stopping potential for this situation? (d) What is the cutoff wavelength for aluminum? SSM

••20 The wavelength associated with the cutoff frequency for silver is 325 nm. Find the maximum kinetic energy of electrons ejected from a silver surface by ultraviolet light of wavelength 254 nm.

••21 The stopping potential for electrons emitted from a surface illuminated by light of wavelength 491 nm is 0.710 V. When the incident wavelength is changed to a new value, the stopping potential is 1.43 V. (a) What is this new wavelength? (b) What is the work function for the surface?

••22 In a photoelectric experiment using a sodium surface, you find a stopping potential of 1.85 V for a wavelength of 300 nm and a stopping potential of 0.820 V for a wavelength of 400 nm. From these data find (a) a value for the Planck constant, (b) the work function Φ for sodium, and (c) the cutoff wavelength λ_0 for sodium.

••23 X rays with a wavelength of 71 pm are directed onto a gold foil and eject tightly bound electrons from the gold atoms. The ejected electrons then move in circular paths of radius r in a region of uniform magnetic field $\vec{B}$. For the fastest of the ejected electrons, the product Br is equal to $1.88 \times 10^{-4}\ T \cdot m$. Find (a) the maximum kinetic energy of those electrons and (b) the work done in removing them from the gold atoms.

••24 Suppose the *fractional efficiency* of a cesium surface (with work function 1.80 eV) is 1.0×10^{-16}; that is, on average one electron is ejected for every 10^{16} photons that reach the surface. What would be the current of electrons ejected from such a surface if it were illuminated with 600 nm light from a 2.00 mW laser and all the ejected electrons took part in the charge flow?

sec. 38-4 Photons Have Momentum

•25 Light of wavelength 2.40 pm is directed onto a target

containing free electrons. (a) Find the wavelength of light scattered at 30.0° from the incident direction. (b) Do the same for a scattering angle of 120°. SSM

•26 (a) In MeV/c, what is the magnitude of the momentum associated with a photon having an energy equal to the electron rest energy? What are the (b) wavelength and (c) frequency of the corresponding radiation?

•27 What (a) frequency, (b) photon energy, and (c) photon momentum magnitude (in keV/c) are associated with x rays having wavelength 35.0 pm?

••28 X rays of wavelength 0.0100 nm are directed in the positive direction of an x axis onto a target containing loosely bound electrons. For Compton scattering from one of those electrons, at an angle of 180°, what are (a) the Compton shift, (b) the corresponding change in photon energy, (c) the kinetic energy of the recoiling electron, and (d) the angle between the positive direction of the x axis and the electron's direction of motion?

••29 Calculate the Compton wavelength for (a) an electron and (b) a proton. What is the photon energy for an electromagnetic wave with a wavelength equal to the Compton wavelength of (c) the electron and (d) the proton? SSM

••30 Gamma rays of photon energy 0.511 MeV are directed onto an aluminum target and are scattered in various directions by loosely bound electrons there. (a) What is the wavelength of the incident gamma rays? (b) What is the wavelength of gamma rays scattered at 90.0° to the incident beam? (c) What is the photon energy of the rays scattered in this direction?

••31 Calculate the percentage change in photon energy during a collision like that in Fig. 38-5 for $\phi = 90°$ and for radiation in (a) the microwave range, with $\lambda = 3.0$ cm; (b) the visible range, with $\lambda = 500$ nm; (c) the x-ray range, with $\lambda = 25$ pm; and (d) the gamma-ray range, with a gamma photon energy of 1.0 MeV. (e) What are your conclusions about the feasibility of detecting the Compton shift in these various regions of the electromagnetic spectrum, judging solely by the criterion of energy loss in a single photon–electron encounter?

••32 What is the maximum wavelength shift for a Compton collision between a photon and a free *proton*?

••33 What percentage increase in wavelength leads to a 75% loss of photon energy in a photon–free electron collision? SSM

••34 What is the maximum kinetic energy of electrons knocked out of a thin copper foil by Compton scattering of an incident beam of 17.5 keV x rays? Assume the work function is negligible.

••35 What are (a) the Compton shift $\Delta\lambda$, (b) the fractional Compton shift $\Delta\lambda/\lambda$, and (c) the change ΔE in photon energy for light of wavelength $\lambda = 590$ nm scattering from a free, initially stationary electron if the scattering is at 90° to the direction of the incident beam? What are (d) $\Delta\lambda$, (e) $\Delta\lambda/\lambda$, and (f) ΔE for 90° scattering for photon energy 50.0 keV (x-ray range)?

••36 Show that when a photon of energy E is scattered from a free electron at rest, the maximum kinetic energy of the recoiling electron is given by

$$K_{max} = \frac{E^2}{E + mc^2/2}.$$

••37 Consider a collision between an x-ray photon of initial energy 50.0 keV and an electron at rest, in which the photon is scattered backward and the electron is knocked forward. (a) What is the energy of the back-scattered photon? (b) What is the kinetic energy of the electron?

••38 Through what angle must a 200 keV photon be scattered by a free electron so that the photon loses 10% of its energy?

sec. 38-6 Electrons and Matter Waves

•39 In an old-fashioned television set, electrons are accelerated through a potential difference of 25.0 kV. What is the de Broglie wavelength of such electrons? (Relativity is not needed.) SSM

•40 Calculate the de Broglie wavelength of (a) a 1.00 keV electron, (b) a 1.00 keV photon, and (c) a 1.00 keV neutron.

••41 The wavelength of the yellow spectral emission line of sodium is 590 nm. At what kinetic energy would an electron have that wavelength as its de Broglie wavelength? SSM

••42 If the de Broglie wavelength of a proton is 100 fm, (a) what is the speed of the proton and (b) through what electric potential would the proton have to be accelerated to acquire this speed?

••43 What is the wavelength of (a) a photon with energy 1.00 eV, (b) an electron with energy 1.00 eV, (c) a photon of energy 1.00 GeV, and (d) an electron with energy 1.00 GeV?

••44 An electron and a photon each have a wavelength of 0.20 nm. What is the momentum (in kg · m/s) of the (a) electron and (b) photon? What is the energy (in eV) of the (c) electron and (d) photon?

••45 Singly charged sodium ions are accelerated through a potential difference of 300 V. (a) What is the momentum acquired by such an ion? (b) What is its de Broglie wavelength? SSM WWW

••46 What are (a) the energy of a photon corresponding to wavelength 1.00 nm, (b) the kinetic energy of an electron with de Broglie wavelength 1.00 nm, (c) the energy of a photon corresponding to wavelength 1.00 fm, and (d) the kinetic energy of an electron with de Broglie wavelength 1.00 fm?

••47 Electrons accelerated to an energy of 50 GeV have a de Broglie wavelength λ small enough for them to probe the structure within a target nucleus by scattering from the structure. Assume that the energy is so large that the extreme relativistic relation $p = E/c$ between momentum magnitude p and energy E applies. (In this extreme situation, the kinetic energy of an electron is much greater than its rest energy.) (a) What is λ? (b) If the target nucleus has radius $R = 5.0$ fm, what is the ratio R/λ?

••48 The existence of the atomic nucleus was discovered in 1911 by Ernest Rutherford, who properly interpreted some experiments in which a beam of alpha particles was scattered from a metal foil of atoms such as gold. (a) If the alpha particles had a kinetic energy of 7.5 MeV, what was their de Broglie wavelength? (b) Explain whether the wave nature of the incident alpha particles should have been taken into account in interpreting these experiments. The mass of an alpha particle is 4.00 u (atomic mass units), and its distance of closest approach to the nuclear center in these experiments was about 30 fm. (The wave nature of matter was not postulated until more than a decade after these crucial experiments were first performed.)

••49 A nonrelativistic particle is moving three times as fast as an electron. The ratio of the de Broglie wavelength of the particle to that of the electron is 1.813×10^{-4}. By calculating its mass, identify the particle. SSM

••50 The highest achievable resolving power of a microscope is limited only by the wavelength used; that is, the smallest item that can be distinguished has dimensions about equal to the wavelength. Suppose one wishes to "see" inside an atom. Assuming the atom to have a diameter of 100 pm, this means that one must be able to resolve a width of, say, 10 pm. (a) If an electron microscope is used, what minimum electron energy is required? (b) If a light microscope is used, what minimum photon energy is required? (c) Which microscope seems more practical? Why?

••51 What accelerating voltage would be required for the electrons of an electron microscope if the microscope is to have the same resolving power as could be obtained using 100 keV gamma rays? (See Problem 50 and assume classical physics holds.) SSM

sec. 38-7 Schrödinger's Equation

••52 (a) Let $n = a + ib$ be a complex number, where a and b are real (positive or negative) numbers. Show that the product nn^* is always a positive real number. (b) Let $m = c + id$ be another complex number. Show that $|nm| = |n|\,|m|$.

••53 Show that Eq. 38-17 is indeed a solution of Eq. 38-16 by substituting $\psi(x)$ and its second derivative into Eq. 38-16 and noting that an identity results.

••54 (a) Write the wave function $\psi(x)$ displayed in Eq. 38-19 in the form $\psi(x) = a + ib$, where a and b are real quantities. (Assume that ψ_0 is real.) (b) Write the time-dependent wave function $\Psi(x, t)$ that corresponds to $\psi(x)$ written in this form.

••55 Show that the angular wave number k for a nonrelativistic free particle of mass m can be written as

$$k = \frac{2\pi\sqrt{2mK}}{h},$$

in which K is the particle's kinetic energy. SSM

••56 Suppose we put $A = 0$ in Eq. 38-17 and relabeled B as ψ_0. (a) What would the resulting wave function then describe? (b) How, if at all, would Fig. 38-12 be altered?

••57 The function $\psi(x)$ displayed in Eq. 38-19 describes a free particle, for which we assumed that $U(x) = 0$ in Schrödinger's equation (Eq. 38-15). Assume now that $U(x) = U_0 =$ a constant in that equation. Show that Eq. 38-19 is still a solution of Schrödinger's equation, with

$$k = \frac{2\pi}{h}\sqrt{2m(E - U_0)}$$

now giving the angular wave number k of the particle. SSM

••58 In Eq. 38-18 keep both terms, putting $A = B = \psi_0$. The equation then describes the superposition of two matter waves of equal amplitude, traveling in opposite directions. (Recall that this is the condition for a standing wave.) (a) Show that $|\Psi(x, t)|^2$ is then given by

$$|\Psi(x, t)|^2 = 2\psi_0^2[1 + \cos 2kx].$$

(b) Plot this function, and demonstrate that it describes the square of the amplitude of a standing matter wave. (c) Show that the nodes of this standing wave are located at

$$x = (2n + 1)(\tfrac{1}{4}\lambda), \qquad \text{where } n = 0, 1, 2, 3, \ldots$$

and λ is the de Broglie wavelength of the particle. (d) Write a similar expression for the most probable locations of the particle.

sec. 38-8 Heisenberg's Uncertainty Principle

•59 The uncertainty in the position of an electron along an x axis is given as 50 pm, which is about equal to the radius of a hydrogen atom. What is the least uncertainty in any simultaneous measurement of the momentum component p_x of this electron?

••60 You will find in Chapter 39 that electrons cannot move in definite orbits within atoms, like the planets in our solar system. To see why, let us try to "observe" such an orbiting electron by using a light microscope to measure the electron's presumed orbital position with a precision of, say, 10 pm (a typical atom has a radius of about 100 pm). The wavelength of the light used in the microscope must then be about 10 pm. (a) What would be the photon energy of this light? (b) How much energy would such a photon impart to an electron in a head-on collision? (c) What do these results tell you about the possibility of "viewing" an atomic electron at two or more points along its presumed orbital path? (*Hint:* The outer electrons of atoms are bound to the atom by energies of only a few electron-volts.)

••61 Figure 38-12 shows a case in which the momentum component p_x of a particle is fixed so that $\Delta p_x = 0$; then, from Heisenberg's uncertainty principle (Eq. 38-20), the position x of the particle is completely unknown. From the same principle it follows that the opposite is also true; that is, if the position of a particle is exactly known ($\Delta x = 0$), the uncertainty in its momentum is infinite.

Consider an intermediate case, in which the position of a particle is measured, not to infinite precision, but to within a distance of $\lambda/2\pi$, where λ is the particle's de Broglie wavelength. Show that the uncertainty in the (simultaneously measured) momentum component is then equal to the component itself; that is, $\Delta p_x = p$. Under these circumstances, would a measured momentum of zero surprise you? What about a measured momentum of $0.5p$? Of $2p$? Of $12p$? SSM

sec. 38-9 Barrier Tunneling

••62 Consider a potential energy barrier like that of Fig. 38-15 but whose height U_b is 6.0 eV and whose thickness L is 0.70 nm. What is the energy of an incident electron whose transmission coefficient is 0.0010?

••63 A 3.0 MeV proton is incident on a potential energy barrier of thickness 10 fm and height 10 MeV. What are (a) the transmission coefficient T, (b) the kinetic energy K_t the proton will have on the other side of the barrier if it tunnels through the barrier, and (c) the kinetic energy K_r it will have if it reflects from the barrier? A 3.0 MeV deuteron (the same charge but twice the mass as a proton) is incident on the same barrier. What are (d) T, (e) K_t, and (f) K_r?

••64 (a) Suppose a beam of 5.0 eV protons strikes a potential energy barrier of height 6.0 eV and thickness 0.70 nm, at a rate equivalent to a current of 1000 A. How long would you have to wait—on average—for one proton to be transmitted? (b) How long would you have to wait if the beam consisted of electrons rather than protons?

••65 Consider the barrier-tunneling situation in Sample Problem 38-7. What percentage change in the transmission coefficient T occurs for a 1.0% change in (a) the barrier height, (b) the barrier thickness, and (c) the kinetic energy of the incident electron? SSM WWW

Additional Problems

66 In about 1916, R. A. Millikan found the following stopping-potential data for lithium in his photoelectric experiments:

Wavelength (nm)	433.9	404.7	365.0	312.5	253.5
Stopping potential (V)	0.55	0.73	1.09	1.67	2.57

Use these data to make a plot like Fig. 38-2 (which is for sodium) and then use the plot to find (a) the Planck constant and (b) the work function for lithium.

67 A spectral emission line is electromagnetic radiation that is emitted in a wavelength range narrow enough to be taken as a single wavelength. One such emission line that is important in astronomy has a wavelength of 21 cm. What is the photon energy in the electromagnetic wave at that wavelength?

68 (a) The smallest amount of energy needed to eject an electron from metallic sodium is 2.28 eV. Does sodium show a photoelectric effect for red light, with $\lambda = 680$ nm? (b) What is the cutoff wavelength for photoelectric emission from sodium? (c) To what color does that wavelength correspond?

69 Consider a balloon filled with helium gas at room temperature and atmospheric pressure. Calculate (a) the average de Broglie wavelength of the helium atoms and (b) the average distance between atoms under these conditions. The average kinetic energy of an atom is equal to $(3/2)kT$, where k is the Boltzmann constant. (c) Can the atoms be treated as particles under these conditions? Explain.

70 Neutrons in thermal equilibrium with matter have an average kinetic energy of $(3/2)kT$, where k is the Boltzmann constant and T, which may be taken to be 300 K, is the temperature of the environment of the neutrons. (a) What is the average kinetic energy of such a neutron? (b) What is the corresponding de Broglie wavelength?

71 Derive Eq. 38-11, the equation for the Compton shift, from Eqs. 38-8, 38-9, and 38-10 by eliminating v and θ.

72 A 1500 kg car moving at 20 m/s approaches a hill that is 24 m high and 30 m long. Although the car and hill are clearly too large to be treated as matter waves, determine what Eq. 38-21 predicts for the transmission coefficient of the car, as if it could tunnel through the hill as a matter wave. Treat the hill as a potential energy barrier where the potential energy is gravitational.

73 Imagine playing baseball in a universe (not ours!) where the Planck constant is 0.60 J · s. What would be the uncertainty in the position of a 0.50 kg baseball that is moving at 20 m/s along an axis if the uncertainty in the speed is 1.0 m/s?

74 Figure 38-12 shows that because of Heisenberg's uncertainty principle, it is not possible to assign an x coordinate to the position of the electron. (a) Can you assign a y or a z coordinate? (*Hint:* The momentum of the electron has no y or z component.) (b) Describe the extent of the matter wave in three dimensions.

75 A bullet of mass 40 g travels at 1000 m/s. Although the bullet is clearly too large to be treated as a matter wave, determine what Eq. 38-13 predicts for the de Broglie wavelength of the bullet at that speed.

76 Using the classical equations for momentum and kinetic energy, show that an electron's de Broglie wavelength in nanometers can be written as $\lambda = 1.226/\sqrt{K}$, in which K is the electron's kinetic energy in electron-volts.

77 Show that $|\psi|^2 = |\Psi|^2$, with ψ and Ψ related as in Eq. 38-14. That is, show that the probability density does not depend on the time variable.

78 Show that $\Delta E/E$, the fractional loss of energy of a photon during a collision with a particle of mass m, is given by

$$\frac{\Delta E}{E} = \frac{hf'}{mc^2}(1 - \cos\phi),$$

where E is the energy of the incident photon, f' is the frequency of the scattered photon, and ϕ is defined as in Fig. 38-5.

79 Show, by analyzing a collision between a photon and a free electron (using relativistic mechanics), that it is impossible for a photon to transfer all its energy to a free electron (and thus for the photon to vanish).

80 An electron of mass m and speed v "collides" with a gamma-ray photon of initial energy hf_0, as measured in the laboratory frame. The photon is scattered in the electron's direction of travel. Verify that the energy of the scattered photon, as measured in the laboratory frame, is

$$E = hf_0\left(1 + \frac{2hf_0}{mc^2}\sqrt{\frac{1 + v/c}{1 - v/c}}\right)^{-1}.$$

81 The work functions for potassium and cesium are 2.25 and 2.14 eV, respectively. (a) For which element will the photoelectric effect occur with incident light of wavelength 565 nm? (b) With light of wavelength 518 nm?

82 Express the Planck constant h in terms of the unit electron-volt-femtoseconds.

83 Show that, for light of wavelength λ in nanometers, the photon energy hf in electron-volts is $1240/\lambda$.

On-Line Simulation Problems

The website http://www.wiley.com/college/halliday has simulation problems about this chapter.

More About Matter Waves

39

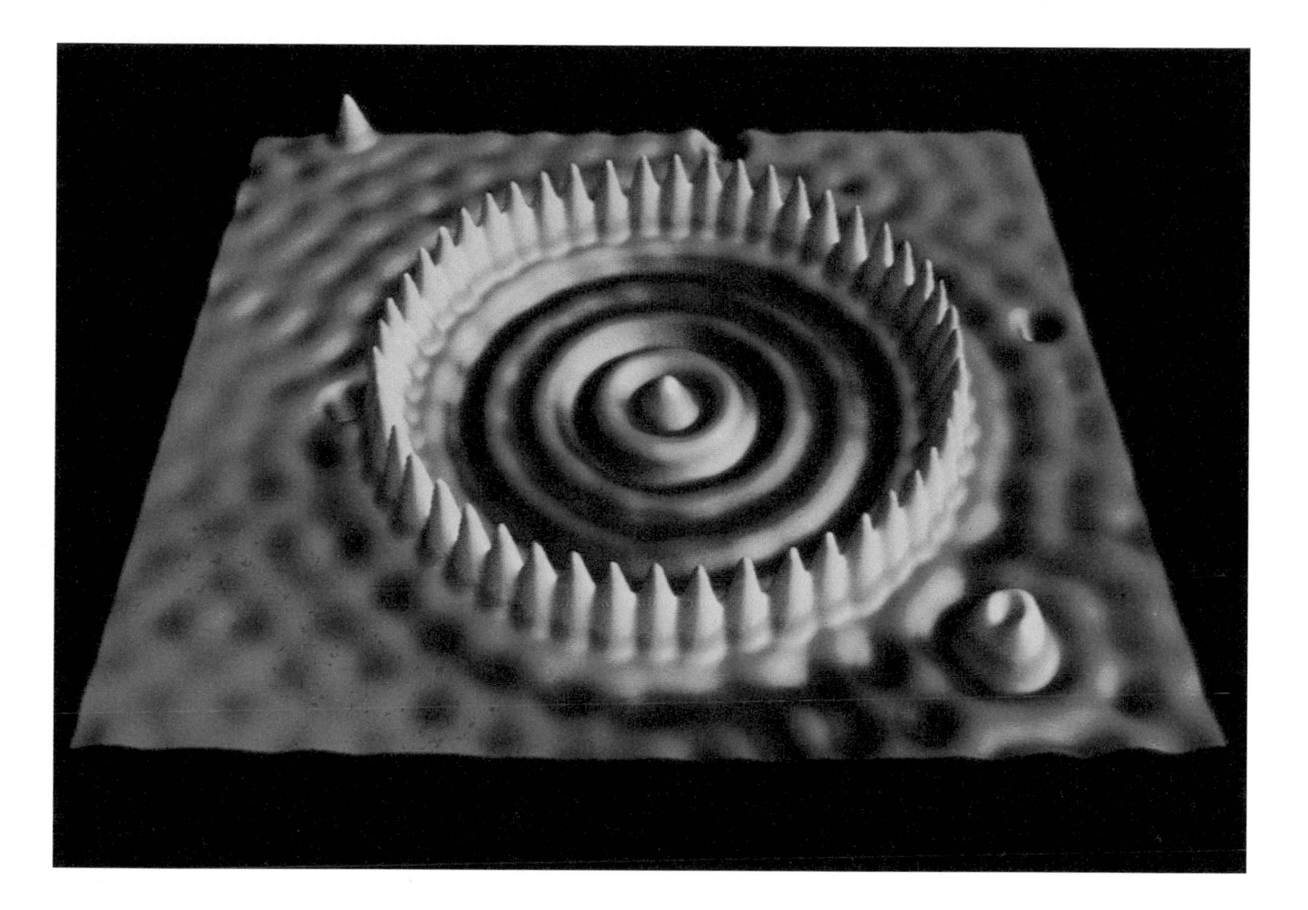

When this image from a scanning tunneling microscope was published by researchers at IBM's Almaden Research Center in California, it stunned many scientists and engineers. The image shows 48 iron atoms that the researchers "dragged" into place to form a circle 14 nm in diameter on a specially prepared copper surface. The arrangement is called a quantum corral *because, as with a ranch corral that fences in livestock, this barrier of iron atoms can fence in something. The ripples are a hint of that something.*

What are *the ripples inside the quantum corral?*

The answer is in this chapter.

39-1 What Is Physics?

One of the long-standing goals of physics has been to understand the nature of atoms. Early in the 20th century nobody knew how the electrons in an atom are arranged, what their motions are, how atoms emit or absorb light, or even why atoms are stable. Without this knowledge it was not possible to understand how atoms combine to form molecules or stack up to form solids. As a consequence, the foundations of chemistry—including biochemistry, which underlies the nature of life itself—were more or less a mystery.

In 1926, all these questions and many others were answered with the development of quantum physics. Its basic premise is that moving electrons, protons, and particles of any kind are best viewed as matter waves, whose motions are governed by Schrödinger's equation. Although quantum theory also applies to massive particles, there is no point in treating baseballs, automobiles, planets, and such objects with quantum theory. For such massive, slow-moving objects, Newtonian physics and quantum physics yield the same answers.

Before we can apply quantum physics to the problem of atomic structure, we need to develop some insights by applying quantum ideas in a few simpler situations. Some of these situations may seem simplistic and unreal, but they allow us to discuss the basic principles of the quantum physics of atoms without having to deal with the often overwhelming complexity of atoms. Besides, with the advances in nanotechnology, situations that were previously found only in textbooks are now being produced in laboratories and put to use in modern electronics and materials science applications. We are on the threshold of being able to use the quantum corrals described in the chapter opening, as well as another type of construction called *quantum dots*, to create "designer atoms" whose properties can be manipulated in the laboratory. For both natural atoms and these artificial ones, the starting point in our discussion is the wave nature of an electron.

39-2 String Waves and Matter Waves

In Chapter 16 we saw that waves of two kinds can be set up on a stretched string. If the string is so long that we can take it to be infinitely long, we can set up a *traveling wave* of essentially any frequency. However, if the stretched string has only a finite length, perhaps because it is rigidly clamped at both ends, we can set up only *standing waves* on it; further, these standing waves can have only discrete frequencies. In other words, confining the wave to a finite region of space leads to *quantization* of the motion—to the existence of discrete *states* for the wave, each state with a sharply defined frequency.

This observation applies to waves of all kinds, including matter waves. For matter waves, however, it is more convenient to deal with the energy E of the associated particle than with the frequency f of the wave. In all that follows we shall focus on the matter wave associated with an electron, but the results apply to any confined matter wave.

Consider the matter wave associated with an electron moving in the positive x direction and subject to no net force—a so-called *free particle.* The energy of such an electron can have any reasonable value, just as a wave traveling along a stretched string of infinite length can have any reasonable frequency.

Consider next the matter wave associated with an atomic electron, perhaps the *valence* (least tightly bound) electron in a sodium atom. Such an electron—held within the atom by the attractive Coulomb force between it and the positively charged nucleus—is *not* a free particle. It can exist only in a set of discrete states, each having a discrete energy E. This sounds much like the discrete states and quantized frequencies that are available to a stretched string of finite length.

For matter waves, then, as for all other kinds of waves, we may state a **confinement principle:**

> Confinement of a wave leads to quantization—that is, to the existence of discrete states with discrete energies. The wave can have only those energies.

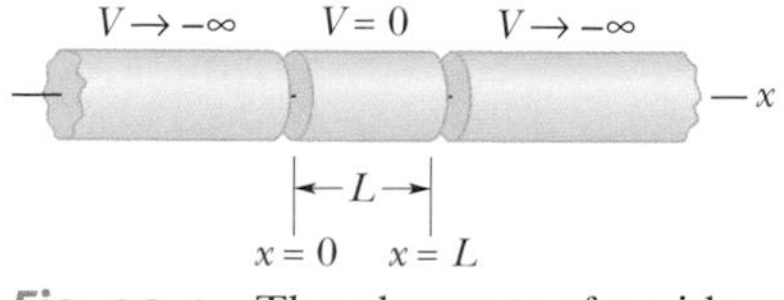

Fig. 39-1 The elements of an idealized "trap" designed to confine an electron to the central cylinder. We take the semi-infinitely long end cylinders to be at an infinitely great negative potential and the central cylinder to be at zero potential.

39-3 Energies of a Trapped Electron

One-Dimensional Traps

Here we examine the matter wave associated with a nonrelativistic electron confined to a limited region of space. We do so by analogy with standing waves on a string of finite length, stretched along an x axis and confined between rigid supports. Because the supports are rigid, the two ends of the string are nodes, or points at which the string is always at rest. There may be other nodes along the string, but these two must always be present, as Fig. 16-23 shows.

The states, or discrete standing wave patterns in which the string can oscillate, are those for which the length L of the string is equal to an integer number of half-wavelengths. That is, the string can occupy only states for which

$$L = \frac{n\lambda}{2}, \qquad \text{for } n = 1, 2, 3, \ldots. \tag{39-1}$$

Each value of n identifies a state of the oscillating string; using the language of quantum physics, we can call the integer n a **quantum number.**

For each state of the string permitted by Eq. 39-1, the transverse displacement of the string at any position x along the string is given by

$$y_n(x) = A \sin\left(\frac{n\pi}{L}x\right), \qquad \text{for } n = 1, 2, 3, \ldots, \tag{39-2}$$

in which the quantum number n identifies the oscillation pattern and A depends on the time at which you inspect the string. (Equation 39-2 is a short version of Eq. 16-60.) We see that for all values of n and for all times, there is a point of zero displacement (a node) at $x = 0$ and at $x = L$, as there must be. Figure 16-22 shows time exposures of such a stretched string for $n = 2$, 3, and 4.

Now let us turn our attention to matter waves. Our first problem is to physically confine an electron that is moving along the x axis so that it remains within a finite segment of that axis. Figure 39-1 shows a conceivable one-dimensional *electron trap.* It consists of two semi-infinitely long cylinders, each of which has an electric potential approaching $-\infty$; between them is a hollow cylinder of length L, which has an electric potential of zero. We put a single electron into this central cylinder to trap it.

The trap of Fig. 39-1 is easy to analyze but is not very practical. Single electrons *can,* however, be trapped in the laboratory with traps that are more complex in design but similar in concept. At the University of Washington, for example, a single electron has been held in a trap for months on end, permitting scientists to make extremely precise measurements of its properties.

Finding the Quantized Energies

Figure 39-2 shows the potential energy of the electron as a function of its position along the x axis of the idealized trap of Fig. 39-1. When the electron is in the central cylinder, its potential energy U $(= -eV)$ is zero because there the potential V is zero. If the electron could get outside this region, its potential energy would be positive and of infinite magnitude because there $V \rightarrow -\infty$. We call the potential energy pattern of Fig. 39-2 an **infinitely deep potential energy well** or, for short, an *infinite potential well.* It is a "well" because an electron placed in

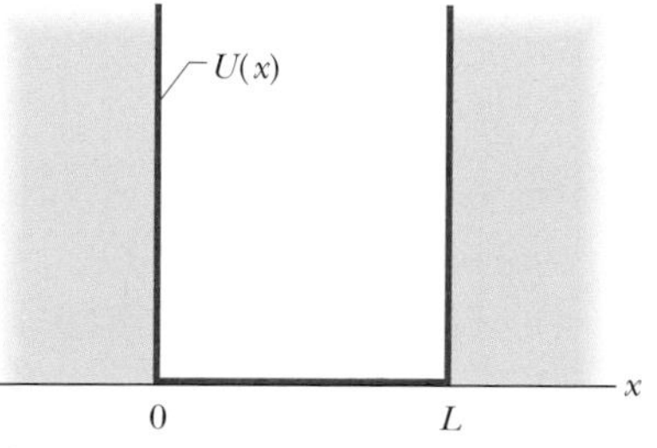

Fig. 39-2 The electric potential energy $U(x)$ of an electron confined to the central cylinder of the idealized trap of Fig. 39-1. We see that $U = 0$ for $0 < x < L$, and $U \rightarrow \infty$ for $x < 0$ and $x > L$.

the central cylinder of Fig. 39-1 cannot escape from it. As the electron approaches either end of the cylinder, a force of essentially infinite magnitude reverses the electron's motion, thus trapping it. Because the electron can move along only a single axis, this trap can be called a *one-dimensional infinite potential well.*

Just like the standing wave in a length of stretched string, the matter wave describing the confined electron must have nodes at $x = 0$ and $x = L$. Moreover, Eq. 39-1 applies to such a matter wave if we interpret λ in that equation as the de Broglie wavelength associated with the moving electron.

The de Broglie wavelength λ is defined in Eq. 38-13 as $\lambda = h/p$, where p is the magnitude of the electron's momentum. Because the electron is nonrelativistic, this momentum magnitude p is related to the electron's kinetic energy K by $p = \sqrt{2mK}$, where m is the mass of the electron. For an electron moving within the central cylinder of Fig. 39-1, where $U = 0$, the total (mechanical) energy E is equal to the kinetic energy. Hence, we can write the de Broglie wavelength of this electron as

$$\lambda = \frac{h}{p} = \frac{h}{\sqrt{2mE}}. \tag{39-3}$$

If we substitute Eq. 39-3 into Eq. 39-1 and solve for the energy E, we find that E depends on n according to

$$E_n = \left(\frac{h^2}{8mL^2}\right)n^2, \quad \text{for } n = 1, 2, 3, \ldots. \tag{39-4}$$

The positive integer n here is the quantum number of the electron's quantum state in the trap.

Equation 39-4 tells us something important: Because the electron is confined to the trap, it can have only the energies given by the equation. It *cannot* have an energy that is, say, halfway between the values for $n = 1$ and $n = 2$. Why this restriction? Because an electron is a matter wave. Were it, instead, a particle as assumed in classical physics, it could have *any* value of energy while it is confined to the trap.

Figure 39-3 is a graph showing the lowest five allowed energy values for an electron in an infinite well with $L = 100$ pm (about the size of a typical atom). The values are called *energy levels,* and they are drawn in Fig. 39-3 as levels, or steps, on a ladder, in an *energy-level diagram.* Energy is plotted vertically; nothing is plotted horizontally.

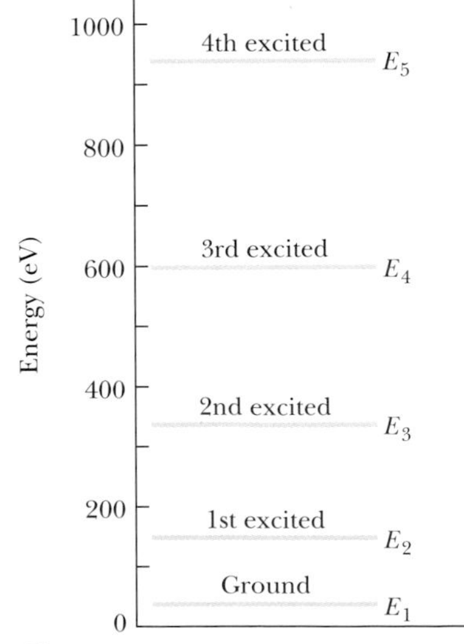

Fig. 39-3 Several of the allowed energies given by Eq. 39-4 for an electron confined to the infinite well of Fig. 39-2. Here width $L = 100$ pm. Such a plot is called an *energy-level diagram.*

The quantum state with the lowest possible energy level E_1 allowed by Eq. 39-4, with quantum number $n = 1$, is called the *ground state* of the electron. The electron tends to be in this lowest energy state. All the quantum states with greater energies (corresponding to quantum numbers $n = 2$ or greater) are called *excited states* of the electron. The state with energy level E_2, for quantum number $n = 2$, is called the *first excited state* because it is the first of the excited states as we move up the energy-level diagram. Similarly, the state with energy level E_3 is called the *second excited state.*

Energy Changes

A trapped electron tends to have the lowest allowed energy and thus to be in its ground state. It can be changed to an excited state (in which it has greater energy) only if an external source provides the additional energy that is required for the change. Let E_{low} be the initial energy of the electron and E_{high} be the greater energy in a state that is higher on its energy-level diagram. Then the amount of energy that is required for the electron's change of state is

$$\Delta E = E_{\text{high}} - E_{\text{low}}. \tag{39-5}$$

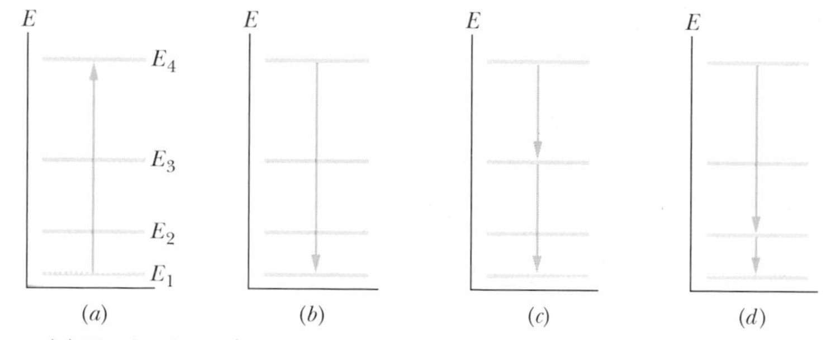

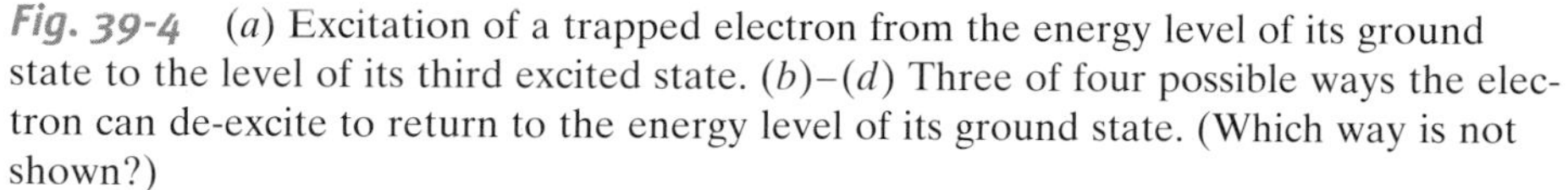

Fig. 39-4 (*a*) Excitation of a trapped electron from the energy level of its ground state to the level of its third excited state. (*b*)–(*d*) Three of four possible ways the electron can de-excite to return to the energy level of its ground state. (Which way is not shown?)

An electron that receives such energy is said to make a *quantum jump* (or *transition*), or to be *excited* from the lower-energy state to the higher-energy state. Figure 39-4*a* represents a quantum jump from the ground state (with energy level E_1) to the third excited state (with energy level E_4). As shown, the jump *must* be from one energy level to another, but it can bypass one or more intermediate energy levels.

One way an electron can gain energy to make a quantum jump up to a greater energy level is to absorb a photon. However, this absorption and quantum jump can occur only if the following condition is met:

> If a confined electron is to absorb a photon, the energy hf of the photon must equal the energy difference ΔE between the initial energy level of the electron and a higher level.

Thus, excitation by the absorption of light requires that

$$hf = \Delta E = E_{\text{high}} - E_{\text{low}}. \qquad (39\text{-}6)$$

When an electron reaches an excited state, it does not stay there but quickly *de-excites* by decreasing its energy. Figures 39-4*b* to *d* represent some of the possible quantum jumps down from the energy level of the third excited state. The electron can reach its ground-state level either with one direct quantum jump (Fig. 39-4*b*) or with shorter jumps via intermediate levels (Figs. 39-4*c* and *d*).

One way in which an electron can decrease its energy is by emitting a photon, but only if the following condition is met:

> If a confined electron emits a photon, the energy hf of that photon must equal the energy difference ΔE between the initial energy level of the electron and a lower level.

Thus, Eq. 39-6 applies to both the absorption and the emission of light by a confined electron. That is, the absorbed or emitted light can have only certain values of hf and thus only certain values of frequency f and wavelength λ.

Aside: Although Eq. 39-6 and what we have discussed about photon absorption and emission can be applied to physical (real) electron traps, they actually cannot be applied to one-dimensional (unreal) electron traps. The reason involves the need to conserve angular momentum in a photon absorption or emission process. In this book, we shall neglect that need and use Eq. 39-6 even for one-dimensional traps.

CHECKPOINT 1 Rank the following pairs of quantum states for an electron confined to an infinite well according to the energy differences between the states, greatest first: (a) $n = 3$ to $n = 1$, (b) $n = 5$ to $n = 4$, (c) $n = 4$ to $n = 3$.

Sample Problem 39-1

An electron is confined to a one-dimensional, infinitely deep potential energy well of width L = 100 pm.

(a) What is the smallest amount of energy the electron can have?

Solution: The **Key Idea** here is that confinement of the electron (a matter wave) to the well leads to quantization of its energy. Because the well is infinitely deep, the allowed energies are given by Eq. 39-4 ($E_n = (h^2/8mL^2)n^2$), with the quantum number n a positive integer. Here, the collection of constants in front of n^2 in Eq. 39-4 is evaluated as

$$\frac{h^2}{8mL^2} = \frac{(6.63 \times 10^{-34}\ \text{J}\cdot\text{s})^2}{(8)(9.11 \times 10^{-31}\ \text{kg})(100 \times 10^{-12}\ \text{m})^2} = 6.031 \times 10^{-18}\ \text{J}. \quad (39\text{-}7)$$

The smallest amount of energy the electron can have corresponds to the lowest quantum number, which is $n = 1$ for the ground state of the electron. Thus, Eqs. 39-4 and 39-7 give us

$$E_1 = \left(\frac{h^2}{8mL^2}\right)n^2 = (6.031 \times 10^{-18}\ \text{J})(1^2) \approx 6.03 \times 10^{-18}\ \text{J} = 37.7\ \text{eV}. \quad \text{(Answer)}$$

(b) How much energy must be transferred to the electron if it is to make a quantum jump from its ground state to its second excited state?

Solution: *First a caution:* Note that, from Fig. 39-3, the *second* excited state corresponds to the *third* energy level, with quantum number n = 3. Then one **Key Idea** is that if the electron is to jump from the n = 1 level to the n = 3 level, the required change in its energy is, from Eq. 39-5,

$$\Delta E_{31} = E_3 - E_1. \quad (39\text{-}8)$$

A second **Key Idea** is that the energies E_3 and E_1 depend on the quantum number n, according to Eq. 39-4. Therefore, substituting that equation into Eq. 39-8 for energies E_3 and E_1 and using Eq. 39-7 lead to

$$\Delta E_{31} = \left(\frac{h^2}{8mL^2}\right)(3)^2 - \left(\frac{h^2}{8mL^2}\right)(1)^2 = \frac{h^2}{8mL^2}(3^2 - 1^2) = (6.031 \times 10^{-18}\ \text{J})(8) = 4.83 \times 10^{-17}\ \text{J} = 302\ \text{eV}. \quad \text{(Answer)}$$

(c) If the electron gains the energy for the jump from energy level E_1 to energy level E_3 by absorbing light, what light wavelength is required?

Solution: One **Key Idea** here is that if light is to transfer energy to the electron, the transfer must be by photon absorption. A second **Key Idea** is that the photon's energy must equal the energy difference ΔE between the initial energy level of the electron and a higher level, according to Eq. 39-6 ($hf = \Delta E$). Otherwise, a photon *cannot* be absorbed. Substituting c/λ for f, we can rewrite Eq. 39-6 as

$$\lambda = \frac{hc}{\Delta E}. \quad (39\text{-}9)$$

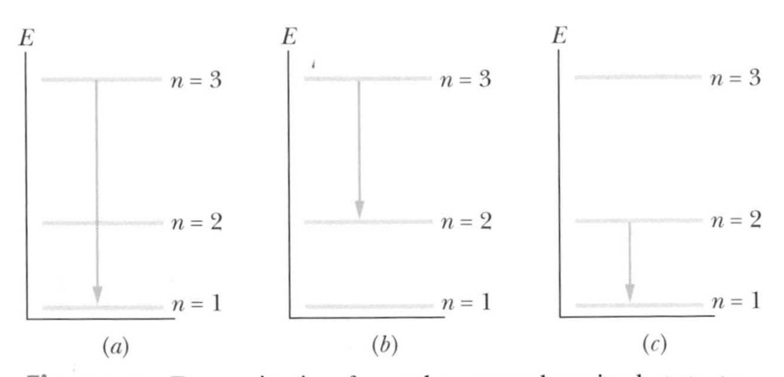

Fig. 39-5 De-excitation from the second excited state to the ground state either directly (*a*) or via the first excited state (*b*, *c*).

For the energy difference ΔE_{31} we found in (b), this equation gives us

$$\lambda = \frac{hc}{\Delta E_{31}} = \frac{(6.63 \times 10^{-34}\ \text{J}\cdot\text{s})(2.998 \times 10^8\ \text{m/s})}{4.83 \times 10^{-17}\ \text{J}} = 4.12 \times 10^{-9}\ \text{m}. \quad \text{(Answer)}$$

(d) Once the electron has been excited to the second excited state, what wavelengths of light can it emit by de-excitation?

Solution: We have three **Key Ideas** here:

1. The electron tends to de-excite, rather than remain in an excited state, until it reaches the ground state (n = 1).
2. If the electron is to de-excite, it must lose just enough energy to jump to a lower energy level.
3. If it is to lose energy by emitting light, then the loss of energy must be by emission of a photon.

Starting in the second excited state (at the n = 3 level), the electron can reach the ground state (n = 1) by *either* making a quantum jump directly to the ground-state energy level (Fig. 39-5*a*) or by making two *separate* jumps by way of the n = 2 level (Figs. 39-5*b* and *c*).

The direct jump involves the same energy difference ΔE_{31} we found in (c). Then the wavelength is the same as we calculated in (c)—except now the wavelength is for light that is emitted, not absorbed. Thus, the electron can jump directly to the ground state by emitting light of wavelength

$$\lambda = 4.12 \times 10^{-9}\ \text{m}. \quad \text{(Answer)}$$

Following the procedure of part (b), you can show that the energy differences for the jumps of Figs. 39-5*b* and *c* are

$$\Delta E_{32} = 3.016 \times 10^{-17}\ \text{J} \quad \text{and} \quad \Delta E_{21} = 1.809 \times 10^{-17}\ \text{J}.$$

From Eq. 39-9, we then find that the wavelength of the light emitted in the first of these jumps (from n = 3 to n = 2) is

$$\lambda = 6.60 \times 10^{-9}\ \text{m}, \quad \text{(Answer)}$$

and the wavelength of the light emitted in the second of these jumps (from n = 2 to n = 1) is

$$\lambda = 1.10 \times 10^{-8}\ \text{m}. \quad \text{(Answer)}$$

39-4 Wave Functions of a Trapped Electron

If we solve Schrödinger's equation for an electron trapped in a one-dimensional infinite potential well of width L, we find that the wave functions for the electron are given by

$$\psi_n(x) = A \sin\left(\frac{n\pi}{L}x\right), \quad \text{for } n = 1, 2, 3, \ldots, \tag{39-10}$$

for $0 \leq x \leq L$ (the wave function is zero outside that range). We shall soon evaluate the amplitude constant A in this equation.

Note that the wave functions $\psi_n(x)$ have the same form as the displacement functions $y_n(x)$ for a standing wave on a string stretched between rigid supports (see Eq. 39-2). We can picture an electron trapped in a one-dimensional well between infinite-potential walls as being a standing matter wave.

Probability of Detection

The wave function $\psi_n(x)$ cannot be detected or directly measured in any way—we cannot simply look inside the well to see the wave the way we can see, say, a wave in a bathtub of water. All we can do is insert a probe of some kind to try to detect the electron. At the instant of detection, the electron would materialize at the point of detection, at some position along the x axis within the well.

If we repeated this detection procedure at many positions throughout the well, we would find that the probability of detecting the electron is related to the probe's position x in the well. In fact, they are related by the *probability density* $\psi_n^2(x)$. Recall from Section 38-7 that in general the probability that a particle can be detected in a specified infinitesimal volume centered on a specified point is proportional to $|\psi_n^2|$. Here, with the electron trapped in a one-dimensional well, we are concerned only with detection of the electron along the x axis. Thus, the probability density $\psi_n^2(x)$ here is a probability per unit length along the x axis. (We can omit the absolute value sign here because $\psi_n(x)$ in Eq. 39-10 is a real quantity, not a complex one.) The probability $p(x)$ that an electron can be detected at position x within the well is

$$\begin{pmatrix}\text{probability } p(x)\\ \text{of detection in width } dx\\ \text{centered on position } x\end{pmatrix} = \begin{pmatrix}\text{probability density } \psi_n^2(x)\\ \text{at position } x\end{pmatrix}(\text{width } dx),$$

or

$$p(x) = \psi_n^2(x)\, dx. \tag{39-11}$$

From Eq. 39-10, we see that the probability density $\psi_n^2(x)$ for the trapped electron is

$$\psi_n^2(x) = A^2 \sin^2\left(\frac{n\pi}{L}x\right), \quad \text{for } n = 1, 2, 3, \ldots, \tag{39-12}$$

for the range $0 \leq x \leq L$ (the probability density is zero outside that range). Figure 39-6 shows $\psi_n^2(x)$ for $n = 1, 2, 3$, and 15 for an electron in an infinite well whose width L is 100 pm.

To find the probability that the electron can be detected in any finite section of the well—say, between point x_1 and point x_2—we must integrate $p(x)$ between those points. Thus, from Eqs. 39-11 and 39-12,

$$\begin{pmatrix}\text{probability of detection}\\ \text{between } x_1 \text{ and } x_2\end{pmatrix} = \int_{x_1}^{x_2} p(x)$$

$$= \int_{x_1}^{x_2} A^2 \sin^2\left(\frac{n\pi}{L}x\right) dx. \tag{39-13}$$

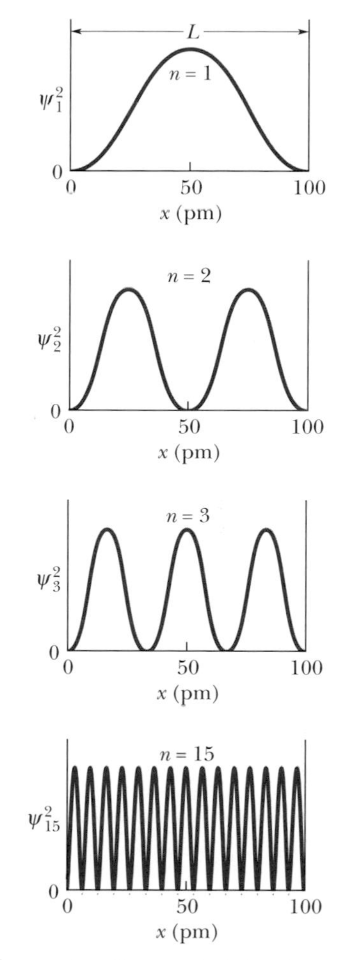

Fig. 39-6 The probability density $\psi_n^2(x)$ for four states of an electron trapped in a one-dimensional infinite well; their quantum numbers are $n = 1, 2, 3$, and 15. The electron is most likely to be found where $\psi_n^2(x)$ is greatest and least likely to be found where $\psi_n^2(x)$ is least.

If classical physics prevailed, we would expect the trapped electron to be detectable with equal probabilities in all parts of the well. From Fig. 39-6 we see that it is not. For example, inspection of that figure or of Eq. 39-12 shows that for the state with $n = 2$, the electron is most likely to be detected near $x = 25$ pm and $x = 75$ pm. It can be detected with near-zero probability near $x = 0$, $x = 50$ pm, and $x = 100$ pm.

The case of $n = 15$ in Fig. 39-6 suggests that as n increases, the probability of detection becomes more and more uniform across the well. This result is an instance of a general principle called the **correspondence principle:**

> At large enough quantum numbers, the predictions of quantum physics merge smoothly with those of classical physics.

This principle, first advanced by Danish physicist Niels Bohr, holds for all quantum predictions.

CHECKPOINT 2 The figure shows three infinite potential wells of widths L, $2L$, and $3L$; each contains an electron in the state for which $n = 10$. Rank the wells according to (a) the number of maxima for the probability density of the electron and (b) the energy of the electron, greatest first.

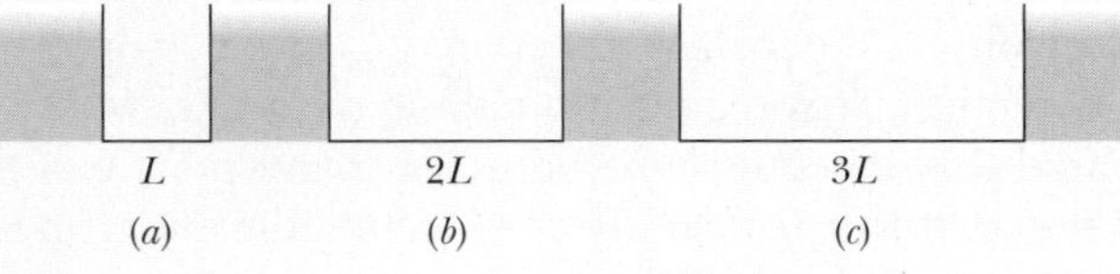

Normalization

The product $\psi_n^2(x)\,dx$ gives the probability that an electron in an infinite well can be detected in the interval of the x axis that lies between x and $x + dx$. We know that the electron must be *somewhere* in the infinite well; so it must be true that

$$\int_{-\infty}^{+\infty} \psi_n^2(x)\,dx = 1 \quad \text{(normalization equation)}, \tag{39-14}$$

because the probability 1 corresponds to certainty. Although the integral is taken over the entire x axis, only the region from $x = 0$ to $x = L$ makes any contribution to the probability. Graphically, the integral in Eq. 39-14 represents the area under each of the plots of Fig. 39-6.

In Sample Problem 39-2 we shall see that if we substitute $\psi_n^2(x)$ from Eq. 39-12 into Eq. 39-14, it is possible to assign a specific value to the amplitude constant A that appears in Eq. 39-12; namely, $A = \sqrt{2/L}$. This process of using Eq. 39-14 to evaluate the amplitude of a wave function is called **normalizing** the wave function. The process applies to *all* one-dimensional wave functions.

Zero-Point Energy

Substituting $n = 1$ in Eq. 39-4 defines the state of lowest energy for an electron in an infinite potential well, the ground state. That is the state the confined electron will occupy unless energy is supplied to it to raise it to an excited state.

The question arises: Why can't we include $n = 0$ among the possibilities listed for n in Eq. 39-4? Putting $n = 0$ in this equation would indeed yield a ground-state energy of zero. However, putting $n = 0$ in Eq. 39-12 would also yield $\psi_n^2(x) = 0$ for all x, which we can interpret only to mean that there is no electron in the well. We know that there is; so $n = 0$ is not a possible quantum number.

It is an important conclusion of quantum physics that confined systems cannot exist in states with zero energy. They must always have a certain minimum energy called the **zero-point energy.**

We can make the zero-point energy as small as we like by making the infinite well wider—that is, by increasing L in Eq. 39-4 for $n = 1$. In the limit as $L \to \infty$, the zero-point energy E_1 approaches zero. In this limit, however, with an infinitely wide well, the electron is a free particle, no longer confined in the x direction. Also, because the energy of a free particle is not quantized, that energy can have any value, including zero. Only a confined particle must have a finite zero-point energy and can never be at rest.

CHECKPOINT 3 Each of the following particles is confined to an infinite well, and all four wells have the same width: (a) an electron, (b) a proton, (c) a deuteron, and (d) an alpha particle. Rank their zero-point energies, greatest first. The particles are listed in order of increasing mass.

Sample Problem 39-2

Evaluate the amplitude constant A in Eq. 39-10 for an infinite potential well extending from $x = 0$ to $x = L$.

Solution: The **Key Idea** here is that the wave functions of Eq. 39-10 must satisfy the normalization requirement of Eq. 39-14, which states that the probability that the electron can be detected somewhere along the x axis is 1. Substituting Eq. 39-10 into Eq. 39-14 and taking the constant A outside the integral yield

$$A^2 \int_0^L \sin^2\left(\frac{n\pi}{L}x\right) dx = 1. \qquad (39\text{-}15)$$

We have changed the limits of the integral from $-\infty$ and $+\infty$ to 0 and L because the wave function is zero outside these new limits (so there's no need to integrate out there).

We can simplify the indicated integration by changing the variable from x to the dimensionless variable y, where

$$y = \frac{n\pi}{L}x, \qquad (39\text{-}16)$$

hence

$$dx = \frac{L}{n\pi}dy.$$

When we change the variable, we must also change the integration limits (again). Equation 39-16 tells us that $y = 0$ when $x = 0$ and that $y = n\pi$ when $x = L$; thus 0 and $n\pi$ are our new limits. With all these substitutions, Eq. 39-15 becomes

$$A^2 \frac{L}{n\pi} \int_0^{n\pi} (\sin^2 y)\, dy = 1.$$

We can use integral 11 in Appendix E to evaluate the integral, obtaining the equation

$$\frac{A^2 L}{n\pi}\left[\frac{y}{2} - \frac{\sin 2y}{4}\right]_0^{n\pi} = 1.$$

Evaluating at the limits yields

$$\frac{A^2 L}{n\pi}\frac{n\pi}{2} = 1;$$

thus

$$A = \sqrt{\frac{2}{L}}. \qquad \text{(Answer)} \quad (39\text{-}17)$$

This result tells us that the dimension for A^2, and thus for $\psi_n^2(x)$, is an inverse length. This is appropriate because the probability density of Eq. 39-12 is a probability *per unit length.*

Sample Problem 39-3

A ground-state electron is trapped in the one-dimensional infinite potential well of Fig. 39-2, with width $L = 100$ pm.

(a) What is the probability that the electron can be detected in the left one-third of the well ($x_1 = 0$ to $x_2 = L/3$)?

Solution: One **Key Idea** here is that if we probe the left one-third of the well, there is no guarantee that we will detect the electron. However, we can calculate the probability of detecting it with the integral of Eq. 39-13. A second **Key Idea** is that the probability very much depends on which state the electron is in—that is, the value of quantum number n. Because here the electron is in the ground state, we set $n = 1$ in Eq. 39-13.

We also set the limits of integration as the positions $x_1 = 0$ and $x_2 = L/3$ and, from Sample Problem 39-2, set the amplitude constant A as $\sqrt{2/L}$. We then see that

$$\begin{pmatrix}\text{probability of detection}\\ \text{in left one-third}\end{pmatrix} = \int_0^{L/3} \frac{2}{L}\sin^2\left(\frac{1\pi}{L}x\right) dx.$$

We could find this probability by substituting 100×10^{-12} m for L and then using a graphing calculator or a computer math package to evaluate the integral. Instead, we shall follow the steps of Sample Problem 39-2. From Eq. 39-16, we obtain for the new integration variable y,

$$y = \frac{\pi}{L}x \quad \text{and} \quad dx = \frac{L}{\pi}dy.$$

From the first of these equations, we find the new limits of integration to be $y_1 = 0$ for $x_1 = 0$ and $y_2 = \pi/3$ for $x_2 = L/3$. We then must evaluate

$$\text{probability} = \left(\frac{2}{L}\right)\left(\frac{L}{\pi}\right) \int_0^{\pi/3} (\sin^2 y)\, dy.$$

Using integral 11 in Appendix E, we then find

$$\text{probability} = \frac{2}{\pi}\left(\frac{y}{2} - \frac{\sin 2y}{4}\right)_0^{\pi/3} = 0.20.$$

Thus, we have

$$\begin{pmatrix}\text{probability of detection}\\ \text{in left one-third}\end{pmatrix} = 0.20. \qquad \text{(Answer)}$$

That is, if we repeatedly probe the left one-third of the well, then on average we can detect the electron with 20% of the probes.

(b) What is the probability that the electron can be detected in the middle one-third of the well (between $x_1 = L/3$ and $x_2 = 2L/3$)?

Solution: We now know that the probability of detection in the left one-third of the well is 0.20. A **Key Idea** here is that by symmetry, the probability of detection in the right one-third of the well is also 0.20. A second **Key Idea** is that because the electron is certainly in the well, the probability of detection in the entire well is 1. Thus, the probability of detection in the middle one-third of the well is

$$\begin{pmatrix}\text{probability of detection}\\ \text{in middle one-third}\end{pmatrix} = 1 - 0.20 - 0.20$$
$$= 0.60. \qquad \text{(Answer)}$$

39-5 An Electron in a Finite Well

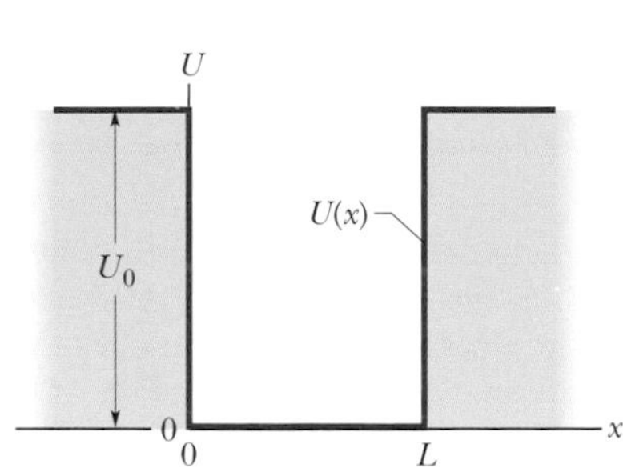

Fig. 39-7 A *finite* potential energy well. The depth of the well is U_0 and its width is L. As in the infinite potential well of Fig. 39-2, the motion of the trapped electron is re-

A potential energy well of infinite depth is an idealization. Figure 39-7 shows a realizable potential energy well—one in which the potential energy of an electron outside the well is not infinitely great but has a finite positive value U_0, called the **well depth.** The analogy between waves on a stretched string and matter waves fails us for wells of finite depth because we can no longer be sure that matter wave nodes exist at $x = 0$ and at $x = L$. (As we shall see, they don't.)

To find the wave functions describing the quantum states of an electron in the finite well of Fig. 39-7, we *must* resort to Schrödinger's equation, the basic equation of quantum physics. From Section 38-7 recall that, for motion in one dimension, we use Schrödinger's equation in the form of Eq. 38-15:

$$\frac{d^2\psi}{dx^2} + \frac{8\pi^2 m}{h^2}[E - U(x)]\psi = 0. \qquad (39\text{-}18)$$

Rather than attempting to solve this equation for the finite well, we simply state the results for particular numerical values of U_0 and L. Figure 39-8 shows these results as graphs of $\psi_n^2(x)$, the probability density, for a well with $U_0 = 450$ eV and $L = 100$ pm.

The probability density $\psi_n^2(x)$ for each graph in Fig. 39-8 satisfies Eq. 39-14, the normalization equation; so we know that the areas under all three probability density plots are numerically equal to 1.

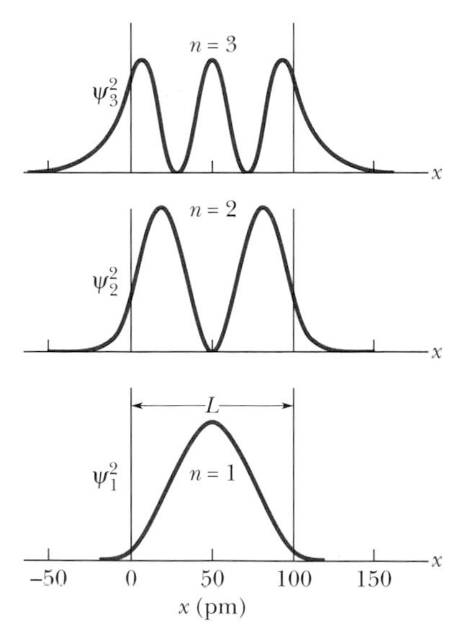

Fig. 39-8 Probability densities $\psi_n^2(x)$ for an electron confined to a finite potential well of depth $U_0 =$ 450 eV and width $L = 100$ pm. The only quantum states the electron can have in this well are those that have quantum numbers $n = 1, 2,$ and 3.

If you compare Fig. 39-8 for a finite well with Fig. 39-6 for an infinite well, you will see one striking difference: For a finite well, the electron matter wave penetrates the walls of the well—into a region in which Newtonian mechanics says the electron cannot exist. This penetration should not be surprising because we saw in Section 38-9 that an electron can tunnel through a potential energy barrier. "Leaking" into the walls of a finite potential energy well is a similar phenomenon. From the plots of ψ^2 in Fig. 39-8, we see that the leakage is greater for greater values of quantum number n.

Because a matter wave *does* leak into the walls of a finite well, the wavelength λ for any given quantum state is greater when the electron is trapped in a finite well than when it is trapped in an infinite well. Equation 39-3 ($\lambda = h/\sqrt{2mE}$) then tells us that the energy E for an electron in any given state is less in the finite well than in the infinite well.

That fact allows us to approximate the energy-level diagram for an electron trapped in a finite well. As an example, we can approximate the diagram for the finitc well of Fig. 39-8, which has width $L = 100$ pm and depth $U_0 = 450$ eV. The energy-level diagram for an *infinite* well of that width is shown in Fig. 39-3. First we remove the portion of Fig. 39-3 above 450 eV. Then we shift the remaining three energy levels down, shifting the level for $n = 3$ the most because the wave leakage into the walls is greatest for $n = 3$. The result is approximately the energy-level diagram for the finite well. The actual diagram is Fig. 39-9.

In that figure, an electron with an energy greater than U_0 (= 450 eV) has too much energy to be trapped in the finite well. Thus, it is not confined, and its energy is not quantized; that is, its energy is not restricted to certain values. To

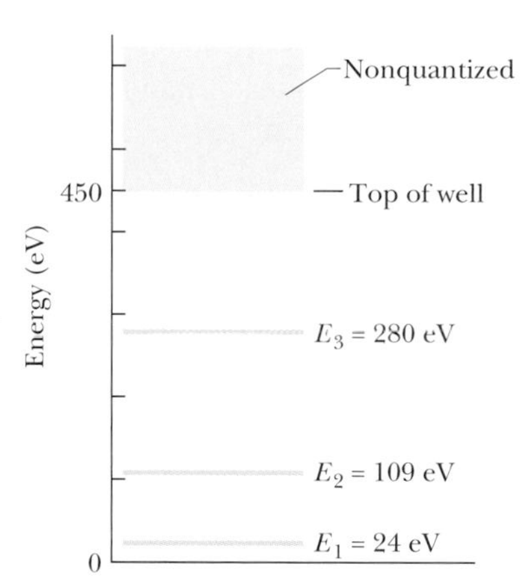

Fig. 39-9 The energy-level diagram corresponding to the probability densities of Fig. 39-8. If an electron is trapped in the finite potential well, it can have only the energies corresponding to $n = 1, 2$, and 3. If it has an energy of 450 eV or greater, it is not trapped and its energy is not quantized.

reach this *nonquantized* portion of the energy-level diagram and thus to be free, a trapped electron must somehow obtain enough energy to have a mechanical energy of 450 eV or greater.

Sample Problem 39-4

Suppose a finite well with $U_0 = 450$ eV and $L = 100$ pm confines a single electron in its ground state.

(a) What wavelength of light is needed to barely free the electron from the potential well if the electron absorbs a single photon from the light?

Solution: One **Key Idea** here is that for the electron to escape from the potential well, it must receive enough energy to put it into the nonquantized energy region of Fig. 39-9. Thus, it must end up with an energy of at least U_0 (= 450 eV). A second **Key Idea** is that the electron is initially in its ground state, with an energy of $E_1 = 24$ eV. Thus, to barely become free, it must receive an energy of

$$U_0 - E_1 = 450 \text{ eV} - 24 \text{ eV} = 426 \text{ eV}.$$

If it receives this energy from light, then it must absorb a photon with that much energy. From Eq. 39-6 ($hf = E_{\text{high}} - E_{\text{low}}$), with c/λ substituted for f, we can then write

$$\frac{hc}{\lambda} = U_0 - E_1,$$

from which we find

$$\lambda = \frac{hc}{U_0 - E_1} = \frac{(6.63 \times 10^{-34} \text{ J}\cdot\text{s})(3.00 \times 10^8 \text{ m/s})}{(426 \text{ eV})(1.60 \times 10^{-19} \text{ J/eV})}$$
$$= 2.92 \times 10^{-9} \text{ m} = 2.92 \text{ nm}. \quad \text{(Answer)}$$

Thus, if the electron absorbs a photon from light of wavelength 2.92 nm, it just barely escapes the potential well.

(b) Can the electron, initially in the ground state, absorb light with a wavelength of 2.00 nm? If so, what then is the electron's energy?

Solution: The **Key Ideas** here are these:

1. In (a) we found that light of 2.92 nm will just barely free the electron from the potential well.
2. We are now considering light with a shorter wavelength of 2.00 nm and thus a greater energy per photon ($hf = hc/\lambda$).
3. Hence, the electron *can* absorb a photon of this light. The energy transfer will not only free the electron but will also provide it with more kinetic energy. Further, because the electron is then no longer trapped, its energy is not quantized and thus there is no restriction on its kinetic energy.

The energy transferred to the electron is the photon energy:

$$hf = h\frac{c}{\lambda} = \frac{(6.63 \times 10^{-34} \text{ J}\cdot\text{s})(3.00 \times 10^8 \text{ m/s})}{2.00 \times 10^{-9} \text{ m}}$$
$$= 9.95 \times 10^{-17} \text{ J} = 622 \text{ eV}.$$

From (a), the energy required to just barely free the electron from the potential well is $U_0 - E_1$ (= 426 eV). The remainder of the 622 eV goes to kinetic energy. Thus, the kinetic energy of the freed electron is

$$K = hf - (U_0 - E_1)$$
$$= 622 \text{ eV} - 426 \text{ eV} = 196 \text{ eV}. \quad \text{(Answer)}$$

39-6 More Electron Traps

Here we discuss three types of artificial electron traps.

Nanocrystallites

Perhaps the most direct way to construct a potential energy well in the laboratory is to prepare a sample of a semiconducting material in the form of a powder

Fig. 39-10 Two samples of powdered cadmium selenide, a semiconductor, differing only in the size of their granules. Each granule serves as an electron trap. The lower sample has the larger granules and consequently the smaller spacing between energy levels and the lower photon energy threshold for the absorption of light. Light not absorbed is scattered, causing the sample to scatter light of greater wavelength and appear red. The upper sample, because of its smaller granules, and consequently its larger level spacing and its larger energy threshold for absorption, appears yellow.

whose granules are small—in the nanometer range—and of uniform size. Each such granule—each **nanocrystallite**—acts as a potential well for the electrons trapped within it.

Equation 39-4 ($E = (h^2/8mL^2)n^2$) shows that we can increase the energy-level values of an electron trapped in an infinite well by reducing the width L of the well. This would also shift the photon energies that the well can absorb to higher values and thus shift the corresponding wavelengths to shorter values.

These general results are also true for a well formed by a nanocrystallite. A given nanocrystallite can absorb photons with an energy above a certain threshold energy E_t ($= hf_t$) and thus wavelengths below a corresponding threshold wavelength

$$\lambda_t = \frac{c}{f_t} = \frac{ch}{E_t}. \tag{39-19}$$

Light with any wavelength longer than λ_t is scattered by the nanocrystallite instead of being absorbed. The color we attribute to the nanocrystallite is then determined by the wavelength composition of the scattered light we intercept.

If we reduce the size of the nanocrystallite, the value of E_t is increased, the value of λ_t is decreased, and the light that is scattered to us changes in its wavelength composition. Thus, the color we attribute to the nanocrystallite changes. As an example, Fig. 39-10 shows two samples of the semiconductor cadmium selenide, each consisting of a powder of nanocrystallites of uniform size. The lower sample scatters light at the red end of the spectrum. The upper sample differs from the lower sample *only* in that the upper sample is composed of smaller nanocrystallites. For this reason its threshold energy E_t is greater and, from Eq. 39-19, its threshold wavelength λ_t is shorter, in the green range of visible light. Thus, the sample now scatters both red and yellow. Because the yellow component happens to be brighter, the sample's color is now dominated by the yellow. The striking contrast in color between the two samples is compelling evidence of the quantization of the energies of trapped electrons and the dependence of these energies on the size of the electron trap.

Quantum Dots

The highly developed techniques used to fabricate computer chips can be used to construct, atom by atom, individual potential energy wells that behave, in many respects, like artificial atoms. These **quantum dots,** as they are usually called, have promising applications in electron optics and computer technology.

In one such arrangement, a "sandwich" is fabricated in which a thin layer of a semiconducting material, shown in purple in Fig. 39-11*a*, is deposited between two insulating layers, one of which is much thinner than the other. Metal end caps with conducting leads are added at both ends. The materials are chosen to

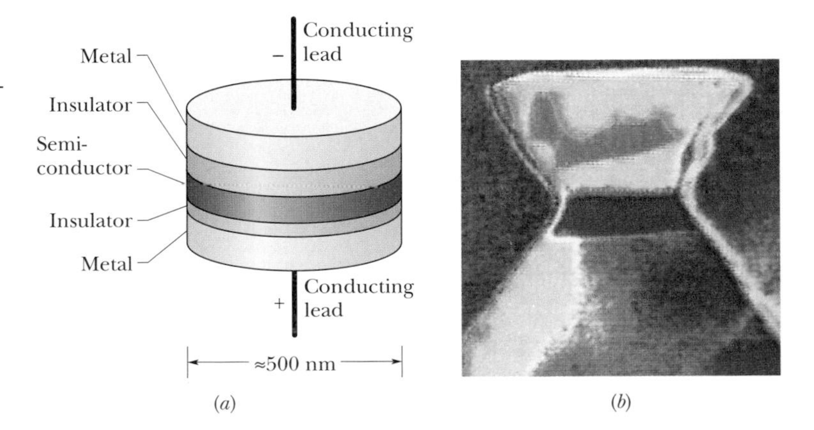

Fig. 39-11 A quantum dot, or "artificial atom." (*a*) A central semiconducting layer forms a potential energy well in which electrons are trapped. The lower insulating layer is thin enough to allow electrons to be added to or removed from the central layer by barrier tunneling if an appropriate voltage is applied between the leads. (*b*) A photograph of an actual quantum dot. The central purple band is the electron confinement region.

ensure that the potential energy of an electron in the central layer is less than it is in the two insulating layers, causing the central layer to act as a potential energy well. Figure 39-11*b* is a photograph of an actual quantum dot; the well in which individual electrons can be trapped is the purple region.

The lower (but not the upper) insulating layer in Fig. 39-11*a* is thin enough to permit electrons to tunnel through it if an appropriate potential difference is applied between the leads. In this way the number of electrons confined to the well can be controlled. The arrangement does indeed behave like an artificial atom with the property that the number of electrons it contains can be controlled. Quantum dots can be constructed in two-dimensional arrays that could well form the basis for computing systems of great speed and storage capacity.

Quantum Corrals

When a scanning tunneling microscope (described in Section 38-9) is in operation, its tip exerts a small force on isolated atoms that may be located on an otherwise smooth surface. By careful manipulation of the position of the tip, such isolated atoms can be "dragged" across the surface and deposited at another location. Using this technique, scientists at IBM's Almaden Research Center moved iron atoms across a carefully prepared copper surface, forming the atoms into a circle (Fig. 39-12), which they named a **quantum corral.** The result is shown in the photograph that opens this chapter. Each iron atom in the circle is nestled in a hollow in the copper surface, equidistant from three nearest-neighbor copper atoms. The corral was fabricated at a low temperature (about 4 K) to minimize the tendency of the iron atoms to move randomly about on the surface because of their thermal energy.

The ripples within the corral are due to matter waves associated with electrons that can move over the copper surface but are largely trapped in the potential well of the corral. The dimensions of the ripples are in excellent agreement with the predictions of quantum theory.

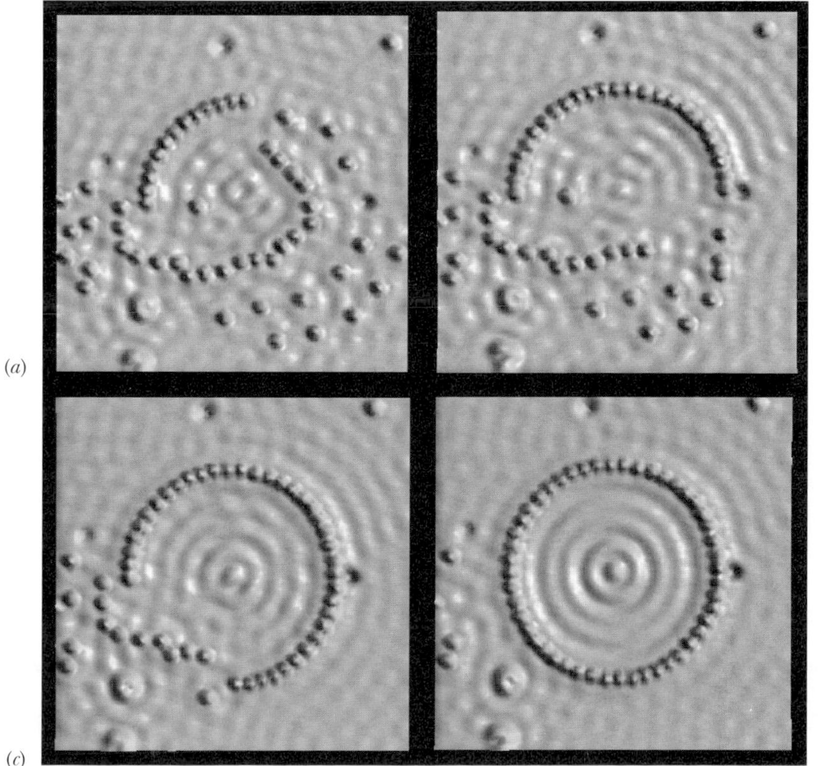

Fig. 39-12 A quantum corral during four stages of construction. Note the appearance of ripples caused by electrons trapped in the corral when it is almost complete.

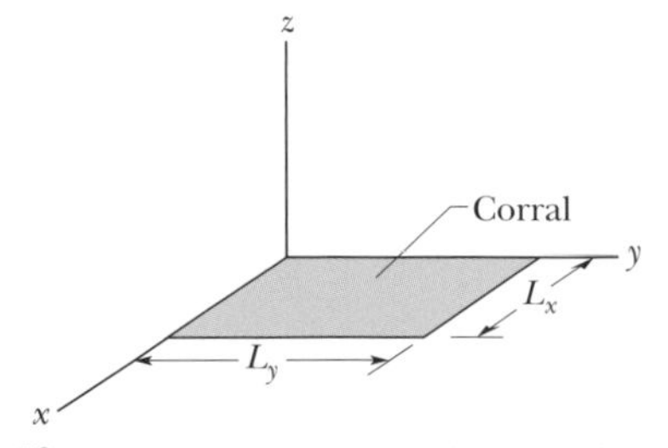

Fig. 39-13 A rectangular corral—a two-dimensional version of the infinite potential well of Fig. 39-2—with widths L_x and L_y.

39-7 Two- and Three-Dimensional Electron Traps

In the next section, we shall discuss the hydrogen atom as being a three-dimensional finite potential well. As a warm-up for the hydrogen atom, let us extend our discussion of infinite potential wells to two and three dimensions.

Rectangular Corral

Figure 39-13 shows the rectangular area to which an electron can be confined by the two-dimensional version of Fig. 39-2—a two-dimensional infinite potential well of widths L_x and L_y that forms a rectangular corral. The corral might be on the surface of a body that somehow prevents the electron from moving parallel to the z axis and thus from leaving the surface. You have to imagine infinite potential energy functions (like $U(x)$ in Fig. 39-2) along each side of the corral, keeping the electron within the corral.

Solution of Schrödinger's equation for the rectangular corral of Fig. 39-13 shows that, for the electron to be trapped, its matter wave must fit into each of the two widths separately, just as the matter wave of a trapped electron must fit into a one-dimensional infinite well. This means the wave is separately quantized in width L_x and in width L_y. Let n_x be the quantum number for which the matter wave fits into width L_x, and let n_y be the quantum number for which the matter wave fits into width L_y. As with a one-dimensional potential well, these quantum numbers can be only positive integers.

The energy of the electron depends on both quantum numbers and is the sum of the energy the electron would have if it were confined along the x axis alone and the energy it would have if it were confined along the y axis alone. From Eq. 39-4, we can write this sum as

$$E_{nx,ny} = \left(\frac{h^2}{8mL_x^2}\right)n_x^2 + \left(\frac{h^2}{8mL_y^2}\right)n_y^2 = \frac{h^2}{8m}\left(\frac{n_x^2}{L_x^2} + \frac{n_y^2}{L_y^2}\right). \tag{39-20}$$

Excitation of the electron by photon absorption and de-excitation of the electron by photon emission have the same requirements as for one-dimensional traps. The only major difference for the two-dimensional corral is that the energy of any given state depends on two quantum numbers (n_x and n_y) instead of just one (n). In general, different states (with different pairs of values for n_x and n_y) have different energies. However, in some situations, different states can have the same energy. Such states (and their energy levels) are said to be *degenerate.* Degenerate states cannot occur in a one-dimensional well.

Rectangular Box

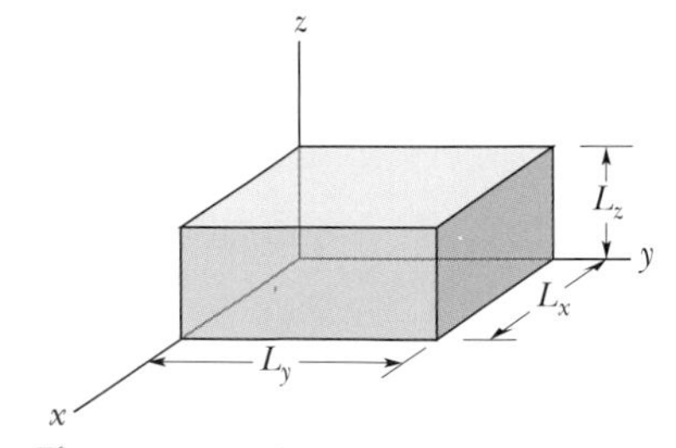

Fig. 39-14 A rectangular box—a three-dimensional version of the infinite potential well of Fig. 39-2—with widths L_x, L_y, and L_z.

An electron can also be trapped in a three-dimensional infinite potential well—a *box.* If the box is rectangular as in Fig. 39-14, then Schrödinger's equation shows us that we can write the energy of the electron as

$$E_{nx,ny,nz} = \frac{h^2}{8m}\left(\frac{n_x^2}{L_x^2} + \frac{n_y^2}{L_y^2} + \frac{n_z^2}{L_z^2}\right). \tag{39-21}$$

Here n_z is a third quantum number, for fitting the matter wave into width L_z.

CHECKPOINT 4 In the notation of Eq. 39-20, is $E_{0,0}$, $E_{1,0}$, $E_{0,1}$, or $E_{1,1}$ the ground-state energy of an electron in a (two-dimensional) rectangular corral?

Sample Problem 39-5

An electron is trapped in a square corral that is a two-dimensional infinite potential well (Fig. 39-13) with widths $L_x = L_y$.

(a) Find the energies of the lowest five possible energy levels for this trapped electron, and construct the corresponding energy-level diagram.

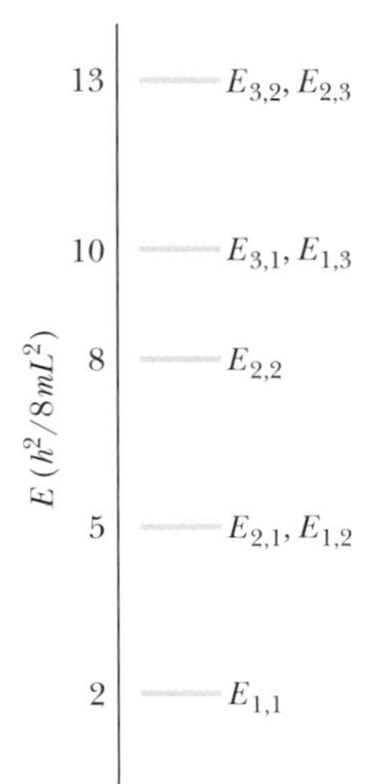

Fig. 39-15 Energy-level diagram for an electron trapped in a square corral.

Solution: The **Key Idea** here is that because the electron is trapped in a two-dimensional well that is rectangular, the electron's energy depends on two quantum numbers, n_x and n_y, according to Eq. 39-20. Because the well is square, we can let the widths be $L_x = L_y = L$. Then Eq. 39-20 simplifies to

$$E_{nx,ny} = \frac{h^2}{8mL^2}(n_x^2 + n_y^2). \tag{39-22}$$

The lowest energy states correspond to low values of the quantum numbers n_x and n_y, which are the positive integers $1, 2, \ldots, \infty$. Substituting those integers for n_x and n_y in Eq. 39-22, starting with the lowest value 1, we can obtain the energy values as listed in Table 39-1. There we can see that several of the pairs of quantum numbers (n_x, n_y) give the same energy. For example, the (1, 2) and (2, 1) states both have an energy of $5(h^2/8mL^2)$. Each such pair is associated with degenerate energy levels. Note also that, perhaps surprisingly, the (4, 1) and (1, 4) states have less energy than the (3, 3) state.

From Table 39-1 (carefully keeping track of degenerate levels), we can construct the energy-level diagram of Fig. 39-15.

(b) As a multiple of $h^2/8mL^2$, what is the energy difference between the ground state and the third excited state of the electron?

Solution: From Fig. 39-15, we see that the ground state is the (1, 1) state, with an energy of $2(h^2/8mL^2)$. We also see that the third excited state (the third state up from the ground state in the energy-level diagram) is the degenerate (1, 3) and (3, 1) states, with an energy of $10(h^2/8mL^2)$. Thus, the difference ΔE between these two states is

$$\Delta E = 10\left(\frac{h^2}{8mL^2}\right) - 2\left(\frac{h^2}{8mL^2}\right) = 8\left(\frac{h^2}{8mL^2}\right). \quad \text{(Answer)}$$

TABLE 39-1 Energy Levels

n_x	n_y	Energy*	n_x	n_y	Energy*
1	3	10	2	4	20
3	1	10	4	2	20
2	2	8	3	3	18
1	2	5	1	4	17
2	1	5	4	1	17
1	1	2	2	3	13
			3	2	13

*In multiples of $h^2/8mL^2$

39-8 The Bohr Model of the Hydrogen Atom

We now move from artificial and fictitious electron traps to natural ones—atoms. In this chapter we focus on the simplest example, a hydrogen atom. This atom consists of a single electron electrically bound to a proton, which is the only constituent of the atom's nucleus at the atom's center. Here we do not consider anything about the nucleus. Rather, we simply use the fact that the negatively charged electron is attracted by the Coulomb force to the positively charged proton. Because the proton mass is much greater than the electron mass, we assume that the proton is fixed in place and that the electron is confined to being near the proton. That is, the electron is trapped.

We have now discussed at length that confinement of an electron means that the electron's energy E is quantized and thus so is any change ΔE in its energy. In this section we want to calculate the quantized energies of the electron in a hydrogen atom. Before we apply the wave approach we used in infinite and finite potential wells, however, let's explore the hydrogen atom at the dawn of quantum physics, when physicists first discovered that atoms are quantized systems.

By the early 1900s, scientists understood that matter came in tiny pieces called atoms and that an atom of hydrogen contained positive charge $+e$ at its center and negative charge $-e$ (an electron) outside that center. However, no one understood why the electrical attraction between the electron and the positive charge did not simply cause the two to collapse together.

One clue came from the fact that a hydrogen atom cannot emit and absorb all wavelengths of visible light. Rather, it can emit and absorb only four particular

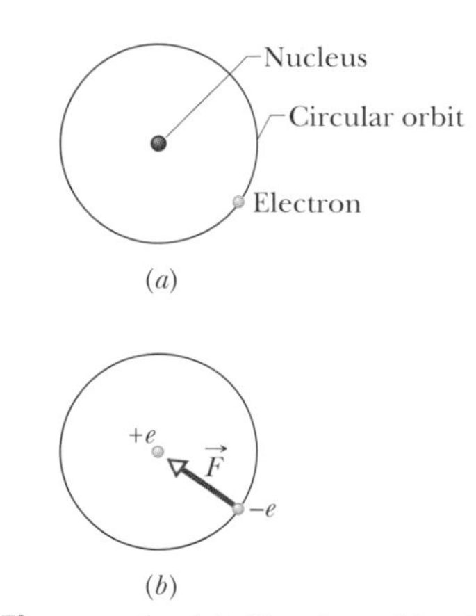

Fig. 39-16 (*a*) Circular orbit of an electron in the Bohr model of the hydrogen atom. (*b*) The Coulomb force $\vec{F}$ on the electron is directed radially inward toward the nucleus.

wavelengths from the visible range. By guesswork, Johann Balmer devised a formula that gave those wavelengths:

$$\frac{1}{\lambda} = R\left(\frac{1}{2^2} - \frac{1}{n^2}\right), \qquad \text{for } n = 3, 4, 5, \text{ and } 6, \tag{39-23}$$

where R is a constant. However, neither Balmer nor anyone else knew why the visible emission and absorption by a hydrogen atom are limited to the four wavelengths given by this formula. Why doesn't a hydrogen atom emit and absorb light at *any* visible wavelength?

No one know until 1913, when Bohr saw Balmer's equation and quickly realized that he could derive it if he made several bold (completely unjustified) assumptions: (1) The electron in a hydrogen atom orbits the nucleus in a circle much like Earth orbits the Sun (Fig. 39-16*a*). (2) The magnitude of the angular momentum $\vec{L}$ of the electron in its orbit is restricted to the values

$$L = n\hbar, \qquad \text{for } n = 1, 2, 3, \ldots, \tag{39-24}$$

where $\hbar$ (h-bar) is $h/2\pi$ and n is a quantum number. Let's see what results from Bohr's two assumptions.

The Orbital Radius Is Quantized in the Bohr Model

Let's examine the orbital motion of the electron in the Bohr model. The force holding the electron in an orbit of radius r is the Coulomb force. From Eq. 21-1, we know that the magnitude of this force is

$$F = k\frac{|q_1||q_2|}{r^2},$$

with $k = 1/4\pi\varepsilon_0$. Here q_1 is the charge $-e$ of the electron and q_2 is the charge $+e$ of the nucleus (the proton). The electron's acceleration is the centripetal acceleration, with a magnitude given by $a = v^2/r$, where v is the electron's speed. Both force $\vec{F}$ and acceleration $\vec{a}$ are radially inward (the negative direction on a radial axis), toward the nucleus (Fig. 39-16*b*). Thus, we can write Newton's second law ($F = ma$) for a radial axis as

$$-\frac{1}{4\pi\varepsilon_0}\frac{e^2}{r^2} = m\left(-\frac{v^2}{r}\right), \tag{39-25}$$

where m is the electron mass.

We next introduce quantization by using Bohr's assumption expressed in Eq. 39-24. From Eq. 11-19, the magnitude ℓ of the angular momentum of a particle of mass m and speed v moving in a circle of radius r is $\ell = rmv \sin\phi$, where ϕ (the angle between $\vec{r}$ and $\vec{v}$) is 90°. Replacing L in Eq. 39-24 with $rmv \sin 90°$ gives us

$$rmv = n\hbar,$$

or

$$v = \frac{n\hbar}{rm}. \tag{39-26}$$

Substituting this equation into Eq. 39-25, replacing $\hbar$ with $h/2\pi$, and rearranging, we find

$$r = \frac{h^2\varepsilon_0}{\pi m e^2}n^2, \qquad \text{for } n = 1, 2, 3, \ldots. \tag{39-27}$$

We can rewrite this as

$$r = an^2, \qquad \text{for } n = 1, 2, 3, \ldots, \tag{39-28}$$

where

$$a = \frac{h^2\varepsilon_0}{\pi m e^2} = 5.291\,772 \times 10^{-11}\text{ m} \approx 52.92\text{ pm}. \tag{39-29}$$

These last three equations tell us that, in the *Bohr model of the hydrogen atom*, the electron's orbital radius r is quantized and the smallest possible orbital radius (for $n = 1$) is a, which is now called the *Bohr radius*. According to the Bohr model, the electron cannot get any closer to the nucleus than orbital radius a, and that is why the attraction between electron and nucleus does not simply collapse them together.

Orbital Energy Is Quantized

Let's next find the energy of the hydrogen atom according to the Bohr model. The electron has kinetic energy $K = \frac{1}{2}mv^2$, and the electron–nucleus system has electric potential energy $U = q_1q_2/4\pi\varepsilon_0 r$ (Eq. 24-43). Again, let q_1 be the electron's charge $-e$ and q_2 be the nuclear charge $+e$. Then the mechanical energy is

$$E = K + U = \tfrac{1}{2}mv^2 + \left(-\frac{1}{4\pi\varepsilon_0}\frac{e^2}{r}\right). \tag{39-30}$$

Solving Eq. 39-25 for mv^2 and substituting the result in Eq. 39-30 lead to

$$E = -\frac{1}{8\pi\varepsilon_0}\frac{e^2}{r}. \tag{39-31}$$

Next, replacing r with its equivalent from Eq. 39-27, we have

$$E_n = -\frac{me^4}{8\varepsilon_0^2 h^2}\frac{1}{n^2}, \qquad \text{for } n = 1, 2, 3, \ldots, \tag{39-32}$$

where the subscript n on E signals that we have now quantized the energy. Evaluating the constants in Eq. 39-32 gives us

$$E_n = -\frac{2.180 \times 10^{-18}\ \text{J}}{n^2} = -\frac{13.60\ \text{eV}}{n^2}, \qquad \text{for } n = 1, 2, 3, \ldots, \tag{39-33}$$

This equation tells us that the energy E_n of the hydrogen atom is quantized; that is, E_n is restricted by its dependence on the quantum number n. Because the nucleus is assumed to be fixed in place and only the electron has motion, we can assign the energy values of Eq. 39-33 either to the atom as a whole or to the electron alone.

Energy Changes

The energy of a hydrogen atom (or, equivalently, of its electron) changes when the atom emits or absorbs light. As we have seen several times since Eq. 39-6, emission and absorption involve a quantum of light according to

$$hf = \Delta E = E_{\text{high}} - E_{\text{low}}. \tag{39-34}$$

Let's make three changes to Eq. 39-34. On the left side, we substitute c/λ for f. On the right side, we use Eq. 39-32 twice to replace the energy terms. Then, with a simple rearrangement, we have

$$\frac{1}{\lambda} = -\frac{me^4}{8\varepsilon_0^2 h^3 c}\left(\frac{1}{n_{\text{high}}^2} - \frac{1}{n_{\text{low}}^2}\right). \tag{39-35}$$

We can rewrite this as

$$\frac{1}{\lambda} = R\left(\frac{1}{n_{\text{low}}^2} - \frac{1}{n_{\text{high}}^2}\right), \tag{39-36}$$

in which

$$R = \frac{me^4}{8\varepsilon_0^2 h^3 c} = 1.097\,373 \times 10^7\ \mathrm{m^{-1}} \tag{39-37}$$

is now known as the *Rydberg constant.*

Compare Eq. 39-36 from the Bohr model with Eq. 39-23 from Balmer's work. In Eq. 39-36, if we replace n_{low} with 2 and then restrict n_{high} to be 3, 4, 5, and 6, we have Balmer's equation. This match was a triumph for Bohr and ushered in the quantum physics of atoms. The triumph was short-lived, however, because even though the Bohr model gives the correct emission and absorption wavelengths for the hydrogen atom, the model is *not* correct because the electron does *not* orbit the nucleus like a planet orbiting the Sun. Indeed, researchers found little success in extending the Bohr model to atoms more complicated than hydrogen. The reason for this lack of success is that an electron trapped in any atom is a matter wave confined to a potential well, and to find the resulting quantized energy values we must apply Schrödinger's equation to the electron.

39-9 Schrödinger's Equation and the Hydrogen Atom

The potential well of a hydrogen atom depends on the electrical potential energy function

$$U(r) = \frac{-e^2}{4\pi\varepsilon_0 r}. \tag{39-38}$$

Because this well is three-dimensional, it is more complex than our previous one- and two-dimensional wells. Because this well is finite, it is more complex than the three-dimensional well of Fig. 39-14. Moreover, it does not have sharply defined walls. Rather, its walls vary in depth with radial distance r. Figure 39-17 is probably the best we can do in drawing the hydrogen potential well, but even that drawing takes much effort to interpret.

Energy Levels and Spectra of the Hydrogen Atom

Although we shall not do so here, we can apply Schrödinger's equation for an electron trapped in the potential well given by Eq. 39-38. In doing so, we would find that the energy values are quantized and that, amazingly, those values are given by Eq. 39-33 just as for the (incorrect) Bohr model. Thus, changes ΔE in energy due to emission or absorption of light are given by Eq. 39-34, and the wavelengths corresponding to ΔE are given by Eq. 39-36. Let's explore these results.

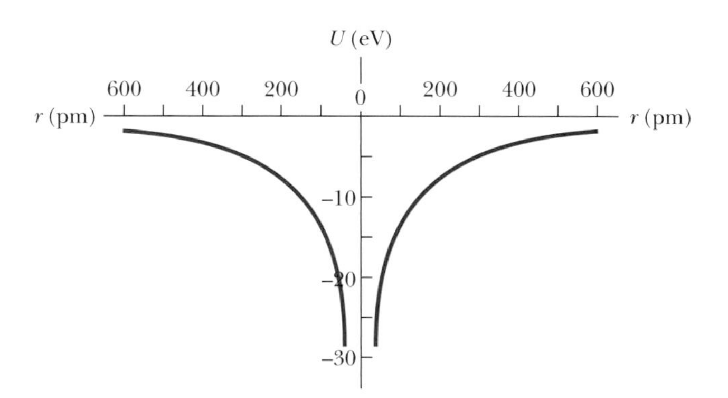

Fig. 39-17 The potential energy U of a hydrogen atom as a function of the separation r between the electron and the central proton. The plot is shown twice (on the left and on the right) to suggest the three-dimensional spherically symmetric trap in which the electron is confined.

Figure 39-18 shows the energy levels corresponding to various values of n in Eq. 39-33. The lowest level, for $n = 1$, is the ground state of hydrogen. Higher levels correspond to excited states, just as we saw for our simpler potential traps. Note several differences, however. (1) The energy levels now have negative values rather than the positive values we previously chose in, for instance, Figs. 39-3 and 39-9. (2) The levels now become progressively closer as we move to higher levels. (3) The energy for the greatest value of n—namely, $n = \infty$—is now $E_\infty = 0$. For any energy greater than $E_\infty = 0$, the electron and proton are not bound together (there is no hydrogen atom), and the $E > 0$ region in Fig. 39-18 is like the nonquantized region for the finite well of Fig. 39-9.

A hydrogen atom can jump between quantized energy levels by emitting or absorbing light at the wavelengths given by Eq. 39-36. Any such wavelength is often called a *line* because of the way it is detected with a spectroscope; thus, a hydrogen atom has *absorption lines* and *emission lines.* A collection of such lines, such as in those in the visible range, is called a **spectrum** of the hydrogen atom.

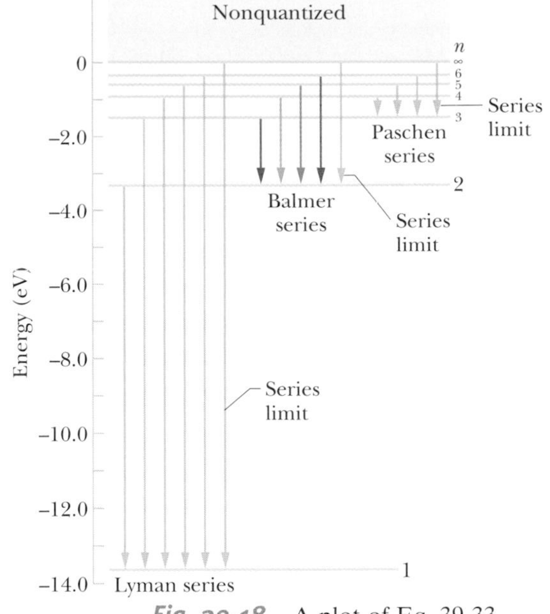

Fig. 39-18 A plot of Eq. 39-33, showing a few of the energy levels of the hydrogen atom. The transitions are grouped into series, each labeled with the name of a person.

The lines for hydrogen are said to be grouped into *series,* according to the level at which upward jumps start and downward jumps end. For example, the emission and absorption lines for all possible jumps up from the $n = 1$ level and down to the $n = 1$ level are said to be in the *Lyman series,* named after the person who first studied those lines. Further, we can say that the Lyman series has a *home-base level* of $n = 1$. Similarly, the *Balmer series* has a home-base level of $n = 2$, and the *Paschen series* has a home-base level of $n = 3$.

Some of the downward quantum jumps for these three series are shown in Fig. 39-18. Four lines in the Balmer series are in the visible range and are the ones Balmer studied. They are represented in Fig. 39-18 with arrows corresponding to their colors. The shortest of those arrows represents the shortest jump in the series, from the $n = 3$ level to the $n = 2$ level. Thus, that jump involves the smallest change in the electron's energy and the smallest amount of emitted photon energy for the series. The emitted light is red. The next jump in the series, from $n = 4$ to $n = 2$, is longer, the photon energy is greater, the wavelength of the emitted light is shorter, and the light is green. The third, fourth, and fifth arrows represent longer jumps and shorter wavelengths. For the fifth jump, the emitted light is in the ultraviolet range and thus is not visible.

The *series limit* of a series is the line produced by the jump between the home-base level and the highest energy level, which is the level with quantum number $n = \infty$. Thus, the series limit is the shortest wavelength in the series. Figure 39-19 is a photograph of the Balmer emission lines taken with a spectroscope (as in Figs. 36-23 and 36-24). The series limit for the series is marked with a small triangle.

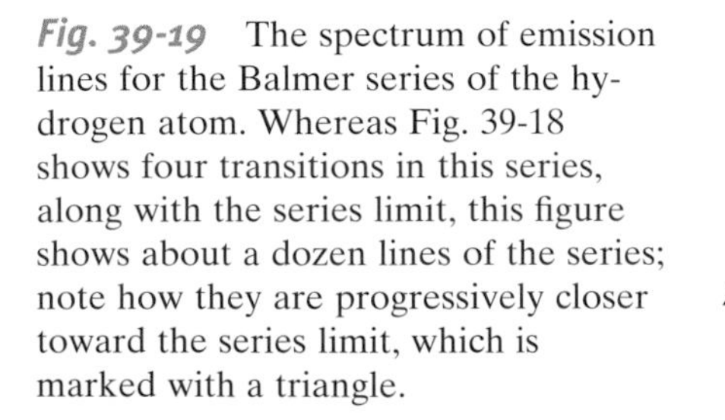

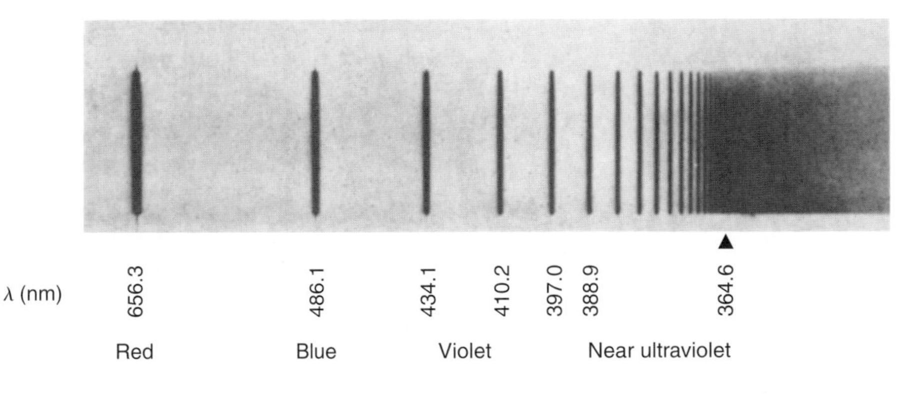

Fig. 39-19 The spectrum of emission lines for the Balmer series of the hydrogen atom. Whereas Fig. 39-18 shows four transitions in this series, along with the series limit, this figure shows about a dozen lines of the series; note how they are progressively closer toward the series limit, which is marked with a triangle.

If a jump is upward into the nonquantized portion of Fig. 39-18, the electron's energy is no longer given by Eq. 39-33 because the electron is no longer trapped in the atom. That is, the hydrogen atom has been *ionized*, meaning that the electron has been removed to a distance so great that the Coulomb force on it from the nucleus is negligible. The atom can be ionized if it absorbs any wavelength greater than the series limit. The free electron then has only kinetic energy K ($= \frac{1}{2}mv^2$, assuming a nonrelativistic situation).

Quantum Numbers for the Hydrogen Atom

Although the energies of the hydrogen atom states can be described by the single quantum number n, the wave functions describing these states require three quantum numbers, corresponding to the three dimensions in which the electron can move. The three quantum numbers, along with their names and the values that they may have, are shown in Table 39-2.

Each set of quantum numbers (n, ℓ, m_ℓ) identifies the wave function of a particular quantum state. The quantum number n, called the **principal quantum number,** appears in Eq. 39-33 for the energy of the state. The **orbital quantum number** ℓ is a measure of the magnitude of the angular momentum associated with the quantum state. The **orbital magnetic quantum number** m_ℓ is related to the orientation in space of this angular momentum vector. The restrictions on the values of the quantum numbers for the hydrogen atom, as listed in Table 39-2, are not arbitrary but come out of the solution to Schrödinger's equation. Note that for the ground state ($n = 1$), the restrictions require that $\ell = 0$ and $m_\ell = 0$. That is, the hydrogen atom in its ground state has zero angular momentum, which is not predicted by Eq. 39-24 in the Bohr model.

CHECKPOINT 5 (a) A group of quantum states of the hydrogen atom has $n = 5$. How many values of ℓ are possible for states in this group? (b) A subgroup of hydrogen atom states in the $n = 5$ group has $\ell = 3$. How many values of m_ℓ are possible for states in this subgroup?

The Wave Function of the Hydrogen Atom's Ground State

The wave function for the ground state of the hydrogen atom, as obtained by solving the three-dimensional Schrödinger equation and normalizing the result, is

$$\psi(r) = \frac{1}{\sqrt{\pi}a^{3/2}} e^{-r/a} \quad \text{(ground state)}, \tag{39-39}$$

where a is the Bohr radius (Eq. 39-29). This radius is loosely taken to be the effective radius of a hydrogen atom and turns out to be a convenient unit of length for other situations involving atomic dimensions.

As with other wave functions, $\psi(r)$ in Eq. 39-39 does not have physical meaning but $\psi^2(r)$ does, being the probability density—the probability per unit volume—that the electron can be detected. Specifically, $\psi^2(r)\,dV$ is the probability that the electron can be detected in any given (infinitesimal) volume element dV

TABLE 39-2
Quantum Numbers for the Hydrogen Atom

Symbol	Name	Allowed Values
n	Principal quantum number	$1, 2, 3, \ldots$
ℓ	Orbital quantum number	$0, 1, 2, \ldots, n - 1$
m_ℓ	Orbital magnetic quantum number	$-\ell, -(\ell - 1), \ldots, +(\ell - 1), +\ell$

located at radius r from the center of the atom:

$$\begin{pmatrix}\text{probability of detection}\\ \text{in volume } dV\\ \text{at radius } r\end{pmatrix} = \begin{pmatrix}\text{volume probability}\\ \text{density } \psi^2(r)\\ \text{at radius } r\end{pmatrix}(\text{volume } dV). \quad (39\text{-}40)$$

Because $\psi^2(r)$ here depends only on r, it makes sense to choose, as a volume element dV, the volume between two concentric spherical shells whose radii are r and $r + dr$. That is, we take the volume element dV to be

$$dV = (4\pi r^2)\,dr, \quad (39\text{-}41)$$

in which $4\pi r^2$ is the surface area of the inner shell and dr is the radial distance between the two shells. Then, combining Eqs. 39-39, 39-40, and 39-41 gives us

$$\begin{pmatrix}\text{probability of detection}\\ \text{in volume } dV\\ \text{at radius } r\end{pmatrix} = \psi^2(r)\,dV = \frac{4}{a^3}e^{-2r/a}r^2\,dr. \quad (39\text{-}42)$$

Describing the probability of detecting an electron is easier if we work with a **radial probability density** $P(r)$ instead of a volume probability density $\psi^2(r)$. This $P(r)$ is a linear probability density such that

$$\begin{pmatrix}\text{radial probability}\\ \text{density } P(r)\\ \text{at radius } r\end{pmatrix}\begin{pmatrix}\text{radial}\\ \text{width } dr\end{pmatrix} = \begin{pmatrix}\text{volume probability}\\ \text{density } \psi^2(r)\\ \text{at radius } r\end{pmatrix}(\text{volume } dV)$$

or

$$P(r)\,dr = \psi^2(r)\,dV. \quad (39\text{-}43)$$

Substituting for $\psi^2(r)\,dV$ from Eq. 39-42, we obtain

$$P(r) = \frac{4}{a^3}r^2e^{-2r/a} \quad \text{(radial probability density, hydrogen atom ground state).} \quad (39\text{-}44)$$

Figure 39-20 is a plot of Eq. 39-44. The area under the plot is unity; that is,

$$\int_0^\infty P(r)\,dr = 1. \quad (39\text{-}45)$$

This equation states that in a hydrogen atom, the electron must be *somewhere* in the space surrounding the nucleus.

The triangular marker on the horizontal axis of Fig. 39-20 is located one Bohr radius from the origin. The graph tells us that in the ground state of the hydrogen atom, the electron is most likely to be found at about this distance from the center of the atom.

Figure 39-20 conflicts sharply with the popular view that electrons in atoms follow well-defined orbits like planets moving around the Sun. *This popular view, however familiar, is incorrect.* Figure 39-20 shows us all that we can ever know about the location of the electron in the ground state of the hydrogen atom. The appropriate question is not "When will the electron arrive at such-and-such a point?" but "What are the odds that the electron will be detected in a small volume centered on such-and-such a point?" Figure 39-21, which we call a dot plot, suggests the probabilistic nature of the wave function and provides a useful mental model of the hydrogen atom in its ground state. Think of the atom in this state as a fuzzy ball with no sharply defined boundary and no hint of orbits.

It is not easy for a beginner to envision subatomic particles in this probabilistic way. The difficulty is our natural impulse to regard an electron as something like a tiny jelly bean, located at certain places at certain times and following a well-defined path. Electrons and other subatomic particles simply do not behave in this way.

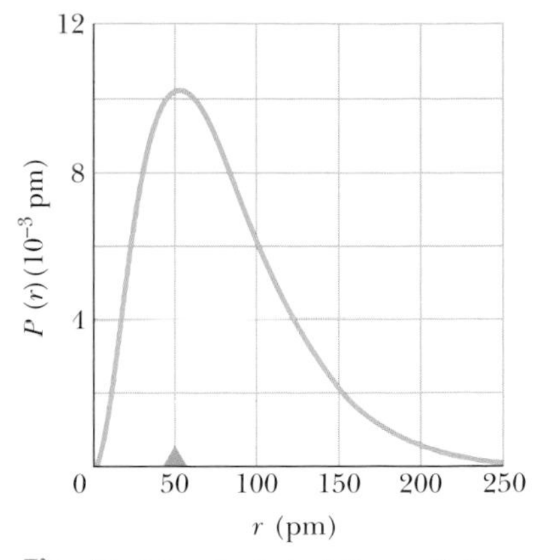

Fig. 39-20 A plot of the radial probability density $P(r)$ for the ground state of the hydrogen atom. The triangular marker is located at one Bohr radius from the origin, and the origin represents the center of the atom.

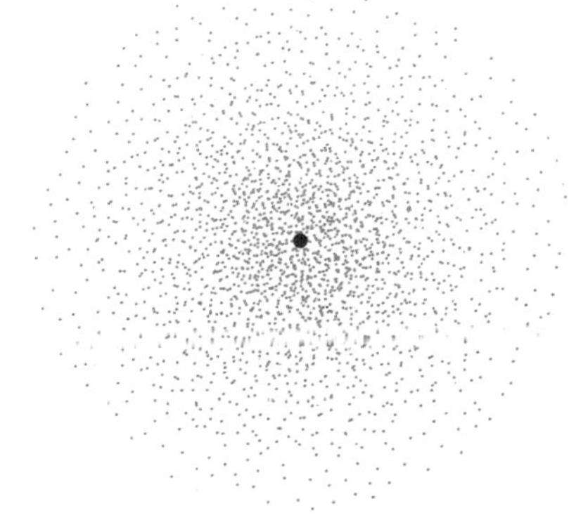

Fig. 39-21 A "dot plot" showing the volume probability density $\psi^2(r)$—not the *radial* probability density $P(r)$—for the ground state of the hydrogen atom. The density of dots drops exponentially with increasing distance from the nucleus, which is represented here by a red spot. Such dot plots provide a mental image of the "electron cloud" of an atom.

The energy of the ground state, found by putting $n = 1$ in Eq. 39-33, is $E_1 = -13.60$ eV. The wave function of Eq. 39-39 results if you solve Schrödinger's equation with this value of the energy. Actually, you can find a solution of Schrödinger's equation for *any* value of the energy—say, $E = -11.6$ eV or -14.3 eV. This may suggest that the energies of the hydrogen atom states are not quantized—but we know that they are.

The puzzle was solved when physicists realized that such solutions of Schrödinger's equation are not physically acceptable because they yield increasingly large values as $r \to \infty$. These "wave functions" tell us that the electron is more likely to be found very far from the nucleus rather than closer to it, which makes no sense. We get rid of these unwanted solutions by imposing what is called a **boundary condition,** in which we agree to accept only solutions of Schrödinger's equation for which $\psi(r) \to 0$ as $r \to \infty$; that is, we agree to deal only with *confined* electrons. With this restriction, the solutions of Schrödinger's equation form a discrete set, with quantized energies given by Eq. 39-33.

Sample Problem 39-6

(a) What is the wavelength of light for the least energetic photon emitted in the Lyman series of the hydrogen atom spectrum lines?

Solution: One **Key Idea** here is that for any series, the transition that produces the least energetic photon is the transition between the home-base level that defines the series and the level immediately above it. A second **Key Idea** is that for the Lyman series, the home-base level is at $n = 1$ (Fig. 39-18). Thus, the transition that produces the least energetic photon is the transition from the $n = 2$ level to the $n = 1$ level. From Eq. 39-33 the energy difference is

$$\Delta E = E_2 - E_1 = -(13.60 \text{ eV})\left(\frac{1}{2^2} - \frac{1}{1^2}\right) = 10.20 \text{ eV}.$$

Then from Eq. 39-6 ($\Delta E = hf$), with c/λ replacing f, we have

$$\lambda = \frac{hc}{\Delta E} = \frac{(6.63 \times 10^{-34} \text{ J}\cdot\text{s})(3.00 \times 10^8 \text{ m/s})}{(10.20 \text{ eV})(1.60 \times 10^{-19} \text{ J/eV})}$$
$$= 1.22 \times 10^{-7} \text{ m} = 122 \text{ nm}. \qquad \text{(Answer)}$$

Light with this wavelength is in the ultraviolet range.

(b) What is the wavelength of the series limit for the Lyman series?

Solution: The **Key Idea** here is that the series limit corresponds to a jump between the home-base level ($n = 1$ for the Lyman series) and the level at the limit $n = \infty$. Now that we have identified the values of n for the transition, we could proceed as in (a) to find the corresponding wavelength λ. Instead, let's use a more direct procedure. From Eq. 39-36, we find

$$\frac{1}{\lambda} = R\left(\frac{1}{n_{\text{low}}^2} - \frac{1}{n_{\text{high}}^2}\right) = 1.097\,373 \times 10^7 \text{ m}^{-1})\left(\frac{1}{1^2} - \frac{1}{\infty^2}\right),$$

which yields

$$\lambda = 9.11 \times 10^{-8} \text{ m} = 91.1 \text{ nm}. \qquad \text{(Answer)}$$

Light with this wavelength is also in the ultraviolet range.

Sample Problem 39-7

Show that the radial probability density for the ground state of the hydrogen atom has a maximum at $r = a$.

Solution: One **Key Idea** here is that the radial probability density for a ground-state hydrogen atom is given by Eq. 39-44,

$$P(r) = \frac{4}{a^3} r^2 e^{-2r/a}.$$

A second **Key Idea** is that to find the maximum (or minimum) of any function, we must differentiate the function and set the result equal to zero. If we differentiate $P(r)$ with respect to r, using derivative 7 of Appendix E and the chain rule for differentiating products, we get

$$\frac{dP}{dr} = \frac{4}{a^3} r^2 \left(\frac{-2}{a}\right) e^{-2r/a} + \frac{4}{a^3} 2r\, e^{-2r/a}$$
$$= \frac{8r}{a^3} e^{-2r/a} - \frac{8r^2}{a^4} e^{-2r/a}$$
$$= \frac{8}{a^4} r(a - r) e^{-2r/a}.$$

If we set the right side equal to zero, we obtain an equation that is true if $r = a$. In other words, dP/dr is equal to zero when $r = a$. (Note that we also have $dP/dr = 0$ at $r = 0$ and at $r = \infty$. However, these conditions correspond to a *minimum* in $P(r)$, as you can see in Fig. 39-20.)

Sample Problem 39-8

It can be shown that the probability $p(r)$ that the electron in the ground state of the hydrogen atom will be detected inside a sphere of radius r is given by

$$p(r) = 1 - e^{-2x}(1 + 2x + 2x^2),$$

in which x, a dimensionless quantity, is equal to r/a. Find r for $p(r) = 0.90$.

Solution: The **Key Idea** here is that there is no guarantee of detecting the electron at any particular radial distance r from the center of the hydrogen atom. However, with the given function, we can calculate the probability that the electron will be detected *somewhere* within a sphere of radius r. We seek the radius of a sphere for which $p(r) = 0.90$. Substituting that value in the expression for $p(r)$, we have

$$0.90 = 1 - e^{-2x}(1 + 2x + 2x^2)$$

or

$$10e^{-2x}(1 + 2x + 2x^2) = 1.$$

We must find the value of x that satisfies this equality. It is not possible to solve explicitly for x, but an equation solver on a calculator yields $x = 2.66$. This means that the radius of a sphere within which the electron will be detected 90% of the time is $2.66a$. Mark this position on the horizontal axis of Fig. 39-20—is it a reasonable answer?

Hydrogen Atom States with $n = 2$

According to the requirements of Table 39-2, there are four states of the hydrogen atom with $n = 2$; their quantum numbers are listed in Table 39-3. Consider first the state with $n = 2$ and $\ell = m_\ell = 0$; its probability density is represented by the dot plot of Fig. 39-22. Note that this plot, like the plot for the ground state shown in Fig. 39-21, is spherically symmetric. That is, in a spherical coordinate system like that defined in Fig. 39-23, the probability density is a function of the radial coordinate r only and is independent of the angular coordinates θ and ϕ.

It turns out that all quantum states with $\ell = 0$ have spherically symmetric wave functions. This is reasonable because the quantum number ℓ is a measure of the angular momentum associated with a given state. If $\ell = 0$, the angular momentum is also zero, which requires that the probability density representing the state have no preferred axis of symmetry.

Dot plots of ψ^2 for the three states with $n = 2$ and $\ell = 1$ are shown in Fig. 39-24. The probability densities for the states with $m_\ell = +1$ and $m_\ell = -1$ are identical. Although these plots are symmetric about the z axis, they are *not* spherically symmetric. That is, the probability densities for these three states are functions of both r and the angular coordinate θ.

Here is a puzzle: What is there about the hydrogen atom that establishes the axis of symmetry that is so obvious in Fig. 39-24? The answer: *absolutely nothing.*

TABLE 39-3
Quantum Numbers for Hydrogen Atom States with $n = 2$

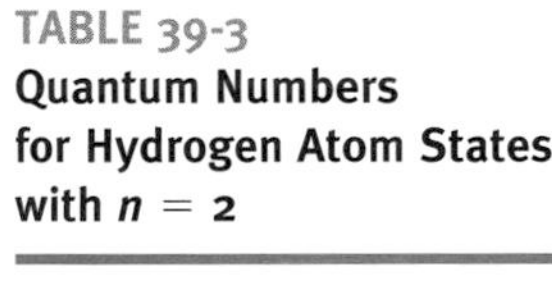

n	ℓ	m_ℓ
2	0	0
2	1	+1
2	1	0
2	1	−1

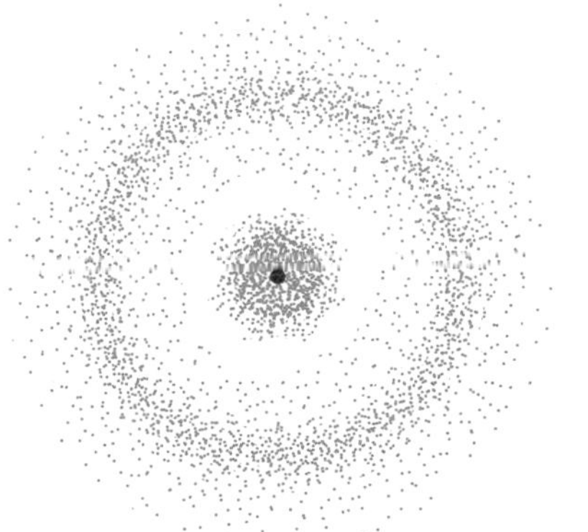

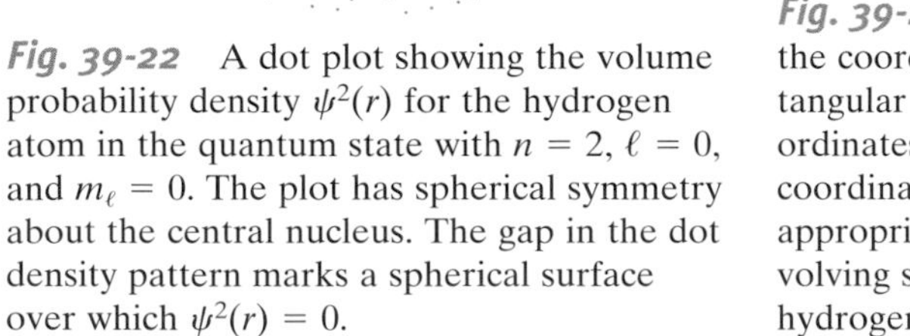

Fig. 39-22 A dot plot showing the volume probability density $\psi^2(r)$ for the hydrogen atom in the quantum state with $n = 2$, $\ell = 0$, and $m_\ell = 0$. The plot has spherical symmetry about the central nucleus. The gap in the dot density pattern marks a spherical surface over which $\psi^2(r) = 0$.

Fig. 39-23 The relationship between the coordinates x, y, and z of the rectangular coordinate system and the coordinates r, θ, and ϕ of the spherical coordinate system. The latter are more appropriate for analyzing situations involving spherical symmetry, such as the hydrogen atom.

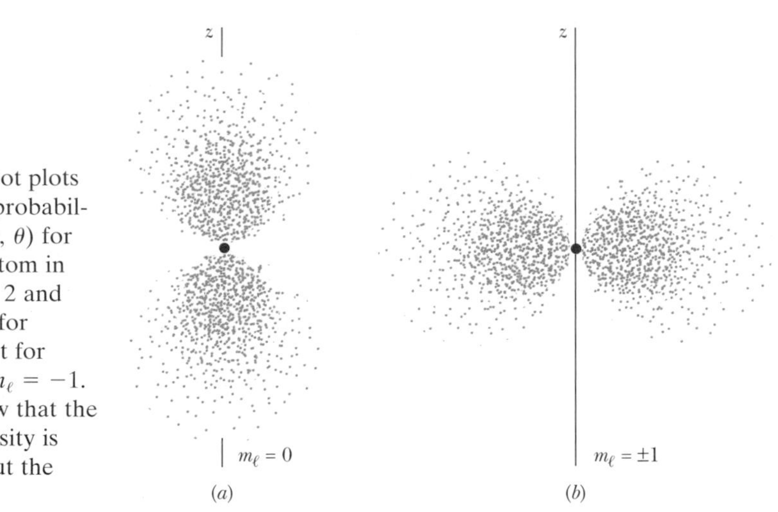

Fig. 39-24 Dot plots of the volume probability density $\psi^2(r, \theta)$ for the hydrogen atom in states with $n = 2$ and $\ell = 1$. (a) Plot for $m_\ell = 0$. (b) Plot for $m_\ell = +1$ and $m_\ell = -1$. Both plots show that the probability density is symmetric about the z axis.

The solution to this puzzle comes about when we realize that all three states shown in Fig. 39-24 have the same energy. Recall that the energy of a state, given by Eq. 39-33, depends only on the principal quantum number n and is independent of ℓ and m_ℓ. In fact, for an *isolated* hydrogen atom there is no way to differentiate experimentally among the three states of Fig. 39-24.

If we add the volume probability densities for the three states for which $n = 2$ and $\ell = 1$, the combined probability density turns out to be spherically symmetrical, with no unique axis. One can, then, think of the electron as spending one-third of its time in each of the three states of Fig. 39-24, and one can think of the weighted sum of the three independent wave functions as defining a spherically symmetric **subshell** specified by the quantum numbers $n = 2$, $\ell = 1$. The individual states will display their separate existence only if we place the hydrogen atom in an external electric or magnetic field. The three states of the $n = 2$, $\ell = 1$ subshell will then have different energies, and the field direction will establish the necessary symmetry axis.

The $n = 2$, $\ell = 0$ state, whose volume probability density is shown in Fig. 39-22, *also* has the same energy as each of the three states of Fig. 39-24. We can view all four states whose quantum numbers are listed in Table 39-3 as forming a spherically symmetric **shell** specified by the single quantum number n. The importance of shells and subshells will become evident in Chapter 40, where we discuss atoms having more than one electron.

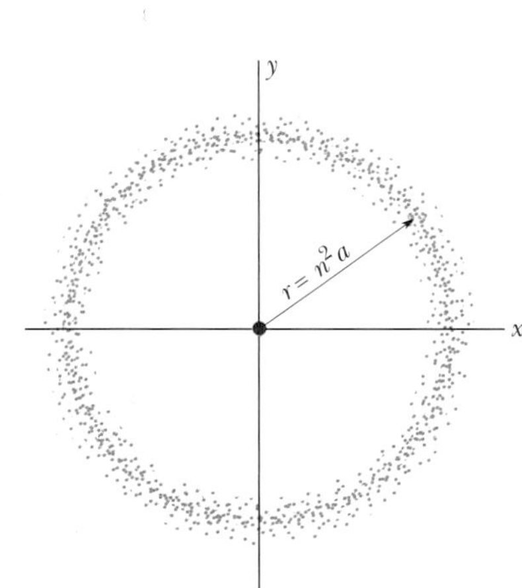

Fig. 39-25 A dot plot of the radial probability density $P(r)$ for the hydrogen atom in a quantum state with a relatively large principal quantum number—namely, $n = 45$—and angular momentum quantum number $\ell = n - 1 = 44$. The dots lie close to the xy plane, the ring of dots suggesting a classical electron orbit.

To round out our picture of the hydrogen atom, we display in Fig. 39-25 a dot plot of the *radial* probability density for a hydrogen atom state with a relatively high quantum number ($n = 45$) and the highest orbital quantum number that the restrictions of Table 39-2 permit ($\ell = n - 1 = 44$). The probability density forms a ring that is symmetrical about the z axis and lies very close to the xy plane. The mean radius of the ring is n^2a, where a is the Bohr radius. This mean radius is more than 2000 times the effective radius of the hydrogen atom in its ground state.

Figure 39-25 suggests the electron orbit of classical physics. Thus, we have another illustration of Bohr's correspondence principle—namely, that at large quantum numbers the predictions of quantum mechanics merge smoothly with those of classical physics. Imagine what a dot plot like that of Figure 39-25 would look like for *really* large values of n and ℓ—say, $n = 1000$ and $\ell = 999$.

Review & Summary

The Confinement Principle The **confinement principle** applies to waves of all kinds, including waves on a string and the matter waves of quantum physics. It states that confinement leads to quantization—that is, to the existence of discrete states with certain energies. States with intermediate values of energy are disallowed.

An Electron in an Infinite Potential Well An infinite potential well is a device for confining an electron. From the confinement principle we expect that the matter wave representing a trapped electron can exist only in a set of discrete states. For a one-dimensional infinite potential well, the energies associated with these *quantum states* are

$$E_n = \left(\frac{h^2}{8mL^2}\right)n^2, \quad \text{for } n = 1, 2, 3, \ldots, \tag{39-4}$$

in which L is the width of the well and n is a **quantum number.** If the electron is to change from one state to another, its energy must change by the amount

$$\Delta E = E_{\text{high}} - E_{\text{low}}, \tag{39-5}$$

where E_{high} is the higher energy and E_{low} is the lower energy. If the change is done by photon absorption or emission, the energy of the photon must be

$$hf = \Delta E = E_{\text{high}} - E_{\text{low}}. \tag{39-6}$$

The **wave functions** associated with the quantum states are

$$\psi_n(x) = A \sin\left(\frac{n\pi}{L}x\right), \quad \text{for } n = 1, 2, 3, \ldots. \tag{39-10}$$

The probability density $\psi_n^2(x)$ for an allowed state has the physical meaning that $\psi_n^2(x)\,dx$ is the probability that the electron will be detected in the interval between x and $x + dx$. For an electron in an infinite well, the probability densities are

$$\psi_n^2(x) = A^2 \sin^2\left(\frac{n\pi}{L}x\right), \quad \text{for } n = 1, 2, 3, \ldots. \tag{39-12}$$

At high quantum numbers n, the electron tends toward classical behavior in that it tends to occupy all parts of the well with equal probability. This fact leads to the **correspondence principle:** At large enough quantum numbers, the predictions of quantum physics merge smoothly with those of classical physics.

Normalization and Zero-Point Energy The amplitude A^2 in Eq. 39-12 can be found from the **normalizing equation,**

$$\int_{-\infty}^{+\infty} \psi_n^2(x)\,dx = 1, \tag{39-14}$$

which asserts that the electron must be *somewhere* within the well because the probability 1 implies certainty.

From Eq. 39-4 we see that the lowest permitted energy for the electron is not zero but the energy that corresponds to $n = 1$. This lowest energy is called the **zero-point energy** of the electron–well system.

An Electron in a Finite Potential Well A finite potential well is one for which the potential energy of an electron inside the well is less than that for one outside the well by a finite amount U_0. The wave function for an electron trapped in such a well extends into the walls of the well.

Two- and Three-Dimensional Electron Traps The quantized energies for an electron trapped in a two-dimensional infinite potential well that forms a rectangular corral are

$$E_{nx,ny} = \frac{h^2}{8m}\left(\frac{n_x^2}{L_x^2} + \frac{n_y^2}{L_y^2}\right), \tag{39-20}$$

where n_x is a quantum number for which the electron's matter wave fits in well width L_x and n_y is a quantum number for which the electron's matter wave fits in well width L_y. Similarly, the energies for an electron trapped in a three-dimensional infinite potential well that forms a rectangular box are

$$E_{nx,ny,nz} = \frac{h^2}{8m}\left(\frac{n_x^2}{L_x^2} + \frac{n_y^2}{L_y^2} + \frac{n_z^2}{L_z^2}\right). \tag{39-21}$$

Here n_z is a third quantum number, one for which the matter wave fits in well width L_z.

The Hydrogen Atom Both the (incorrect) Bohr model of the hydrogen atom and the (correct) application of Schrödinger's equation to this atom give the quantized energy levels of the atom as

$$E_n = -\frac{me^4}{8\varepsilon_0^2 h^2}\frac{1}{n^2} = -\frac{13.60\text{ eV}}{n^2}, \quad \text{for } n = 1, 2, 3, \ldots. \tag{39-32, 39-33}$$

From this we find that if the atom makes a transition between any two energy levels as a result of having emitted or absorbed light, the wavelength of the light is given by

$$\frac{1}{\lambda} = R\left(\frac{1}{n_{\text{low}}^2} - \frac{1}{n_{\text{high}}^2}\right), \tag{39-36}$$

where

$$R = \frac{me^4}{8\varepsilon_0^2 h^3 c} = 1.097\,373 \times 10^7\ \text{m}^{-1} \tag{39-37}$$

is the *Rydberg constant.*

The **radial probability density** $P(r)$ for a state of the hydrogen atom is defined so that $P(r)\,dr$ is the probability that the electron will be detected somewhere in the space between two concentric shells of radii r and $r + dr$ centered on the atom's nucleus. For the hydrogen atom's ground state,

$$P(r) = \frac{4}{a^3}r^2 e^{-2r/a}, \tag{39-44}$$

in which a, the **Bohr radius,** is a length unit equal to 52.92 pm. Figure 39-20 is a plot of $P(r)$ for the ground state.

Figures 39-22 and 39-24 represent the volume probability densities (not the *radial* probability densities) for the four hydrogen atom states with $n = 2$. The plot of Fig. 39-22 ($n = 2$, $\ell = 0$, $m_\ell = 0$) is spherically symmetric. The plots of Fig. 39-24 ($n = 2$, $\ell = 1$, $m_\ell = 0, +1, -1$) are symmetric about the z axis but, when added together, are also spherically symmetric.

All four states with $n = 2$ have the same energy and may be usefully regarded as constituting a **shell,** identified as the $n = 2$ shell. The three states of Fig. 39-24, taken together, may be regarded as constituting the $n = 2$, $\ell = 1$ **subshell.** It is not possible to separate the four $n = 2$ states experimentally unless the hydrogen atom is placed in an electric or magnetic field, which permits the establishment of a definite symmetry axis.

Questions

1 Three electrons are trapped in three different one-dimensional infinite potential wells of widths (a) 50 pm, (b) 200 pm, and (c) 100 pm. Rank the electrons according to their ground-state energies, greatest first.

2 If you double the width of a one-dimensional infinite potential well, (a) is the energy of the ground state of the trapped electron multiplied by 4, 2, $\frac{1}{2}$, $\frac{1}{4}$, or some other number? (b) Are the energies of the higher energy states multiplied by this factor or by some other factor, depending on their quantum number?

3 If you wanted to use the idealized trap of Fig. 39-1 to trap a positron, would you need to change (a) the geometry of the trap, (b) the electric potential of the central cylinder, or (c) the electric potentials of the two semi-infinite end cylinders? (A positron has the same mass as an electron but is positively charged.)

4 Figure 39-26 shows three infinite potential wells, each on an x axis. Without written calculation, determine the wave function ψ for a ground-state electron trapped in each well.

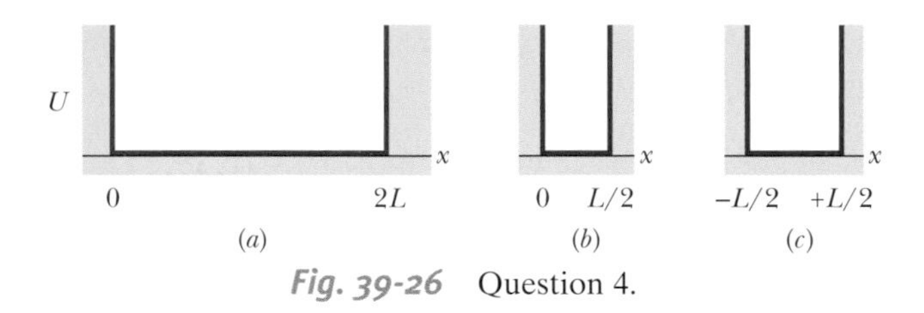

Fig. 39-26 Question 4.

5 An electron is trapped in a one-dimensional infinite potential well in a state with $n = 17$. How many points of (a) zero probability and (b) maximum probability does its matter wave have?

6 Is the ground-state energy of a proton trapped in a one-dimensional infinite potential well greater than, less than, or equal to that of an electron trapped in the same potential well?

7 A proton and an electron are trapped in identical one-dimensional infinite potential wells; each particle is in its ground state. At the center of the wells, is the probability density for the proton greater than, less than, or equal to that of the electron?

8 You want to modify the finite potential well of Fig. 39-7 to allow its trapped electron to exist in more than three quantum states. Could you do so by making the well (a) wider or narrower, (b) deeper or shallower?

9 The table lists the quantum numbers for five proposed hydrogen atom states. Which of them are not possible?

	n	ℓ	m_ℓ
(a)	3	2	0
(b)	2	3	1
(c)	4	3	−4
(d)	5	5	0
(e)	5	3	−2

10 An electron is trapped in a finite potential well that is deep enough to allow the electron to exist in a state with $n = 4$. How many points of (a) zero probability and (b) maximum probability does its matter wave have within the well?

11 An electron that is trapped in a one-dimensional infinite potential well of width L is excited from the ground state to the first excited state. Does the excitation increase, decrease, or have no effect on the probability of detecting the electron in a small length of the x axis (a) at the center of the well and (b) near one of the well walls?

12 Figure 39-27 indicates the lowest energy levels (in electron-volts) for five situations in which an electron is trapped in a one-dimensional infinite potential well. In wells B, C, D, and E, the electron is in the ground state. We shall excite the electron in well A to the fourth excited state (at 25 eV). The electron can then de-excite to the ground state by emitting one or more photons, corresponding to one long jump or several short jumps. Which photon *emission* energies of this de-excitation match a photon *absorption* energy (from the ground state) of the other four electrons? Give the n values.

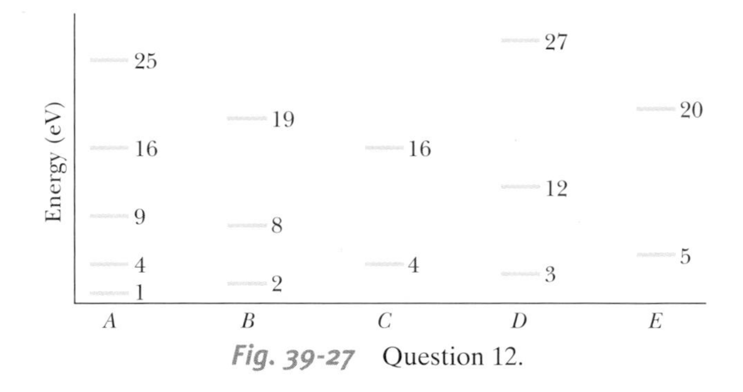

Fig. 39-27 Question 12.

13 From a visual inspection of Fig. 39-8, rank the quantum numbers of the three quantum states according to the de Broglie wavelength of the electron, greatest first.

14 An electron, trapped in a finite potential energy well such as that of Fig. 39-7, is in its state of lowest energy. Are (a) its de Broglie wavelength, (b) the magnitude of its momentum, and (c) its energy greater than, the same as, or less than they would be if the potential well were infinite, as in Fig. 39-2?

15 In 1996, physicists working at an accelerator laboratory succeeded in producing atoms of antihydrogen. Such atoms consist of a positron moving in the electric field of an antiproton. A positron has the same mass as an electron but the opposite charge. An antiproton has the same mass as a proton but the opposite charge. Would you expect the spectrum of antihydrogen to be the same as that of normal hydrogen?

16 (a) From Fig. 39-18, you can show that the photon energy of the second spectral line of the Lyman series is equal to the sum of the photon energies of two other hydrogen lines. What are those lines? (b) The photon energy of the second spectral line of the Lyman series is also equal to the *difference* between the photon energies of two other hydrogen lines. What are *those* lines?

17 A hydrogen atom is in the third excited state. To what state (give the quantum number n) should it jump to (a) emit light with the longest possible wavelength, (b) emit light with the shortest possible wavelength, and (c) absorb light with the longest possible wavelength?

Problems

SSM Solution is in the Student Solutions Manual.
WWW Solution is at http://www.wiley.com/college/halliday
• – ••• Number of dots indicates level of problem difficulty.

sec. 39-3 Energies of a Trapped Electron

•1 An electron in a one-dimensional infinite potential well of length L has ground-state energy E_1. The length is changed to L' so that the new ground-state energy is $E_1' = 0.500E_1$. What is the ratio L'/L?

•2 What is the ground-state energy of (a) an electron and (b) a proton if each is trapped in a one-dimensional infinite potential well that is 200 pm wide?

•3 Consider an atomic nucleus to be equivalent to a one-dimensional infinite potential well with $L = 1.4 \times 10^{-14}$ m, a typical nuclear diameter. What would be the ground-state energy of an electron if it were trapped in such a potential well? (*Note:* Nuclei do not contain electrons.) SSM

•4 A proton is confined to a one-dimensional infinite potential well 100 pm wide. What is its ground-state energy?

•5 What must be the width of a one-dimensional infinite potential well if an electron trapped in it in the $n = 3$ state is to have an energy of 4.7 eV?

•6 An electron, trapped in a one-dimensional infinite potential well 250 pm wide, is in its ground state. How much energy must it absorb if it is to jump up to the state with $n = 4$?

•7 The ground-state energy of an electron trapped in a one-dimensional infinite potential well is 2.6 eV. What will this quantity be if the width of the potential well is doubled?

••8 An electron is trapped in a one-dimensional infinite potential well. For what (a) higher quantum number and (b) lower quantum number is the corresponding energy difference equal to the energy of the $n = 5$ level? (c) Show that no pair of adjacent levels has an energy difference equal to the energy of the $n = 6$ level.

••9 An electron is trapped in a one-dimensional infinite well of width 250 pm and is in its ground state. What are the (a) longest, (b) second longest, and (c) third longest wavelengths of light that can excite the electron from the ground state via a single photon absorption? SSM

••10 An electron is trapped in a one-dimensional infinite potential well. For what (a) higher quantum number and (b) lower quantum number is the corresponding energy difference equal to the energy difference ΔE_{43} between the levels $n = 4$ and $n = 3$? (c) Show that no pair of adjacent levels has an energy difference equal to $2\Delta E_{43}$.

••11 Suppose that an electron trapped in a one-dimensional infinite well of width 250 pm is excited from its first excited state to its third excited state. (a) What energy must be transferred to the electron for this quantum jump? The electron then de-excites back to its ground state by emitting light. In the various possible ways it can do this, what are the (b) shortest, (c) second shortest, (d) longest, and (e) second longest wavelengths that can be emitted? (f) Show the various possible ways on an energy-level diagram. If light of wavelength 29.4 nm happens to be emitted, what are the (g) longest and (h) shortest wavelength that can be emitted afterwards?

••12 An electron is trapped in a one-dimensional infinite well and is in its first excited state. Figure 39-28 indicates the five longest wavelengths of light that the electron could absorb in transitions from this initial state via a single photon absorption. What is the width of the potential well?

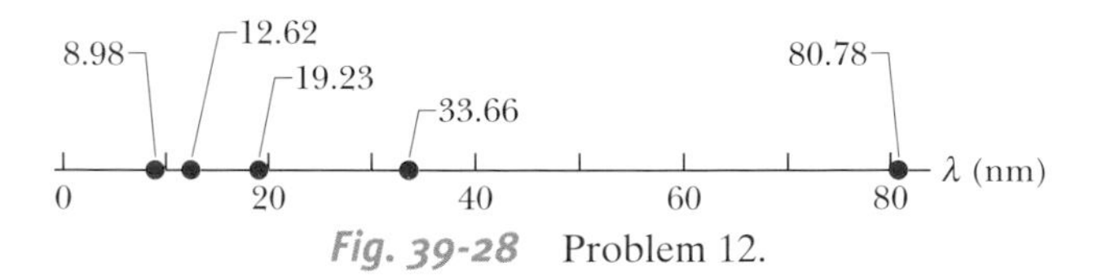

Fig. 39-28 Problem 12.

sec. 39-4 Wave Functions of a Trapped Electron

••13 An electron is trapped in a one-dimensional infinite potential well that is 100 pm wide; the electron is in its ground state. What is the probability that you can detect the electron in an interval of width $\Delta x = 5.0$ pm centered at $x =$ (a) 25 pm, (b) 50 pm, and (c) 90 pm? (*Hint:* The interval Δx is so narrow that you can take the probability density to be constant within it.) SSM WWW

••14 A particle is confined to the one-dimensional infinite potential well of Fig. 39-2. If the particle is in its ground state, what is its probability of detection between (a) $x = 0$ and $x = 0.25L$, (b) $x = 0.75L$ and $x = L$, and (c) $x = 0.25L$ and $x = 0.75L$?

sec. 39-5 An Electron in a Finite Well

•15 An electron in the $n = 2$ state in the finite potential well of Fig. 39-7 absorbs 400 eV of energy from an external source. Using the energy-level diagram of Fig. 39-9, determine the electron's kinetic energy after this absorption, assuming that the electron moves to a position for which $x > L$.

•16 Figure 39-9 gives the energy levels for an electron trapped in a finite potential energy well 450 eV deep. If the electron is in the $n = 3$ state, what is its kinetic energy?

••17 (a) Show that for the region $x > L$ in the finite potential well of Fig. 39-7, $\psi(x) = De^{2kx}$ is a solution of Schrödinger's equation in its one-dimensional form, where D is a constant and k is positive. (b) On what basis do we find this mathematically acceptable solution to be physically unacceptable? SSM

sec. 39-7 Two- and Three-Dimensional Electron Traps

•18 An electron is contained in the rectangular corral of Fig. 39-13, with widths $L_x = 800$ pm and $L_y = 1600$ pm. What is the electron's ground-state energy?

•19 An electron is contained in the rectangular box of Fig. 39-14, with widths $L_x = 800$ pm, $L_y = 1600$ pm, and $L_z = 390$ pm. What is the electron's ground-state energy?

••20 A rectangular corral of widths $L_x = L$ and $L_y = 2L$ contains an electron. What multiple of $h^2/8mL^2$, where m is the electron mass, gives (a) the energy of the electron's ground state, (b) the energy of its first excited state, (c) the energy of its lowest degenerate states, and (d) the difference between the energies of its second and third excited states?

••21 An electron (mass m) is contained in a rectangular corral of widths $L_x = L$ and $L_y = 2L$. (a) How many different frequencies of light could the electron emit or absorb if it makes a transition between a pair of the lowest five energy levels? What multiple of $h/8mL^2$ gives the (b) lowest, (c) second lowest, (d) third lowest, (e) highest, (f) second highest, and (g) third highest frequency? SSM WWW

••22 A cubical box of widths $L_x = L_y = L_z = L$ contains an electron. What multiple of $h^2/8mL^2$, where m is the electron mass, is (a) the energy of the electron's ground state, (b) the energy of its second excited state, and (c) the difference between the energies of its second and third excited states? How many degenerate states have the energy of (d) the first excited state and (e) the fifth excited state?

••23 An electron (mass m) is contained in a cubical box of widths $L_x = L_y = L_z$. (a) How many different frequencies of light could the electron emit or absorb if it makes a transition between a pair of the lowest five energy levels? What multiple of $h/8mL^2$ gives the (b) lowest, (c) second lowest, (d) third lowest, (e) highest, (f) second highest, and (g) third highest frequency?

sec. 39-9 Schrödinger's Equation and the Hydrogen Atom

•24 An atom (not a hydrogen atom) absorbs a photon whose associated wavelength is 375 nm and then immediately emits a photon whose associated wavelength is 580 nm. How much net energy is absorbed by the atom in this process?

•25 What is the ratio of the shortest wavelength of the Balmer series to the shortest wavelength of the Lyman series? SSM

•26 An atom (not a hydrogen atom) absorbs a photon whose associated frequency is 6.2×10^{14} Hz. By what amount does the energy of the atom increase?

•27 What are the (a) energy, (b) magnitude of the momentum, and (c) wavelength of the photon emitted when a hydrogen atom undergoes a transition from a state with $n = 3$ to a state with $n = 1$?

•28 (a) What is the energy E of the hydrogen-atom electron whose probability density is represented by the dot plot of Fig. 39-22? (b) What minimum energy is needed to remove this electron from the atom?

•29 A neutron with a kinetic energy of 6.0 eV collides with a stationary hydrogen atom in its ground state. Explain why the collision must be elastic—that is, why kinetic energy must be conserved. (*Hint:* Show that the hydrogen atom cannot be excited as a result of the collision.) SSM

•30 Calculate the radial probability density $P(r)$ for the hydrogen atom in its ground state at (a) $r = 0$, (b) $r = a$, and (c) $r = 2a$, where a is the Bohr radius.

•31 For the hydrogen atom in its ground state, calculate (a) the probability density $\psi^2(r)$ and (b) the radial probability density $P(r)$ for $r = a$, where a is the Bohr radius.

•32 A hydrogen atom is excited from its ground state to the state with $n = 4$. (a) How much energy must be absorbed by the atom? Consider the photon energies that can be emitted by the atom as it de-excites to the ground state in the several possible ways. (b) How many different energies are possible; what are the (c) highest, (d) second highest, (e) third highest, (f) lowest, (g) second lowest, and (h) third lowest energies?

••33 How much work must be done to pull apart the electron and the proton that make up the hydrogen atom if the atom is initially in (a) its ground state and (b) the state with $n = 2$? SSM

••34 A hydrogen atom, initially at rest in the $n = 4$ quantum state, undergoes a transition to the ground state, emitting a photon in the process. What is the speed of the recoiling hydrogen atom?

••35 In the ground state of the hydrogen atom, the electron has a total energy of -13.6 eV. What are (a) its kinetic energy and (b) its potential energy if the electron is one Bohr radius from the central nucleus?

••36 What are the (a) wavelength range and (b) frequency range of the Lyman series? What are the (c) wavelength range and (d) frequency range of the Balmer series?

••37 Schrödinger's equation for states of the hydrogen atom for which the orbital quantum number ℓ is zero is

$$\frac{1}{r^2}\frac{d}{dr}\left(r^2\frac{d\psi}{dr}\right) + \frac{8\pi^2 m}{h^2}[E - U(r)]\psi = 0.$$

Verify that Eq. 39-39, which describes the ground state of the hydrogen atom, is a solution of this equation. SSM WWW

••38 Light of wavelength 121.6 nm is emitted by a hydrogen atom. What are the (a) higher quantum number and (b) lower quantum number of the transition producing this emission? (c) What is the name of the series that includes the transition?

••39 What is the probability that in the ground state of the hydrogen atom, the electron will be found at a radius greater than the Bohr radius? (*Hint:* See Sample Problem 39-8.)

••40 A hydrogen atom in a state having a *binding energy* (the energy required to remove an electron) of 0.85 eV makes a transition to a state with an *excitation energy* (the difference between the energy of the state and that of the ground state) of 10.2 eV. (a) What is the energy of the photon emitted as a result of the transition? What are the (b) higher quantum number and (c) lower quantum number of the transition producing this emission?

••41 Verify that Eq. 39-44, the radial probability density for the ground state of the hydrogen atom, is normalized. That is, verify that

$$\int_0^\infty P(r)\,dr = 1$$

is true. SSM

••42 Calculate the probability that the electron in the hydrogen atom, in its ground state, will be found between spherical shells whose radii are a and $2a$, where a is the Bohr radius. (*Hint:* See Sample Problem 39-8.)

••43 The wave functions for the three states with the dot plots shown in Fig. 39-24, which have $n = 2$, $\ell = 1$, and $m_\ell = 0, +1$, and -1, are

$$\psi_{210}(r, \theta) = (1/4\sqrt{2\pi})(a^{-3/2})(r/a)e^{-r/2a}\cos\theta,$$
$$\psi_{21+1}(r, \theta) = (1/8\sqrt{\pi})(a^{-3/2})(r/a)e^{-r/2a}(\sin\theta)e^{+i\phi},$$
$$\psi_{21-1}(r, \theta) = (1/8\sqrt{\pi})(a^{-3/2})(r/a)e^{-r/2a}(\sin\theta)e^{-i\phi},$$

in which the subscripts on $\psi(r, \theta)$ give the values of the quantum numbers n, ℓ, m_ℓ and the angles θ and ϕ are defined in Fig. 39-23. Note that the first wave function is real but the others, which involve the imaginary number i, are complex. Find the radial probability density $P(r)$ for (a) ψ_{210} and (b) ψ_{21+1} (same as for ψ_{21-1}). (c) Show that each $P(r)$ is consistent with the corresponding dot plot in Fig. 39-24. (d) Add the radial probability densities for ψ_{210}, ψ_{21+1}, and ψ_{21-1} and then show that the sum is spherically symmetric, depending only on r. SSM

••44 Light of wavelength 102.6 nm is emitted by a hydrogen atom. What are the (a) higher quantum number and (b) lower quantum number of the transition producing this emission? (c) What is the name of the series that includes the transition?

••45 What is the probability that an electron in the ground state of the hydrogen atom will be found between two spherical shells whose radii are r and $r + \Delta r$, (a) if $r = 0.500a$ and $\Delta r = 0.010a$ and (b) if $r = 1.00a$ and $\Delta r = 0.01a$, where a is the Bohr radius? (*Hint:* Δr is small enough to permit the radial probability density to be taken to be constant between r and $r + \Delta r$.) SSM

••46 For what value of the principal quantum number n would the effective radius, as shown in a probability density dot plot for the hydrogen atom, be 1.0 mm? Assume that ℓ has its maximum value of $n - 1$. (*Hint:* See Fig. 39-25.)

•••47 In Sample Problem 39-7 we showed that the radial probability density for the ground state of the hydrogen atom is a maximum when $r = a$, where a is the Bohr radius. Show that the *average* value of r, defined as

$$r_{avg} = \int P(r)\, r\, dr,$$

has the value $1.5a$. In this expression for r_{avg}, each value of $P(r)$ is weighted with the value of r at which it occurs. Note that the average value of r is greater than the value of r for which $P(r)$ is a maximum. SSM

•••48 The wave function for the hydrogen-atom quantum state represented by the dot plot shown in Fig. 39-22, which has $n = 2$ and $\ell = m_\ell = 0$, is

$$\psi_{200}(r) = \frac{1}{4\sqrt{2\pi}}\, a^{-3/2}\left(2 - \frac{r}{a}\right)e^{-r/2a},$$

in which a is the Bohr radius and the subscript on $\psi(r)$ gives the values of the quantum numbers n, ℓ, m_ℓ. (a) Plot $\psi^2_{200}(r)$ and show that your plot is consistent with the dot plot of Fig. 39-22. (b) Show analytically that $\psi^2_{200}(r)$ has a maximum at $r = 4a$. (c) Find the radial probability density $P_{200}(r)$ for this state. (d) Show that

$$\int_0^\infty P_{200}(r)\, dr = 1$$

and thus that the expression above for the wave function $\psi_{200}(r)$ has been properly normalized.

Additional Problems

49 An electron is trapped in a one-dimensional infinite potential well. Show that the energy difference ΔE between its quantum levels n and $n + 2$ is $(h^2/2mL^2)(n + 1)$.

50 An electron is confined to a narrow evacuated tube of length 3.0 m; the tube functions as a one-dimensional infinite potential well. (a) What is the energy difference between the electron's ground state and its first excited state? (b) At what quantum number n would the energy difference between adjacent energy levels be 1.0 eV—which is measurable, unlike the result of (a)? At that quantum number, (c) what multiple of the electron's rest energy would give the electron's total energy and (d) would the electron be relativistic?

51 (a) For a given value of the principal quantum number n, how many values of the orbital quantum number ℓ are possible? (b) For a given value of ℓ, how many values of the orbital magnetic quantum number m_ℓ are possible? (c) For a given value of n, how many values of m_ℓ are possible?

52 As Fig. 39-8 suggests, the probability density for an electron in the region $0 < x < L$ for the finite potential well of Fig. 39-7 is sinusoidal, being given by $\psi^2(x) = B \sin^2 kx$, in which B is a constant. (a) Show that the wave function $\psi(x)$ that may be found from this equation is a solution of Schrödinger's equation in its one-dimensional form. (b) Find an expression for k that makes this true.

53 As Fig. 39-8 suggests, the probability density for the region $x > L$ in the finite potential well of Fig. 39-7 drops off exponentially according to $\psi^2(x) = Ce^{-2kx}$, where C is a constant. (a) Show that the wave function $\psi(x)$ that may be found from this equation is a solution of Schrödinger's equation in its one-dimensional form. (b) Find an expression for k for this to be true.

54 Let ΔE_{adj} be the energy difference between two adjacent energy levels for an electron trapped in a one-dimensional infinite potential well. Let E be the energy of either of the two levels. (a) Show that the ratio $\Delta E_{adj}/E$ approaches the value $2/n$ at large values of the quantum number n. As $n \to \infty$, does (b) ΔE_{adj}, (c) E, or (d) $\Delta E_{adj}/E$ approach zero? (e) What do these results mean in terms of the correspondence principle?

55 (a) Show that the terms in Schrödinger's equation (Eq. 39-18) have the same dimensions. (b) What is the common SI unit for each of these terms?

56 Verify that the combined value of the constants appearing in Eq. 39-32 is 13.6 eV.

57 Light of wavelength 486.1 nm is emitted by a hydrogen atom. What are the (a) higher quantum number and (b) lower quantum number of the transition producing this emission? (c) What is the name of the series that includes the transition?

58 Repeat Sample Problem 39-6 for the Balmer series of the hydrogen atom.

59 Verify the wavelengths given in Fig. 39-19 for the visible spectral lines of the Balmer series.

On-Line Simulation Problems

The website http://www.wiley.com/college/halliday has simulation problems about this chapter.

40 All About Atoms

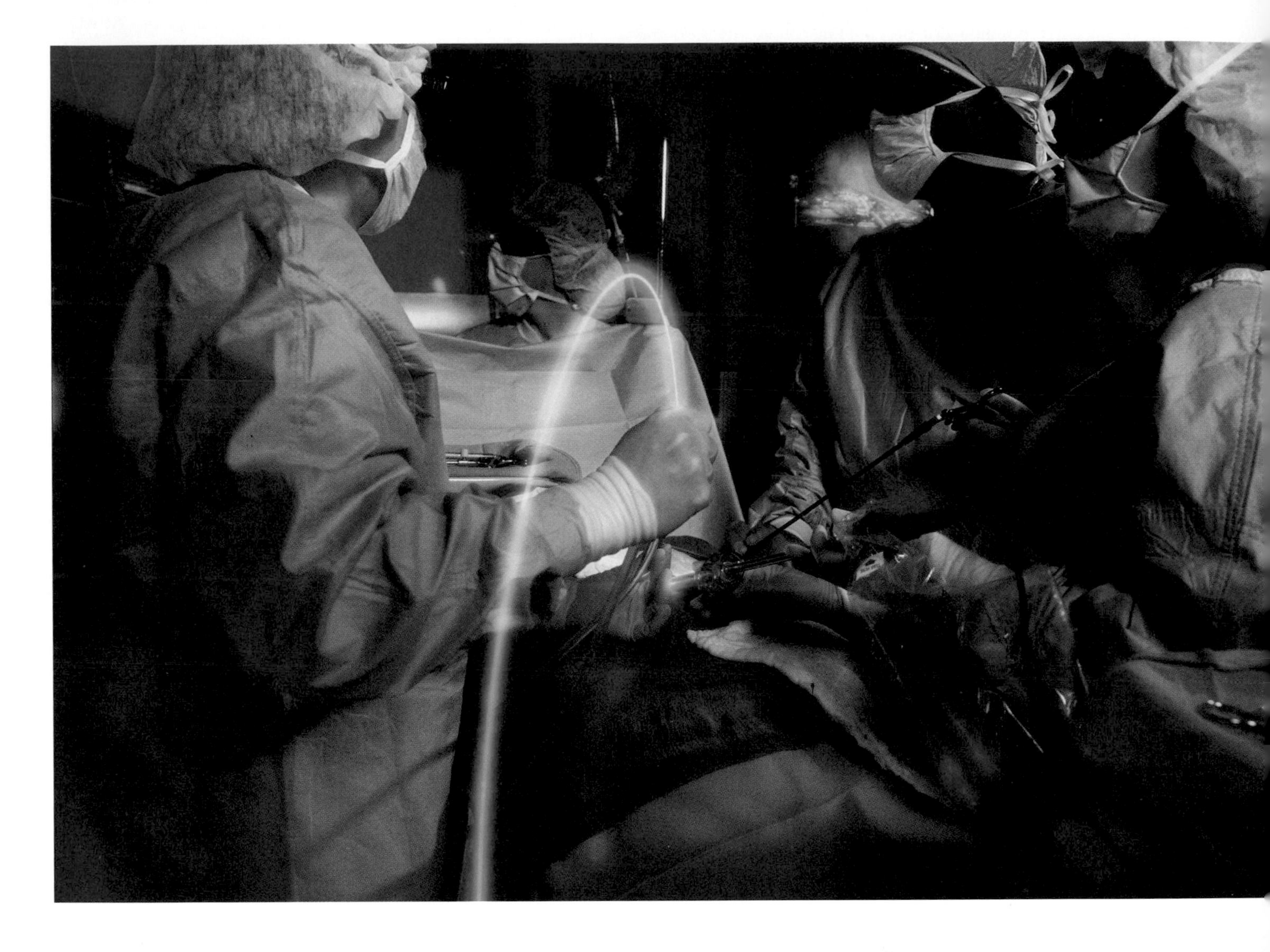

What, then, is so different about the light from a laser?

The answer is in this chapter.

Soon after lasers were invented in the 1960s, they became novel sources of light in research laboratories. Today, lasers are ubiquitous and are found in such diverse applications as voice and data transmission, surveying, welding, and grocery-store price scanning. The photograph shows surgery being performed with laser light transmitted via optical fibers. Light from a laser and light from any other source are both due to emissions by atoms.

40-1 What Is Physics?

In this chapter we continue with a primary goal of physics—discovering and understanding the properties of atoms. About 100 years ago, researchers struggled to find experiments that would prove the existence of atoms. Now we take their existence for granted and even have photographs (scanning tunneling microscope images) of atoms. We can drag them around on surfaces, such as to make the quantum corral shown in the opening photograph of Chapter 39. We can even hold an individual atom indefinitely in a trap (Fig. 40-1) so as to study its properties when it is completely isolated from other atoms.

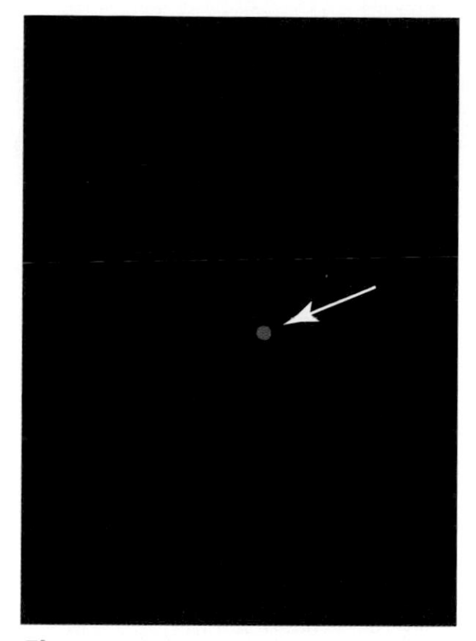

Fig. 40-1 The blue dot is a photograph of the light emitted from a single barium ion held for a long time in a trap at the University of Washington. Special techniques caused the ion to emit light over and over again as it underwent transitions between the same pair of energy levels. The dot represents the cumulative emission of many photons.

40-2 Some Properties of Atoms

You may think the details of atomic physics are remote from your daily life. However, consider how the following properties of atoms—so basic that we rarely think about them—affect the way we live in our world.

Atoms are stable. Essentially all the atoms that form our tangible world have existed without change for billions of years. What would the world be like if atoms continually changed into other forms, perhaps every few weeks or every few years?

Atoms combine with each other. They stick together to form stable molecules and stack up to form rigid solids. An atom is mostly empty space, but you can stand on a floor—made up of atoms—without falling through it.

These basic properties of atoms can be explained by quantum physics, as can the three less apparent properties that follow.

Atoms Are Put Together Systematically

Figure 40-2 shows an example of a repetitive property of the elements as a function of their position in the periodic table (Appendix G). The figure is a plot of the **ionization energy** of the elements; the energy required to remove the most loosely bound electron from a neutral atom is plotted as a function of the position in the periodic table of the element to which the atom belongs. The remarkable similarities in the chemical and physical properties of the elements in each vertical column of the periodic table are evidence enough that the atoms are constructed according to systematic rules.

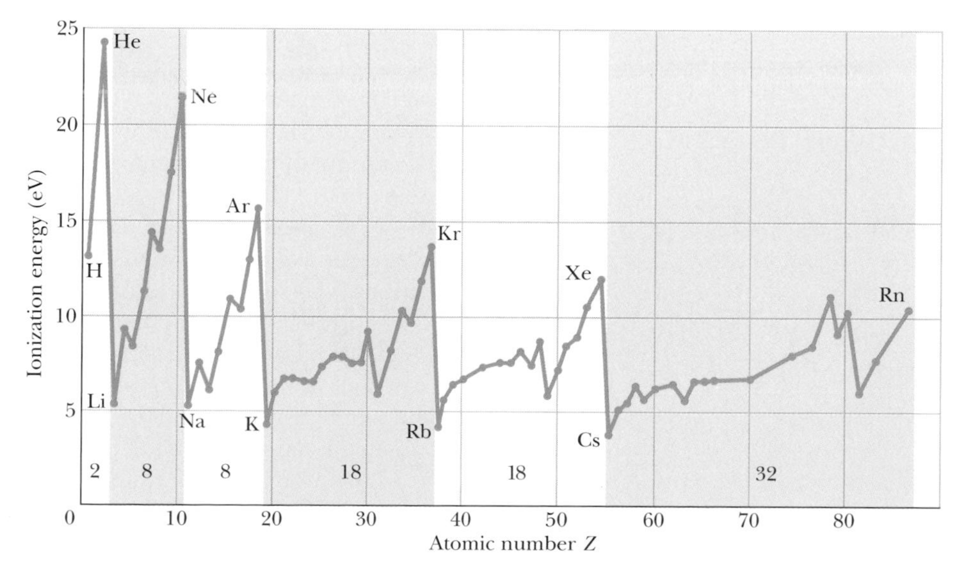

Fig. 40-2 A plot of the ionization energies of the elements as a function of atomic number, showing the periodic repetition of properties through the six complete horizontal periods of the periodic table. The number of elements in each of these periods is indicated.

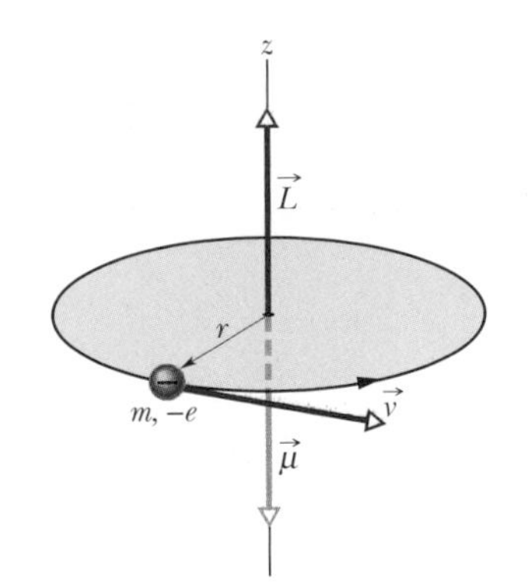

Fig. 40-3 A classical model showing a particle of mass m and charge $-e$ moving with speed v in a circle of radius r. The moving particle has an angular momentum $\vec{L}$ given by $\vec{r} \times \vec{p}$, where $\vec{p}$ is its linear momentum $m\vec{v}$. The particle's motion is equivalent to a current loop that has an associated magnetic moment $\vec{\mu}$ that is directed opposite $\vec{L}$.

The elements are arranged in the periodic table in six horizontal **periods;** except for the first, each period starts at the left with a highly reactive alkali metal (lithium, sodium, potassium, and so on) and ends at the right with a chemically inert noble gas (neon, argon, krypton, and so on). Quantum physics accounts for the chemical properties of these elements. The numbers of elements in the six periods are

$$2, 8, 8, 18, 18, \text{ and } 32.$$

Quantum physics predicts these numbers.

Atoms Emit and Absorb Light

We have already seen that atoms can exist only in discrete quantum states, each state having a certain energy. An atom can make a transition from one state to another by emitting light (to jump to a lower energy level E_{low}) or by absorbing light (to jump to a higher energy level E_{high}). As we first discussed in Section 39-3, the light is emitted or absorbed as a photon with energy

$$hf = E_{\text{high}} - E_{\text{low}}. \tag{40-1}$$

Thus, the problem of finding the frequencies of light emitted or absorbed by an atom reduces to the problem of finding the energies of the quantum states of that atom. Quantum physics allows us—in principle at least—to calculate these energies.

Atoms Have Angular Momentum and Magnetism

Figure 40-3 shows a negatively charged particle moving in a circular orbit around a fixed center. As we discussed in Section 32-7, the orbiting particle has both an angular momentum $\vec{L}$ and (because its path is equivalent to a tiny current loop) a magnetic dipole moment $\vec{\mu}$. As Fig. 40-3 shows, vectors $\vec{L}$ and $\vec{\mu}$ are both perpendicular to the plane of the orbit but, because the charge is negative, they point in opposite directions.

The model of Fig. 40-3 is strictly classical and does not accurately represent an electron in an atom. In quantum physics, the rigid orbit model has been replaced by the probability density model, best visualized as a dot plot. In quantum physics, however, it is still true that in general, each quantum state of an electron in an atom involves an angular momentum $\vec{L}$ and a magnetic dipole moment $\vec{\mu}$ that have opposite directions (those vector quantities are said to be *coupled*).

The Einstein–de Haas Experiment

In 1915, well before the discovery of quantum physics, Albert Einstein and Dutch physicist W. J. de Haas carried out a clever experiment designed to show that the angular momentum and magnetic moment of individual atoms are coupled.

Einstein and de Haas suspended an iron cylinder from a thin fiber, as shown in Fig. 40-4. A solenoid was placed around the cylinder but not touching it. Initially, the magnetic dipole moments $\vec{\mu}$ of the atoms of the cylinder point in random directions, and so their external magnetic effects cancel (Fig. 40-4*a*). However, when a current is switched on in the solenoid (Fig. 40-4*b*) so that a magnetic field $\vec{B}$ is set up parallel to the long axis of the cylinder, the magnetic dipole moments of the atoms of the cylinder reorient themselves, lining up with that field. If the angular momentum $\vec{L}$ of each atom is coupled to its magnetic moment $\vec{\mu}$, then this alignment of the atomic magnetic moments must cause an alignment of the atomic angular momenta opposite the magnetic field.

No external torques initially act on the cylinder; thus, its angular momentum must remain at its initial zero value. However, when $\vec{B}$ is turned on and the atomic angular momenta line up antiparallel to $\vec{B}$, they tend to give a net angular mo-

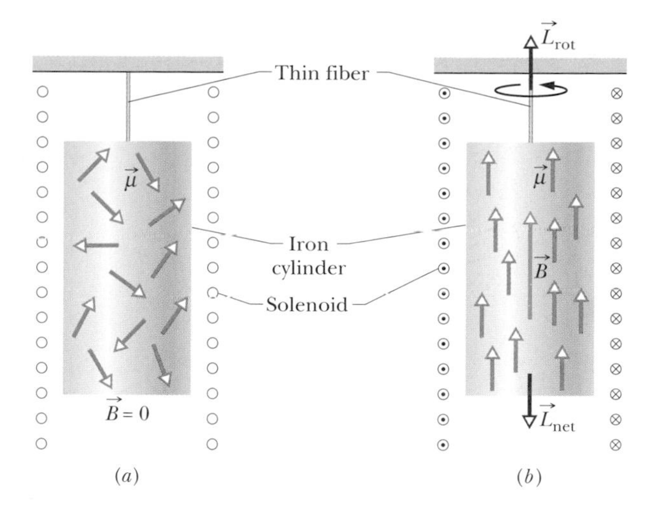

Fig. 40-4 The Einstein–de Haas experimental setup. (*a*) Initially, the magnetic field in the iron cylinder is zero and the magnetic dipole moment vectors $\vec{\mu}$ of its atoms are randomly oriented. The atomic angular momentum vectors (not shown) are directed opposite the magnetic dipole moment vectors and thus are also randomly oriented. (*b*) When a magnetic field $\vec{B}$ is set up along the cylinder's axis, the magnetic dipole moment vectors line up parallel to $\vec{B}$, which means that the angular momentum vectors line up opposite $\vec{B}$. Because the cylinder is initially isolated from external torques, its angular momentum is conserved and the cylinder as a whole must begin to rotate as shown.

mentum $\vec{L}_{net}$ to the cylinder as a whole (directed downward in Fig. 40-4*b*). To maintain zero angular momentum, the cylinder begins to rotate around its central axis to produce an angular momentum $\vec{L}_{rot}$ in the opposite direction (upward in Fig. 40-4*b*).

Were it not for the fiber, the cylinder would continue to rotate for as long as the magnetic field is present. However, the twisting of the fiber quickly produces a torque that momentarily stops the cylinder's rotation and then rotates the cylinder in the opposite direction as the twisting is undone. Thereafter, the fiber will twist and untwist as the cylinder oscillates about its initial orientation in angular simple harmonic motion.

Observation of the cylinder's rotation verified that the angular momentum and the magnetic dipole moment of an atom are coupled in opposite directions. Moreover, it dramatically demonstrated that the angular momenta associated with quantum states of atoms can result in *visible* rotation of an object of everyday size.

40-3 Electron Spin

As we discussed in Section 32-7, whether an electron is *trapped* in an atom or is *free,* it has an intrinsic **spin angular momentum** $\vec{S}$, often called simply **spin.** (Recall that *intrinsic* means that $\vec{S}$ is a basic characteristic of an electron, like its mass and electric charge.) As we shall discuss in the next section, the magnitude of $\vec{S}$ is quantized and depends on a **spin quantum number** s, which is always $\frac{1}{2}$ for electrons (and for protons and neutrons). In addition, the component of $\vec{S}$ measured along any axis is quantized and depends on a **spin magnetic quantum number** m_s, which can have only the value $+\frac{1}{2}$ or $-\frac{1}{2}$.

The existence of electron spin was postulated on an empirical basis by two Dutch graduate students, George Uhlenbeck and Samuel Goudsmit, from their studies of atomic spectra. The quantum physics basis for electron spin was provided a few years later, by British physicist P. A. M. Dirac, who developed (in 1929) a relativistic quantum theory of the electron.

It is tempting to account for electron spin by thinking of the electron as a tiny sphere spinning about an axis. However, that classical model, like the classical model of orbits, does not hold up. In quantum physics, spin angular momentum is best thought of as a measurable intrinsic property of the electron; you simply can't visualize it with a classical model.

In Section 39-9, we briefly discussed the quantum numbers generated by applying Schrödinger's equation to the electron in a hydrogen atom (Table 39-2). We can now extend the list of quantum numbers by including s and m_s, as

TABLE 40-1 Electron States for an Atom

Quantum Number	Symbol	Allowed Values	Related to
Principal	n	$1, 2, 3, \ldots$	Distance from the nucleus
Orbital	ℓ	$0, 1, 2, \ldots, (n-1)$	Orbital angular momentum
Orbital magnetic	m_ℓ	$0, \pm 1, \pm 2, \ldots, \pm \ell$	Orbital angular momentum (z component)
Spin	s	$\frac{1}{2}$	Spin angular momentum
Spin magnetic	m_s	$\pm\frac{1}{2}$	Spin angular momentum (z component)

shown in Table 40-1. This set of five quantum numbers completely specify the quantum state of an electron in a hydrogen atom or any other atom. All states with the same value of n form a **shell.** By counting the allowed values of ℓ and m_ℓ and then doubling the number to account for the two allowed values of m_s, you can verify that a shell defined by quantum number n has $2n^2$ states. All states with the same value of n and ℓ form a **subshell** and have the same energy. You can verify that a subshell defined by quantum number ℓ has $2(2\ell + 1)$ states.

40-4 Angular Momenta and Magnetic Dipole Moments

Every quantum state of an electron in an atom has an associated orbital angular momentum and a corresponding orbital magnetic dipole moment. Every electron, whether trapped in an atom or free, has a spin angular momentum and a corresponding spin magnetic dipole moment. We discuss these quantities separately first, and then in combination.

Orbital Angular Momentum and Magnetism

The magnitude L of the **orbital angular momentum** $\vec{L}$ of an electron *in an atom* is quantized; that is, it can have only certain values. These values are

$$L = \sqrt{\ell(\ell + 1)}\hbar, \tag{40-2}$$

in which ℓ is the orbital quantum number and $\hbar$ is $h/2\pi$. According to Table 40-1, ℓ must be either zero or a positive integer no greater than $n - 1$. For a state with $n = 3$, for example, only $\ell = 2$, $\ell = 1$, and $\ell = 0$ are permitted.

As we discussed in Section 32-7, a magnetic dipole is associated with the orbital angular momentum $\vec{L}$ of an electron in an atom. This magnetic dipole has an **orbital magnetic dipole moment** $\vec{\mu}_{orb}$, which is related to the angular momentum by Eq. 32-28:

$$\vec{\mu}_{orb} = -\frac{e}{2m}\vec{L}. \tag{40-3}$$

The minus sign in this relation means that $\vec{\mu}_{orb}$ is directed opposite $\vec{L}$. Because the magnitude of $\vec{L}$ is quantized (Eq. 40-2), the magnitude of $\vec{\mu}_{orb}$ must also be quantized and given by

$$\mu_{orb} = \frac{e}{2m}\sqrt{\ell(\ell + 1)}\hbar. \tag{40-4}$$

Neither $\vec{\mu}_{orb}$ nor $\vec{L}$ can be measured in any way. However, we *can* measure the components of those two vectors along a given axis. Let us imagine that the atom is located in a magnetic field $\vec{B}$; assume that a z axis extends in the direction of the field lines at the atom's location. Then we can measure the z components of $\vec{\mu}_{orb}$ and $\vec{L}$ along that axis.

The components $\mu_{orb,z}$ of the orbital magnetic dipole moment are quantized and given by

$$\mu_{\text{orb},z} = -m_\ell \mu_B. \tag{40-5}$$

Here m_ℓ is the orbital magnetic quantum number of Table 40-1 and μ_B is the **Bohr magneton:**

$$\mu_B = \frac{eh}{4\pi m} = \frac{e\hbar}{2m} = 9.274 \times 10^{-24}\ \text{J/T} \quad \text{(Bohr magneton)}, \tag{40-6}$$

where m is the electron mass.

The components L_z of the angular momentum are also quantized, and they are given by

$$L_z = m_\ell \hbar. \tag{40-7}$$

Figure 40-5 shows the five quantized components L_z of the orbital angular momentum for an electron with $\ell = 2$, as well as the associated orientations of the angular momentum $\vec{L}$. However, *do not take the figure literally* because we cannot detect $\vec{L}$ in any way. Thus, drawing it in a figure like Fig. 40-5 is merely a visual aide. We can extend that visual aide by saying that $\vec{L}$ makes a certain angle θ with the z axis, such that

$$\cos\theta = \frac{L_z}{L}. \tag{40-8}$$

We can call θ the *semi-classical angle* between vector $\vec{L}$ and the z axis because θ is a classical measurement of something that quantum theory tells us cannot be measured.

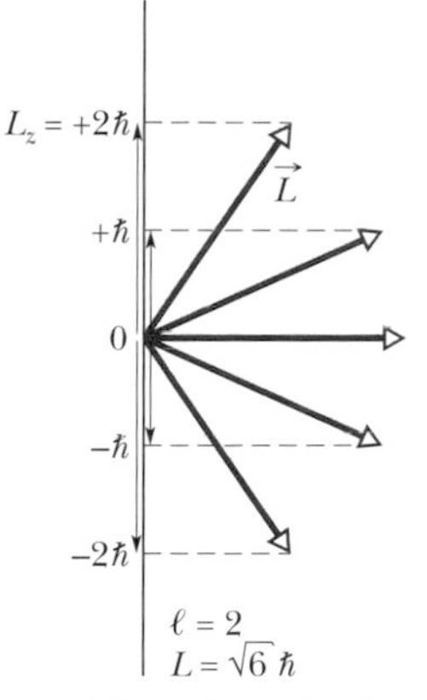

Fig. 40-5 The allowed values of L_z for an electron in a quantum state with $\ell = 2$. For every orbital angular momentum vector $\vec{L}$ in the figure, there is a vector pointing in the opposite direction, representing the magnitude and direction of the orbital magnetic dipole moment $\vec{\mu}_{\text{orb}}$.

Spin Angular Momentum and Spin Magnetic Dipole Moment

The magnitude S of the spin angular momentum $\vec{S}$ of any electron, whether *free or trapped,* has the single value given by

$$\begin{aligned} S &= \sqrt{s(s+1)}\hbar \\ &= \sqrt{(\tfrac{1}{2})(\tfrac{1}{2}+1)}\hbar = 0.866\hbar, \end{aligned} \tag{40-9}$$

where s $(=\frac{1}{2})$ is the spin quantum number of the electron.

As we discussed in Section 32-7, an electron has an intrinsic magnetic dipole that is associated with its spin angular momentum $\vec{S}$, whether the electron is confined to an atom or free. This magnetic dipole has a **spin magnetic dipole moment** $\vec{\mu}_s$, which is related to the spin angular momentum $\vec{S}$ by Eq. 32-22:

$$\vec{\mu}_s = -\frac{e}{m}\vec{S}. \tag{40-10}$$

The minus sign in this relation means that $\vec{\mu}_s$ is directed opposite $\vec{S}$. Because the magnitude of $\vec{S}$ is quantized (Eq. 40-9), the magnitude of $\vec{\mu}_s$ must also be quantized and given by

$$\mu_s = \frac{e}{m}\sqrt{s(s+1)}\hbar. \tag{40-11}$$

Neither $\vec{S}$ nor $\vec{\mu}_s$ can be measured in any way. However, we *can* measure their components along any given axis—call it the z axis. The components S_z of the spin angular momentum are quantized and given by

$$S_z = m_s \hbar, \tag{40-12}$$

where m_s is the spin magnetic quantum number of Table 40-1. That quantum number can have only two values: $m_s = +\frac{1}{2}$ (the electron is said to be *spin up*) and $m_s = -\frac{1}{2}$ (the electron is said to be *spin down*).

The components $\mu_{s,z}$ of the spin magnetic dipole moment are also quantized, and they are given by

$$\mu_{s,z} = -2m_s \mu_B. \tag{40-13}$$

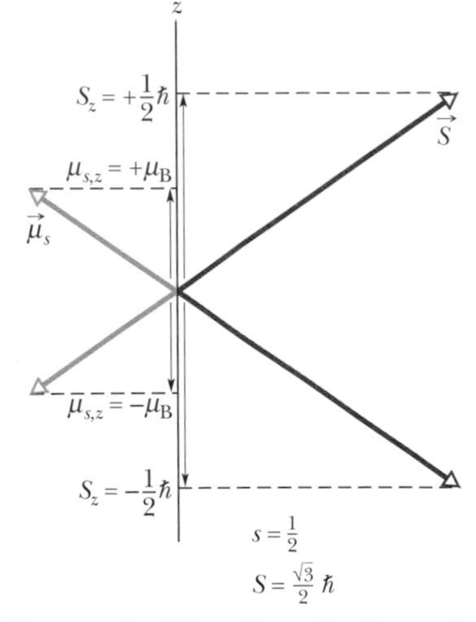

Fig. 40-6 The allowed values of S_z and μ_z for an electron.

Figure 40-6 shows the two quantized components S_z of the spin angular momentum for an electron and the associated orientations of vector $\vec{S}$. It also shows the quantized components $\mu_{s,z}$ of the spin magnetic dipole moment and the associated orientations of $\vec{\mu}_s$.

Orbital and Spin Angular Momenta Combined

For an atom containing more than one electron, we define a total angular momentum $\vec{J}$, which is the vector sum of the angular momenta of the individual electrons—both their orbital and their spin angular momenta. Each element in the periodic table is defined by the number of protons in the nucleus of an atom of the element. This number of protons is defined as being the **atomic number** (or **charge number**) Z of the element. Because an electrically neutral atom contains equal numbers of protons and electrons, Z is also the number of electrons in the neutral atom, and we use this fact to indicate a $\vec{J}$ value for a neutral atom:

$$\vec{J} = (\vec{L}_1 + \vec{L}_2 + \vec{L}_3 + \cdots + \vec{L}_Z) + (\vec{S}_1 + \vec{S}_2 + \vec{S}_3 + \cdots + \vec{S}_Z). \quad (40\text{-}14)$$

Similarly, the total magnetic dipole moment of a multielectron atom is the vector sum of the magnetic dipole moments (both orbital and spin) of its individual electrons. However, because of the factor 2 in Eq. 40-13, the resultant magnetic dipole moment for the atom does not have the direction of vector $-\vec{J}$; instead, it makes a certain angle with that vector. The **effective magnetic dipole moment** $\vec{\mu}_{\text{eff}}$ for the atom is the component of the vector sum of the individual magnetic dipole moments in the direction of $-\vec{J}$ (Fig. 40-7).

As you will see in the next section, in typical atoms the orbital angular momenta and the spin angular momenta of most of the electrons sum vectorially to zero. Then $\vec{J}$ and $\vec{\mu}_{\text{eff}}$ of those atoms are due to a relatively small number of electrons, often only a single valence electron.

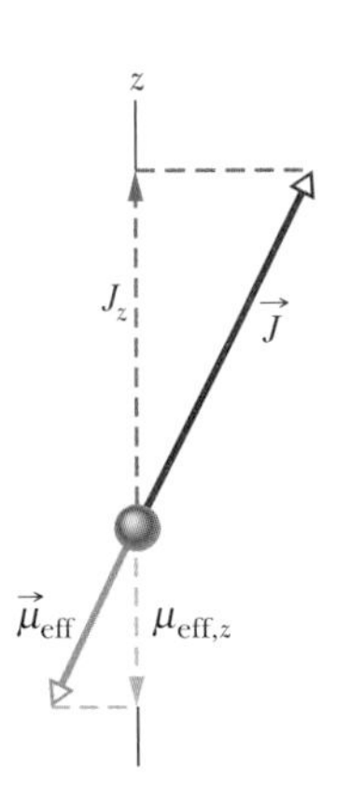

Fig. 40-7 A classical model showing the total angular momentum vector $\vec{J}$ and the effective magnetic moment vector $\vec{\mu}_{\text{eff}}$.

CHECKPOINT 1 An electron is in a quantum state for which the magnitude of the electron's orbital angular momentum $\vec{L}$ is $2\sqrt{3}\hbar$. How many projections of the electron's orbital magnetic dipole moment on a z axis are allowed?

40-5 The Stern–Gerlach Experiment

In 1922, Otto Stern and Walther Gerlach at the University of Hamburg in Germany showed experimentally that the magnetic moment of silver atoms is quantized. In the Stern–Gerlach experiment, as it is now known, silver is vaporized in an oven, and some of the atoms in that vapor escape through a narrow slit in the oven wall and pass into an evacuated tube. Some of those escaping atoms then pass through a second narrow slit, to form a narrow beam of atoms (Fig. 40-8). (The atoms are said to be *collimated*—made into a beam—and the second slit is called a *collimator.*) The beam passes between the poles of an electromagnet and then lands on a glass detector plate where it forms a silver deposit.

When the electromagnet is off, the silver deposit is a narrow spot. However, when the electromagnet is turned on, the silver deposit should be spread vertically. The reason is that silver atoms are magnetic dipoles, and so vertical magnetic forces act on them as they pass through the vertical magnetic field of the electromagnet; these forces deflect them slightly up or down. Thus, by analyzing the silver deposit on the plate, we can determine what deflections the atoms underwent in the magnetic field. When Stern and Gerlach analyzed the pattern of silver on their detector plate, they found a surprise. However, before we discuss that surprise and its quantum implications, let us discuss the magnetic deflecting force acting on the silver atoms.

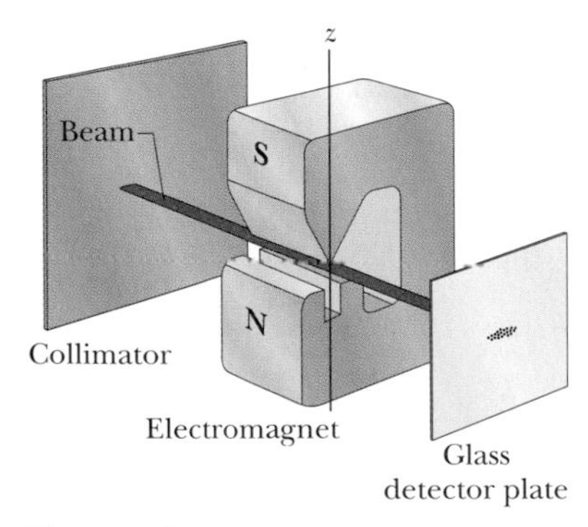

Fig. 40-8 Apparatus used by Stern and Gerlach.

The Magnetic Deflecting Force on a Silver Atom

We have not previously discussed the type of magnetic force that deflects the silver atoms in a Stern–Gerlach experiment. It is *not* the magnetic deflecting force that acts on a moving charged particle, as given by Eq. 28-2 ($\vec{F} = q\vec{v} \times \vec{B}$). The reason is simple: A silver atom is electrically neutral (its net charge q is zero), and thus this type of magnetic force is also zero.

The type of magnetic force we seek is due to an interaction between the magnetic field $\vec{B}$ of the electromagnet and the magnetic dipole of the individual silver atom. We can derive an expression for the force in this interaction by starting with the potential energy U of the dipole in the magnetic field. Equation 28-38 tells us that

$$U = -\vec{\mu} \cdot \vec{B}, \tag{40-15}$$

where $\vec{\mu}$ is the magnetic dipole moment of a silver atom. In Fig. 40-8, the positive direction of the z axis and the direction of $\vec{B}$ are vertically upward. Thus, we can write Eq. 40-15 in terms of the component μ_z of the atom's magnetic dipole moment along the direction of $\vec{B}$:

$$U = -\mu_z B. \qquad 3(40\text{-}16)$$

Then, using Eq. 8-22 ($F = -dU/dx$) for the z axis shown in Fig. 40-8, we obtain

$$F_z = -\frac{dU}{dz} = \mu_z \frac{dB}{dz}. \tag{40-17}$$

This is what we sought—an equation for the magnetic force that deflects a silver atom as the atom passes through a magnetic field.

The term dB/dz in Eq. 40-17 is the *gradient* of the magnetic field along the z axis. If the magnetic field does not change along the z axis (as in a uniform magnetic field or no magnetic field), then $dB/dz = 0$ and a silver atom is not deflected as it moves between the magnet's poles. In the Stern–Gerlach experiment, the poles are designed to maximize the gradient dB/dz, so as to vertically deflect the silver atoms passing between the poles as much as possible, so that their deflections show up in the deposit on the glass plate.

According to classical physics, the components μ_z of silver atoms passing through the magnetic field in Fig. 40-8 should range in value from $-\mu$ (the dipole moment $\vec{\mu}$ is directed straight down the z axis) to $+\mu$ ($\vec{\mu}$ is directed straight up the z axis). Thus, from Eq. 40-17, there should be a range of forces on the atoms, and therefore a range of deflections of the atoms, from a greatest downward deflection to a greatest upward deflection. This means that we should expect the atoms to land along a vertical line on the glass plate, but they *don't*.

The Experimental Surprise

What Stern and Gerlach found was that the atoms formed two distinct spots on the glass plate, one spot above the point where they would have landed with no deflection and the other spot just as far below that point. This two-spot result can be seen in the plots of Fig. 40-9, which shows the outcome of a more recent version of the Stern–Gerlach experiment. In that version, a beam of cesium atoms (magnetic dipoles like the silver atoms in the original Stern–Gerlach experiment) was sent through a magnetic field with a large vertical gradient dB/dz. The field could be turned on and off, and a detector could be moved up and down through the beam.

When the field was turned off, the beam was, of course, undeflected and the detector recorded the central-peak pattern shown in Fig. 40-9. When the field was turned on, the original beam was split vertically by the magnetic field into two smaller beams, one beam higher than the previously undeflected beam and the other beam lower. As the detector moved vertically up through these two smaller beams, it recorded the two-peak pattern shown in Fig. 40-9.

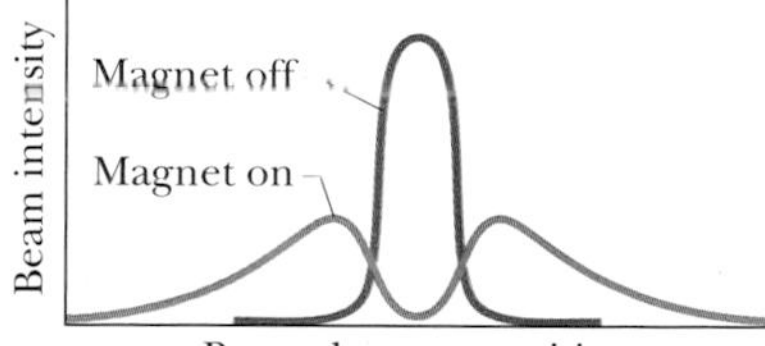

Fig. 40-9 Results of a modern repetition of the Stern–Gerlach experiment. With the electromagnet turned off, there is only a single beam; with the electromagnet turned on, the original beam splits into two subbeams. The two subbeams correspond to parallel and antiparallel alignment of the magnetic moments of cesium atoms with the external magnetic field.

The Meaning of the Results

In the original Stern–Gerlach experiment, two spots of silver were formed on the glass plate, not a vertical line of silver. This means that the component μ_z along $\vec{B}$ (and along z) could not have any value between $-\mu$ and $+\mu$ as classical physics predicts. Instead, μ_z is restricted to only two values, one for each spot on the glass. Thus, the original Stern–Gerlach experiment showed that μ_z is quantized, implying (correctly) that $\vec{\mu}$ is also. Moreover, because the angular momentum $\vec{L}$ of an atom is associated with $\vec{\mu}$, that angular momentum and its component L_z are also quantized.

With modern quantum theory, we can add to the explanation of the two-spot result in the Stern–Gerlach experiment. We now know that a silver atom consists of many electrons, each with a spin magnetic moment and an orbital magnetic moment. We also know that all those moments vectorially cancel out *except* for a single electron, and the orbital dipole moment of that electron is zero. Thus, the combined dipole moment $\vec{\mu}$ of a silver atom is the *spin* magnetic dipole moment of that single electron. According to Eq. 40-13, this means that μ_z can have only two components along the z axis in Fig. 40-8. One component is for quantum number $m_s = +\frac{1}{2}$ (the single electron is spin up), and the other component is for quantum number $m_s = -\frac{1}{2}$ (the single electron is spin down). Substituting into Eq. 40-13 gives us

$$\mu_{s,z} = -2(+\tfrac{1}{2})\mu_B = -\mu_B \quad \text{and} \quad \mu_{s,z} = -2(-\tfrac{1}{2})\mu_B = +\mu_B. \tag{40-18}$$

Then substituting these expressions for μ_z in Eq. 40-17, we find that the force component F_z deflecting the silver atoms as they pass through the magnetic field can have only the two values

$$F_z = -\mu_B\left(\frac{dB}{dz}\right) \quad \text{and} \quad F_z = +\mu_B\left(\frac{dB}{dz}\right), \tag{40-19}$$

which result in the two spots of silver on the glass.

Sample Problem 40-1

In the Stern–Gerlach experiment of Fig. 40-8, a beam of silver atoms passes through a magnetic field gradient dB/dz of magnitude 1.4 T/mm that is set up along the z axis. This region has a length w of 3.5 cm in the direction of the original beam. The speed of the atoms is 750 m/s. By what distance d have the atoms been deflected when they leave the region of the field gradient? The mass M of a silver atom is 1.8×10^{-25} kg.

Solution: One **Key Idea** here is that the deflection of a silver atom in the beam is due to an interaction between the magnetic dipole of the atom and the magnetic field, because of the gradient dB/dz. The deflecting force is directed along the field gradient (along the z axis) and is given by Eqs. 40-19. Let us consider only deflection in the positive direction of z; thus, we shall use $F_z = \mu_B(dB/dz)$ from Eqs. 40-19.

A second **Key Idea** is that we assume the field gradient dB/dz has the same value throughout the region through which the silver atoms travel. Thus, force component F_z is constant in that region, and from Newton's second law, the acceleration a_z of an atom along the z axis due to F_z is also constant and is given by

$$a_z = \frac{F_z}{M} = \frac{\mu_B(dB/dz)}{M}.$$

Because this acceleration is constant, we can use Eq. 2-15 (from Table 2-1) to write the deflection d parallel to the z axis as

$$d = v_{0z}t + \tfrac{1}{2}a_z t^2 = 0t + \tfrac{1}{2}\left(\frac{\mu_B(dB/dz)}{M}\right)t^2. \tag{40-20}$$

Because the deflecting force on the atom acts perpendicular to the atom's original direction of travel, the component v of the atom's velocity along the original direction of travel is not changed by the force. Thus, the atom requires time $t = w/v$ to travel through length w in that direction. Substituting w/v for t into Eq. 40-20, we find

$$\begin{aligned} d &= \tfrac{1}{2}\left(\frac{\mu_B(dB/dz)}{M}\right)\left(\frac{w}{v}\right)^2 = \frac{\mu_B(dB/dz)w^2}{2Mv^2} \\ &= (9.27 \times 10^{-24}\ \text{J/T})(1.4 \times 10^3\ \text{T/m}) \\ &\quad \times \frac{(3.5 \times 10^{-2}\ \text{m})^2}{(2)(1.8 \times 10^{-25}\ \text{kg})(750\ \text{m/s})^2} \\ &= 7.85 \times 10^{-5}\ \text{m} \approx 0.08\ \text{mm}. \end{aligned} \quad \text{(Answer)}$$

The separation between the two subbeams is twice this, or 0.16 mm. This separation is not large but is easily measured.

40-6 Magnetic Resonance

As we discussed briefly in Section 32-7, a proton has a spin magnetic dipole moment $\vec{\mu}$ that is associated with the proton's intrinsic spin angular momentum $\vec{S}$. The two vectors are said to be coupled together and, because the proton is positively charged, they are in the same direction. Suppose a proton is located in a magnetic field $\vec{B}$ that is directed along the positive direction of a z axis. Then $\vec{\mu}$ has two possible quantized components along that axis: the component can be $+\mu_z$ if the vector is in the direction of $\vec{B}$ (Fig. 40-10*a*) or $-\mu_z$ if it is opposite the direction of $\vec{B}$ (Fig. 40-10*b*).

From Eq. 28-38, ($U(\theta) = -\vec{\mu} \cdot \vec{B}$), recall that a potential energy is associated with the orientation of any magnetic dipole moment $\vec{\mu}$ located in an external magnetic field $\vec{B}$. Thus, energy is associated with the two orientations shown in Figs. 40-10*a* and *b*. The orientation in Fig. 40-10*a* is the lower-energy state and is called the *spin-up state* because the proton's spin component S_z (not shown) is also aligned with $\vec{B}$. The orientation in Fig. 40-10*b* (the *spin-down state*) is the higher-energy state ($\mu_z B$). Thus, the energy difference between these two states is

$$\Delta E = \mu_z B - (-\mu_z B) = 2\mu_z B. \tag{40-21}$$

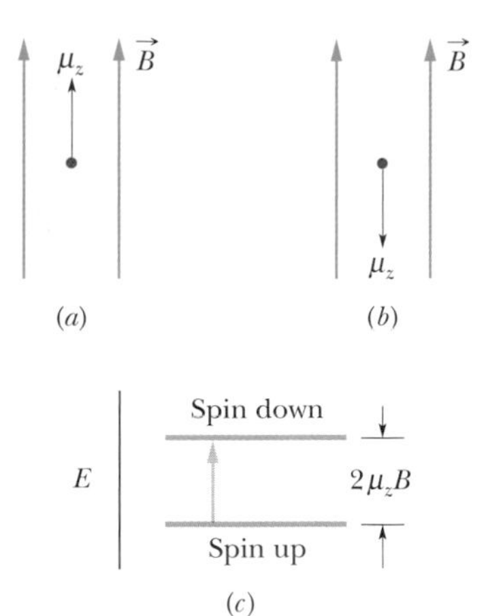

Fig. 40-10 The z component of $\vec{\mu}$ for a proton in the (a) lower-energy (spin-up) and (b) higher-energy (spin-down) state. (c) An energy-level diagram for the states, showing the upward quantum jump the proton makes when its spin flips from up to down.

If we place a sample of water in a magnetic field $\vec{B}$, the protons in the hydrogen portions of each water molecule tend to be in the lower-energy state. (We shall not consider the oxygen portions.) Any one of these protons can jump to the higher-energy state by absorbing a photon with an energy hf equal to ΔE. That is, the proton can jump by absorbing a photon of energy

$$hf = 2\mu_z B. \tag{40-22}$$

Such absorption is called **magnetic resonance** or, as originally, **nuclear magnetic resonance** (NMR), and the consequent reversal of the spin component S_z is called *spin-flipping*.

In practice, the photons required for magnetic resonance have an associated frequency in the radio-frequency (RF) range and are provided by a small coil wrapped around the sample undergoing resonance. An electromagnetic oscillator called an *RF source* drives a sinusoidal current in the coil at frequency f. The electromagnetic (EM) field set up within the coil and sample also oscillates at frequency f. If f meets the requirement of Eq. 40-22, the oscillating EM field can transfer a quantum of energy to a proton in the sample via a photon absorption, spin-flipping the proton.

The magnetic field magnitude B that appears in Eq. 40-22 is actually the magnitude of the net magnetic field $\vec{B}$ at the site where a given proton undergoes spin-flipping. That net field is the vector sum of the external field $\vec{B}_{\text{ext}}$ set up by the magnetic resonance equipment (primarily a large magnet) and the internal field $\vec{B}_{\text{int}}$ set up by the magnetic dipole moments of the atoms and nuclei near the given proton. For practical reasons we do not discuss here, magnetic resonance is usually detected by sweeping the magnitude B_{ext} through a range of values while the frequency f of the RF source is kept at a predetermined value and the energy of the RF source is monitored. A graph of the energy loss of the RF source versus B_{ext} shows a *resonance peak* when B_{ext} sweeps through the value at which spin-flipping occurs. Such a graph is called a *nuclear magnetic resonance spectrum*, or *NMR spectrum*.

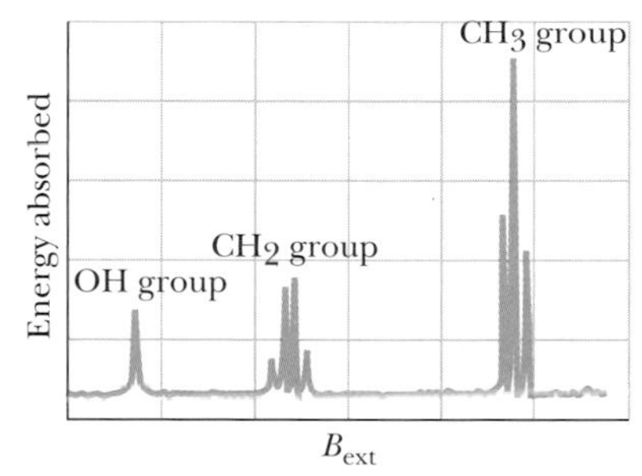

Fig. 40-11 A nuclear magnetic resonance spectrum for ethanol, CH_3CH_2OH. The spectral lines represent the absorption of energy associated with spin-flips of protons. The three groups of lines correspond, as indicated, to protons in the OH group, the CH_2 group, and the CH_3 group of the ethanol molecule. Note that the two protons in the CH_2 group occupy four different local environments. The entire horizontal axis covers less than 10^{-4} T.

Figure 40-11 shows the NMR spectrum of ethanol, which is a molecule consisting of three groups of atoms: CH_3, CH_2, and OH. Protons in each group can undergo magnetic resonance, but each group has its own unique magnetic-resonance value of B_{ext} because the groups lie in different internal fields $\vec{B}_{\text{int}}$ due to their arrangement within the CH_3CH_2OH molecule. Thus, the resonance peaks

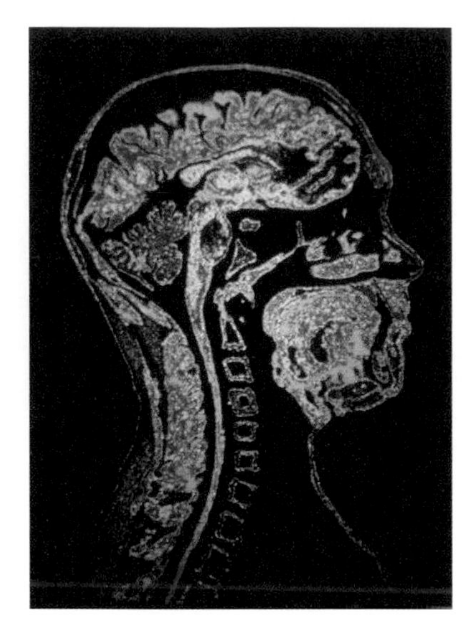

Fig. 40-12 A cross-sectional view of a human head and neck produced by magnetic resonance imaging. Some of the details visible here would not show up on an x-ray image, even with a modern computerized axial tomography scanner (CAT scanner).

in the spectrum of Fig. 40-11 form a unique NMR signature by which ethanol can be indentified.

Because many substances have unique NMR signatures, magnetic resonance is used to identify unknown substances, such as in forensic work of a criminal investigation. Additionally, a procedure called **magnetic resonance imaging** (MRI) has been applied to medical diagnostics with great success. The protons in the various tissues of the human body are situated in many different internal magnetic environments. When the body, or part of it, is immersed in a strong external magnetic field, these environmental differences can be detected by spin-flip techniques and translated by computer processing into an image resembling those produced by x rays. Figure 40-12, for example, shows a cross section of a human head imaged by this method.

Sample Problem 40-2

In an NMR experiment, a drop of water is suspended in a uniform external magnetic field $\vec{B}_{ext}$. Assume the internal field $\vec{B}_{int}$ is negligible. The magnitude of μ_z for a proton in the hydrogen atoms of the water molecules is 1.41×10^{-26} J/T. Magnetic resonance occurs when $B_{ext} = 1.80$ T. What is the frequency f of the RF source causing the protons to spin-flip, and what wavelength λ is associated with a photon absorbed in the spin-flipping?

Solution: One **Key Idea** here is that when a proton is located in an external magnetic field $\vec{B}_{ext}$, it has a potential energy because it is a magnetic dipole. A second **Key Idea** is that this potential energy is restricted to two values, with a difference of $2\mu_z B$. The third **Key Idea** is that if the proton is to jump between these two energies (spin-flip), the photon energy hf must be equal to the energy difference $2\mu_z B$, according to Eq. 40-22. From that equation with $B = B_{ext} = 1.80$ T, we then find

$$f = \frac{2\mu_z B}{h} = \frac{(2)(1.41 \times 10^{-26}\ \text{J/T})(1.80\ \text{T})}{6.63 \times 10^{-34}\ \text{J}\cdot\text{s}}$$
$$= 7.66 \times 10^{7}\ \text{Hz} = 76.6\ \text{MHz}. \qquad \text{(Answer)}$$

This is the frequency associated with the photons absorbed in the spin-flipping; it is also the frequency of the RF source and thus of the oscillating electromagnetic fields set up by that source. The wavelength associated with a photon absorbed in the spin-flipping is

$$\lambda = \frac{c}{f} = \frac{3.00 \times 10^{8}\ \text{m/s}}{7.66 \times 10^{7}\ \text{Hz}} = 3.92\ \text{m}. \qquad \text{(Answer)}$$

40-7 The Pauli Exclusion Principle

In Chapter 39 we considered a variety of electron traps, from fictional one-dimensional traps to the real three-dimensional trap of a hydrogen atom. In all those examples, we trapped only one electron. However, when we discuss traps containing two or more electrons (as we shall in the next two sections), we must consider a principle that governs any particle whose spin quantum number s is not zero or an integer. This principle applies not only to electrons but also to protons and neutrons, all of which have $s = \frac{1}{2}$. The principle is known as the **Pauli exclusion principle** after Wolfgang Pauli, who formulated it in 1925. For electrons, it states that

> No two electrons confined to the same trap can have the same set of values for their quantum numbers.

As we shall discuss in Section 40-9, this principle means that no two electrons in an atom can have the same four values for the quantum numbers n, ℓ, m_ℓ, and m_s. All electrons have the same quantum number $s = \frac{1}{2}$. Thus, any two electrons in an atom must differ in at least one of these other quantum numbers. Were this not true, atoms would collapse, and thus you and the world could not exist.

40-8 Multiple Electrons in Rectangular Traps

To prepare for our discussion of multiple electrons in atoms, let us discuss two electrons confined to the rectangular traps of Chapter 39. We shall again use the quantum numbers we found for those traps when only one electron was confined. However, here we shall also include the spin angular momenta of the two electrons. To do this, we assume that the traps are located in a uniform magnetic field. Then according to Eq. 40-12 ($S_z = m_s\hbar$), an electron can be either spin up with $m_s = \frac{1}{2}$ or spin down with $m_s = -\frac{1}{2}$. (We assume that the field is very weak so that we can neglect the potential energies of the electrons due to it.)

As we confine the two electrons to one of the traps, we must keep the Pauli exclusion principle in mind; that is, the electrons cannot have the same set of values for their quantum numbers.

1. *One-dimensional trap.* In the one-dimensional trap of Fig. 39-2, fitting an electron wave to the trap's width L requires the single quantum number n. Therefore, any electron confined to the trap must have a certain value of n, and its quantum number m_s can be either $+\frac{1}{2}$ or $-\frac{1}{2}$. The two electrons could have different values of n, or they could have the same value of n if one of them is spin up and the other is spin down.
2. *Rectangular corral.* In the rectangular corral of Fig. 39-13, fitting an electron wave to the corral's widths L_x and L_y requires the two quantum numbers n_x and n_y. Thus, any electron confined to the trap must have certain values for those two quantum numbers, and its quantum number m_s can be either $+\frac{1}{2}$ or $-\frac{1}{2}$; so now there are three quantum numbers. According to the Pauli exclusion principle, two electrons confined to the trap must have different values for at least one of those three quantum numbers.
3. *Rectangular box.* In the rectangular box of Fig. 39-14, fitting an electron wave to the box's widths L_x, L_y, and L_z requires the three quantum numbers n_x, n_y, and n_z. Thus, any electron confined to the trap must have certain values for these three quantum numbers, and its quantum number m_s can be either $+\frac{1}{2}$ or $-\frac{1}{2}$; so now there are four quantum numbers. According to the Pauli exclusion principle, two electrons confined to the trap must have different values for at least one of those four quantum numbers.

Suppose we add more than two electrons, one by one, to a rectangular trap in the preceding list. The first electrons naturally go into the lowest possible energy level—they are said to *occupy* that level. However, eventually the Pauli exclusion principle disallows any more electrons from occupying that lowest energy level, and the next electron must occupy the next higher level. When an energy level cannot be occupied by more electrons because of the Pauli exclusion principle, we say that level is **full** or **fully occupied.** In contrast, a level that is not occupied by any electrons is **empty** or **unoccupied.** For intermediate situations, the level is **partially occupied.** The *electron configuration* of a system of trapped electrons is a listing or drawing either of the energy levels the electrons occupy or of the set of the quantum numbers of the electrons.

Finding the Total Energy

We shall later want to find the energy of a *system* of two or more electrons confined to a rectangular trap. That is, we shall want to find the total energy for any configuration of the trapped electrons.

For simplicity, we shall assume that the electrons do not electrically interact with one another; that is, we shall neglect the electric potential energies of pairs of electrons. In that case, we can calculate the total energy for any electron configuration by calculating the energy of each electron as we did in Chapter 39, and then summing those energies. (In Sample Problem 40-3 we do so for seven electrons confined to a rectangular corral.)

A good way to organize the energy values of a given system of electrons is with an energy-level diagram *for the system,* just as we did for a single electron in the traps of Chapter 39. The lowest level, with energy E_{gr}, corresponds to the ground state of the system. The next higher level, with energy E_{fe}, corresponds to the first excited state of the system. The next level, with energy E_{se}, corresponds to the second excited state of the system, and so on.

Sample Problem 40-3

Seven electrons are confined to the square corral of Sample Problem 39-5, where the corral is a two-dimensional infinite potential well with widths $L_x = L_y = L$ (Fig. 39-13). Assume that the electrons do not electrically interact with one another.

(a) What is the electron configuration for the ground state of the system of seven electrons?

Solution: We can determine the electron configuration of the system by placing the seven electrons in the corral one by one, to build up the system. One **Key Idea** here is that because we assume the electrons do not electrically interact with one another, we can use the energy-level diagram for a single trapped electron in order to keep track of how we place the seven electrons in the corral. That *one-electron energy-level diagram* is given in Fig. 39-15 and partially reproduced here as Fig. 40-13*a*. Recall that the levels are labeled as $E_{nx,ny}$ for their associated energy. For example, the lowest level is for energy $E_{1,1}$, where quantum number n_x is 1 and quantum number n_y is 1.

A second **Key Idea** here is that the trapped electrons must obey the Pauli exclusion principle; that is, no two electrons can have the same set of values for their quantum numbers n_x, n_y, and m_s.

The first electron goes into energy level $E_{1,1}$ and can have $m_s = \frac{1}{2}$ or $m_s = -\frac{1}{2}$. We arbitrarily choose the latter and draw a down arrow (to represent spin down) on the $E_{1,1}$ level in Fig. 40-13*a*. The second electron also goes into the $E_{1,1}$ level but must have $m_s = +\frac{1}{2}$ so that one of its quantum numbers differs from those of the first electron. We represent this second electron with an up arrow (for spin up) on the $E_{1,1}$ level in Fig. 40-13*b*.

Another **Key Idea** now comes into play: The level for energy $E_{1,1}$ is fully occupied, and thus the third electron cannot have that energy. Therefore, the third electron goes into the next higher level, which is for the equal energies $E_{2,1}$ and $E_{1,2}$ (the level is degenerate). This third electron can have quantum numbers n_x and n_y of either 1 and 2 or 2 and 1, respectively. It can also have a quantum number m_s of either $+\frac{1}{2}$ or $-\frac{1}{2}$. Let us arbitrarily assign it the quantum numbers $n_x = 2$, $n_y = 1$, and $m_s = -\frac{1}{2}$. We then represent it with a down arrow on the level for $E_{1,2}$ and $E_{2,1}$ in Fig. 40-13*c*.

You can show that the next three electrons can also go into the level for energies $E_{2,1}$ and $E_{1,2}$, provided that no set of three quantum numbers is completely duplicated. That level then contains four electrons, with quantum numbers (n_x, n_y, m_s) of

$$(2, 1, -\tfrac{1}{2}),\ (2, 1, +\tfrac{1}{2}),\ (1, 2, -\tfrac{1}{2}),\ (1, 2, +\tfrac{1}{2}),$$

and the level is fully occupied. Thus, the seventh electron goes into the next higher level, which is the $E_{2,2}$ level. Let us arbitrarily assume this electron is spin down, with $m_s = -\frac{1}{2}$.

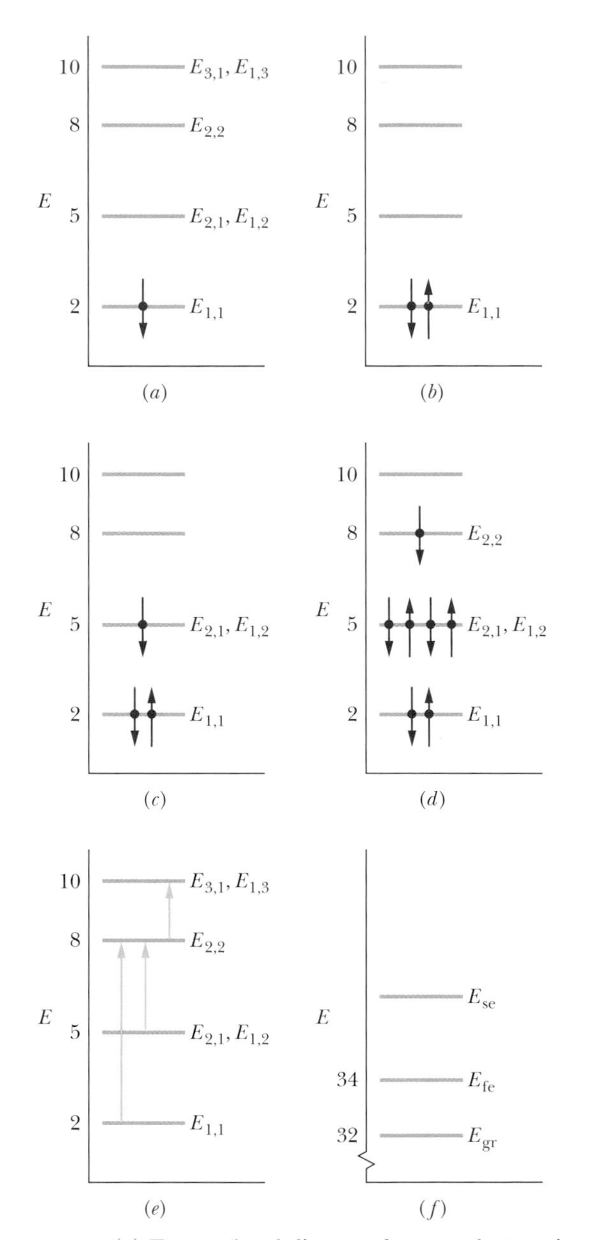

Fig. 40-13 (*a*) Energy-level diagram for one electron in a square corral of widths L. (Energy E is in multiples of $h^2/8mL^2$.) A spin-down electron occupies the lowest level. (*b*) Two electrons (one spin down, the other spin up) occupy the lowest level of the one-electron energy-level diagram. (*c*) A third electron occupies the next energy level. (*d*) The system's ground-state configuration, for all seven electrons. (*e*) Three transitions to consider as possibly taking the seven-electron system to its first excited state. (*f*) The system's energy-level diagram, for the lowest three total energies of the system (in multiples of $h^2/8mL^2$).

Figure 40-13*d* shows all seven electrons on a one-electron energy-level diagram. We now have seven electrons in the corral, and they are in the configuration with the lowest energy that satisfies the Pauli exclusion principle. Thus, the ground-state configuration of the system is that shown in Fig. 40-13*d* and listed in Table 40-2.

(b) What is the total energy of the seven-electron system in its ground state, as a multiple of $h^2/8mL^2$?

Solution: The **Key Idea** here is that the total energy E_{gr} of the system in its ground state is the sum of the energies of the individual electrons in the system's ground-state configuration. The energy of each electron can be read from Table 39-1, which is partially reproduced in Table 40-2, or from Fig. 40-13*d*. Because there are two electrons in the first (lowest) level, four in the second level, and one in the third level, we have

$$E_{gr} = 2\left(2\frac{h^2}{8mL^2}\right) + 4\left(5\frac{h^2}{8mL^2}\right) + 1\left(8\frac{h^2}{8mL^2}\right)$$

$$= 32\frac{h^2}{8mL^2}. \qquad \text{(Answer)}$$

(c) How much energy must be transferred to the system for it to jump to its first excited state, and what is the energy of that state?

Solution: The **Key Ideas** here are these:

1. If the system is to be excited, one of the seven electrons must make a quantum jump up the one-electron energy-level diagram of Fig. 40-13*d*.
2. If that jump is to occur, the energy change ΔE of the electron (and thus of the system) must be $\Delta E = E_{high} - E_{low}$ (Eq. 39-5), where E_{low} is the energy of the level where the jump begins and E_{high} is the energy of the level where the jump ends.
3. The Pauli exclusion principle must still apply; in particular, an electron *cannot* jump to a level that is fully occupied.

Let us consider the three jumps shown in Fig. 40-13*e*; all are allowed by the Pauli exclusion principle because they are jumps to either empty or partially occupied states. In one of those possible jumps, an electron jumps from the $E_{1,1}$ level to the partially occupied $E_{2,2}$ level. The change in the energy is

$$\Delta E = E_{2,2} - E_{1,1} = 8\frac{h^2}{8mL^2} - 2\frac{h^2}{8mL^2} = 6\frac{h^2}{8mL^2}.$$

(We shall assume that the spin orientation of the electron making the jump can change as needed.)

TABLE 40-2
Ground-State Configuration and Energies

n_x	n_y	m_s	Energy*
2	2	$-\frac{1}{2}$	8
2	1	$+\frac{1}{2}$	5
2	1	$-\frac{1}{2}$	5
1	2	$+\frac{1}{2}$	5
1	2	$-\frac{1}{2}$	5
1	1	$+\frac{1}{2}$	2
1	1	$-\frac{1}{2}$	2
			Total 32

*In multiples of $h^2/8mL^2$

In another of the possible jumps in Fig. 40-13*e*, an electron jumps from the degenerate level of $E_{2,1}$ and $E_{1,2}$ to the partially occupied $E_{2,2}$ level. The change in the energy is

$$\Delta E = E_{2,2} - E_{2,1} = 8\frac{h^2}{8mL^2} - 5\frac{h^2}{8mL^2} = 3\frac{h^2}{8mL^2}.$$

In the third possible jump in Fig. 40-13*e*, the electron in the $E_{2,2}$ level jumps to the unoccupied, degenerate level of $E_{1,3}$ and $E_{3,1}$. The change in energy is

$$\Delta E = E_{1,3} - E_{2,2} = 10\frac{h^2}{8mL^2} - 8\frac{h^2}{8mL^2} = 2\frac{h^2}{8mL^2}.$$

Of these three possible jumps, the one requiring the least energy change ΔE is the last one. We could consider even more possible jumps, but none would require less energy. Thus, for the system to jump from its ground state to its first excited state, the electron in the $E_{2,2}$ level must jump to the unoccupied, degenerate level of $E_{1,3}$ and $E_{3,1}$, and the required energy is

$$\Delta E = 2\frac{h^2}{8mL^2}. \qquad \text{(Answer)}$$

The energy E_{fe} of the first excited state of the system is then

$$E_{fe} = E_{gr} + \Delta E$$

$$= 32\frac{h^2}{8mL^2} + 2\frac{h^2}{8mL^2} = 34\frac{h^2}{8mL^2}. \qquad \text{(Answer)}$$

We can represent this energy and the energy E_{gr} for the ground state of the system on an energy-level diagram *for the system*, as shown in Fig. 40-13*f*.

40-9 Building the Periodic Table

The four quantum numbers n, ℓ, m_ℓ, and m_s identify the quantum states of individual electrons in a multielectron atom. The wave functions for these states, however, are not the same as the wave functions for the corresponding states of the hydrogen atom because, in multielectron atoms, the potential energy associated with a given electron is determined not only by the charge and position of the atom's nucleus but also by the charges and positions of all the other electrons in the atom. Solutions of Schrödinger's equation for multielectron atoms can be carried out numerically—in principle at least—using a computer.

As we discussed in Sections 39-9 and 40-3, all states with the same values of the quantum numbers n and ℓ form a subshell. For a given value of ℓ, there are

$2\ell + 1$ possible values of the magnetic quantum number m_ℓ and, for each m_ℓ, there are two possible values for the spin quantum number m_s. Thus, there are $2(2\ell + 1)$ states in a subshell. It turns out that *all states in a given subshell have the same energy,* its value being determined primarily by the value of n and to a lesser extent by the value of ℓ.

For the purpose of labeling subshells, the values of ℓ are represented by letters:

$$\begin{array}{llllllll} \ell = 0 & 1 & 2 & 3 & 4 & 5 & \ldots \\ s & p & d & f & g & h & \ldots . \end{array}$$

For example, the $n = 3$, $\ell = 2$ subshell would be labeled the $3d$ subshell.

When we assign electrons to states in a multielectron atom, we must be guided by the Pauli exclusion principle of Section 40-7; that is, no two electrons in an atom can have the same set of the quantum numbers n, ℓ, m_ℓ, and m_s. If this important principle did not hold, *all* the electrons in any atom could jump to the atom's lowest energy level, which would eliminate the chemistry of atoms and molecules, and thus also eliminate biochemistry and us. Let us examine the atoms of a few elements to see how the Pauli exclusion principle operates in the building up of the periodic table.

Neon

The neon atom has 10 electrons. Only two of them fit into the lowest-energy subshell, the $1s$ subshell. These two electrons both have $n = 1$, $\ell = 0$, and $m_\ell = 0$, but one has $m_s = +\frac{1}{2}$ and the other has $m_s = -\frac{1}{2}$. The $1s$ subshell contains $2[2(0) + 1] = 2$ states. Because this subshell then contains all the electrons permitted by the Pauli principle, it is said to be **closed.**

Two of the remaining eight electrons fill the next lowest energy subshell, the $2s$ subshell. The last six electrons just fill the $2p$ subshell which, with $\ell = 1$, holds $2[2(1) + 1] = 6$ states.

In a closed subshell, all allowed z projections of the orbital angular momentum vector $\vec{L}$ are present and, as you can verify from Fig. 40-5, these projections cancel for the subshell as a whole; for every positive projection there is a corresponding negative projection of the same magnitude. Similarly, the z projections of the spin angular momenta also cancel. Thus, a closed subshell has no angular momentum and no magnetic moment of any kind. Furthermore, its probability density is spherically symmetric. Then neon with its three closed subshells ($1s$, $2s$, and $2p$) has no "loosely dangling electrons" to encourage chemical interaction with other atoms. Neon, like the other **noble gases** that form the right-hand column of the periodic table, is almost chemically inert.

Sodium

Next after neon in the periodic table comes sodium, with 11 electrons. Ten of them form a closed neon-like core, which, as we have seen, has zero angular momentum. The remaining electron is largely outside this inert core, in the $3s$ subshell—the next lowest energy subshell. Because this **valence electron** of sodium is in a state with $\ell = 0$ (that is, an s state using the lettering system above), the sodium atom's angular momentum and magnetic dipole moment must be due entirely to the spin of this single electron.

Sodium readily combines with other atoms that have a "vacancy" into which sodium's loosely bound valence electron can fit. Sodium, like the other **alkali metals** that form the left-hand column of the periodic table, is chemically active.

Chlorine

The chlorine atom, which has 17 electrons, has a closed 10-electron, neon-like core, with 7 electrons left over. Two of them fill the $3s$ subshell, leaving five to be assigned to the $3p$ subshell, which is the subshell next lowest in energy. This subshell, which has $\ell = 1$, can hold $2[2\ell(0) + 1] = 6$ electrons, and so there is a vacancy, or a "hole," in this subshell.

Chlorine is receptive to interacting with other atoms that have a valence electron that might fill this hole. Sodium chloride (NaCl), for example, is a very stable compound. Chlorine, like the other **halogens** that form column VIIA of the periodic table, is chemically active.

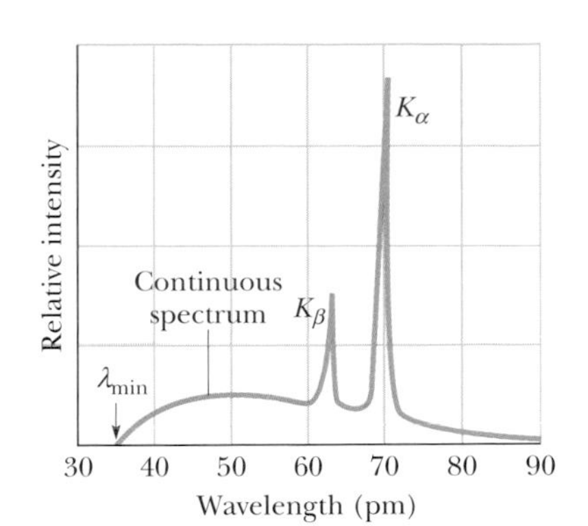

Fig. 40-14 The distribution by wavelength of the x rays produced when 35 keV electrons strike a molybdenum target. The sharp peaks and the continuous spectrum from which they rise are produced by different mechanisms.

Iron

The arrangement of the 26 electrons of the iron atom can be represented as follows:

$$\underbrace{1s^2 \;\; 2s^2\, 2p^6 \;\; 3s^2\, 3p^6}\; 3d^6 \;\; 4s^2.$$

The subshells are listed in numerical order and, following convention, a superscript gives the number of electrons in each subshell. From Table 40-1 we can see that an s subshell ($\ell = 0$) can hold 2 electrons, a p subshell ($\ell = 1$) can hold 6, and a d subshell ($\ell = 2$) can hold 10. Thus, iron's first 18 electrons form the five filled subshells that are marked off by the bracket, leaving 8 electrons to be accounted for. Six of the eight go into the $3d$ subshell, and the remaining two go into the $4s$ subshell.

The reason the last two electrons do not also go into the $3d$ subshell (which can hold 10 electrons) is that the $3d^6\, 4s^2$ configuration results in a lower-energy state for the atom as a whole than would the $3d^8$ configuration. An iron atom with 8 electrons (rather than 6) in the $3d$ subshell would quickly make a transition to the $3d^6\, 4s^2$ configuration, emitting electromagnetic radiation in the process. The lesson here is that except for the simplest elements, the states may not be filled in what we might think of as their "logical" sequence.

40-10 X Rays and the Ordering of the Elements

When a solid target, such as solid copper or tungsten, is bombarded with electrons whose kinetic energies are in the kiloelectron-volt range, electromagnetic radiation called **x rays** is emitted. Our concern here is what these rays—whose medical, dental, and industrial usefulness is so well known and widespread—can teach us about the atoms that absorb or emit them. Figure 40-14 shows the wavelength spectrum of the x rays produced when a beam of 35 keV electrons falls on a molybdenum target. We see a broad, continuous spectrum of radiation on which are superimposed two peaks of sharply defined wavelengths. The continuous spectrum and the peaks arise in different ways, which we next discuss separately.

The Continuous X-Ray Spectrum

Here we examine the continuous x-ray spectrum of Fig. 40-14, ignoring for the time being the two prominent peaks that rise from it. Consider an electron of initial kinetic energy K_0 that collides (interacts) with one of the target atoms, as in Fig. 40-15. The electron may lose an amount of energy ΔK, which will appear as the energy of an x-ray photon that is radiated away from the site of the collision. (Very little energy is transferred to the recoiling atom because of the relatively large mass of the atom; here we neglect that transfer.)

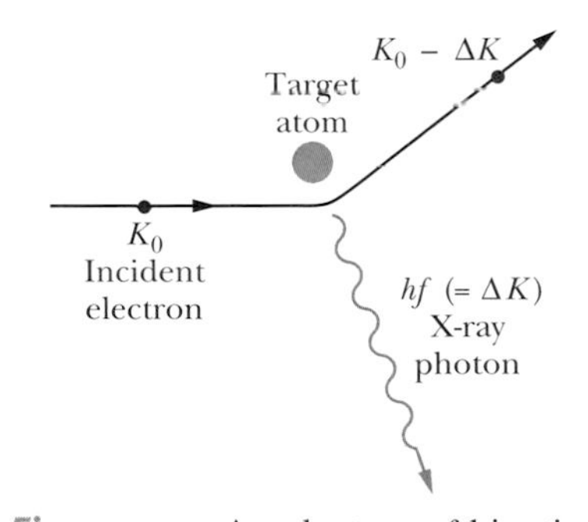

Fig. 40-15 An electron of kinetic energy K_0 passing near an atom in the target may generate an x-ray photon, the electron losing part of its energy in the process. The continuous x-ray spectrum arises in this way.

The scattered electron in Fig. 40-15, whose energy is now less than K_0, may have a second collision with a target atom, generating a second photon, whose energy will in general be different from the energy of the photon produced in the first collision. This electron-scattering process can continue until the electron is approximately stationary. All the photons generated by these collisions form part of the continuous x-ray spectrum.

A prominent feature of that spectrum in Fig. 40-14 is the sharply defined **cutoff wavelength** λ_{min}, below which the continuous spectrum does not exist. This minimum wavelength corresponds to a collision in which an incident electron loses *all* its initial kinetic energy K_0 in a single head-on collision with a target atom. Essentially all this energy appears as the energy of a single photon, whose associated wavelength—the minimum possible x-ray wavelength—is found from

$$K_0 = hf = \frac{hc}{\lambda_{min}},$$

or

$$\lambda_{min} = \frac{hc}{K_0} \quad \text{(cutoff wavelength).} \tag{40-23}$$

The cutoff wavelength is totally independent of the target material. If we were to switch from a molybdenum target to a copper target, for example, all features of the x-ray spectrum of Fig. 40-14 would change *except* the cutoff wavelength.

CHECKPOINT 2 Does the cutoff wavelength λ_{min} of the continuous x-ray spectrum increase, decrease, or remain the same if you (a) increase the kinetic energy of the electrons that strike the x-ray target, (b) allow the electrons to strike a thin foil rather than a thick block of the target material, (c) change the target to an element of higher atomic number?

Sample Problem 40-4

A beam of 35.0 keV electrons strikes a molybdenum target, generating the x rays whose spectrum is shown in Fig. 40-14. What is the cutoff wavelength?

Solution: The **Key Idea** here is that the cutoff wavelength λ_{min} corresponds to an electron transferring (approximately) all of its energy to an x-ray photon, thus producing a photon with the greatest possible frequency and least possible wavelength. From Eq. 40-23, we have

$$\lambda_{min} = \frac{hc}{K_0} = \frac{(4.14 \times 10^{-15}\ \text{eV}\cdot\text{s})(3.00 \times 10^8\ \text{m/s})}{35.0 \times 10^3\ \text{eV}}$$
$$= 3.55 \times 10^{-11}\ \text{m} = 35.5\ \text{pm}. \quad \text{(Answer)}$$

The Characteristic X-Ray Spectrum

We now turn our attention to the two peaks of Fig. 40-14, labeled K_α and K_β. These (and other peaks that appear at wavelengths beyond the range displayed in Fig. 40-14) form the **characteristic x-ray spectrum** of the target material.

The peaks arise in a two-part process. (1) An energetic electron strikes an atom in the target and, while it is being scattered, the incident electron knocks out one of the atom's deep-lying (low n value) electrons. If the deep-lying electron is in the shell defined by $n = 1$ (called, for historical reasons, the K shell), there remains a vacancy, or *hole,* in this shell. (2) An electron in one of the shells with a higher energy jumps to the K shell, filling the hole in this shell. During this jump, the atom emits a characteristic x-ray photon. If the electron that fills the K-shell vacancy jumps from the shell with $n = 2$ (called the L shell), the emitted radiation is the K_α line of Fig. 40-14; if it jumps from the shell with $n = 3$ (called the M shell), it produces the K_β line, and so on. The hole left in either the L or M shell will be filled by an electron from still farther out in the atom.

In studying x rays, it is more convenient to keep track of where a hole is created deep in the atom's "electron cloud" than to record the changes in the quantum state of the electrons that jump to fill that hole. Figure 40-16 does exactly that; it is an energy-level diagram for molybdenum, the element to which Fig. 40-14 refers. The baseline ($E = 0$) represents the neutral atom in its ground state. The level marked K (at $E = 20$ keV) represents the energy of the molybdenum atom with a hole in its K shell, the level marked L (at $E = 2.7$ keV) represents the atom with a hole in its L shell, and so on.

The transitions marked K_α and K_β in Fig. 40-16 are the ones that produce the two x-ray peaks in Fig. 40-14. The K_α spectral line, for example, originates when an electron from the L shell fills a hole in the K shell. To state this transition in terms of what the arrows in Fig. 40-16 show, a hole originally in the K shell moves to the L shell.

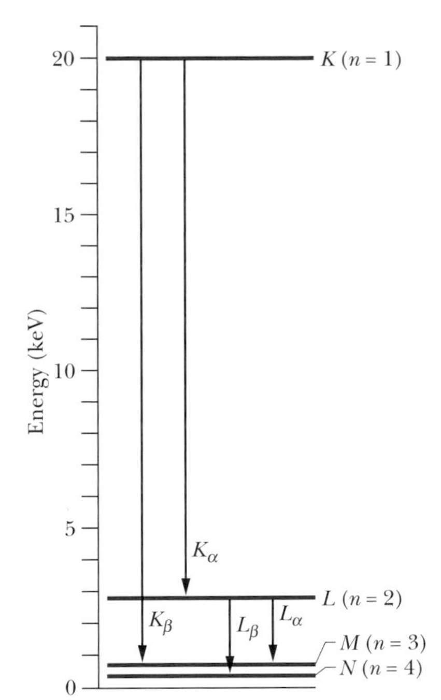

Fig. 40-16 A simplified energy-level diagram for a molybdenum atom, showing the transitions (of holes rather than electrons) that give rise to some of the characteristic x rays of that element. Each horizontal line represents the energy of the atom with a hole (a missing electron) in the shell indicated.

Ordering the Elements

In 1913, British physicist H. G. J. Moseley generated characteristic x rays for as many elements as he could find—he found 38—by using them as targets for electron bombardment in an evacuated tube of his own design. By means of a trolley manipulated by strings, Moseley was able to move the individual targets into the path of an electron beam. He measured the wavelengths of the emitted x rays by the crystal diffraction method described in Section 36-10.

Moseley then sought (and found) regularities in these spectra as he moved from element to element in the periodic table. In particular, he noted that if, for a given spectral line such as K_α, he plotted for each element the square root of the frequency f against the position of the element in the periodic table, a straight line resulted. Figure 40-17 shows a portion of his extensive data. Moseley's conclusion was this:

> We have here a proof that there is in the atom a fundamental quantity, which increases by regular steps as we pass from one element to the next. This quantity can only be the charge on the central nucleus.

As a result of Moseley's work, the characteristic x-ray spectrum became the universally accepted signature of an element, permitting the solution of a number of periodic table puzzles. Prior to that time (1913), the positions of elements in the table were assigned in order of atomic *mass,* although it was necessary to invert this order for several pairs of elements because of compelling chemical evidence; Moseley showed that it is the nuclear charge (that is, atomic number Z) that is the real basis for ordering the elements.

In 1913 the periodic table had several empty squares, and a surprising number of claims for new elements had been advanced. The x-ray spectrum provided a conclusive test of such claims. The lanthanide elements, often called the rare earth elements, had been sorted out only imperfectly because their similar chemical properties made sorting difficult. Once Moseley's work was reported, these elements were properly organized. In more recent times, the identities of some elements beyond uranium were pinned down beyond dispute when the elements became available in quantities large enough to permit a study of their individual x-ray spectra.

It is not hard to see why the characteristic x-ray spectrum shows such impressive regularities from element to element whereas the optical spectrum in

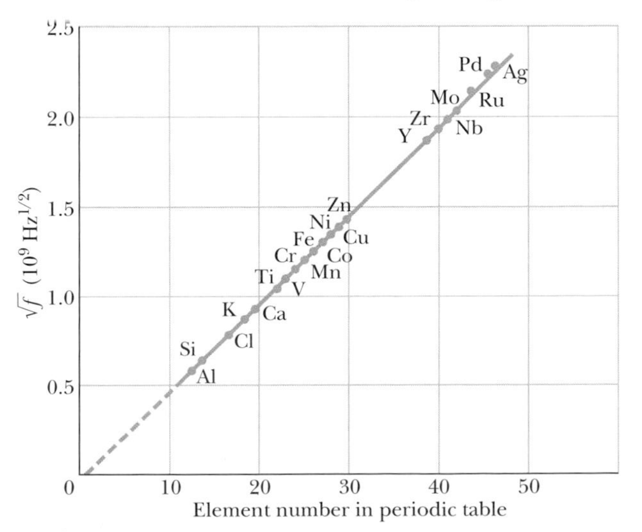

Fig. 40-17 A Moseley plot of the K_α line of the characteristic x-ray spectra of 21 elements. The frequency is calculated from the measured wavelength.

the visible and near-visible region does not: The key to the identity of an element is the charge on its nucleus. Gold, for example, is what it is because its atoms have a nuclear charge of $+79e$ (that is, $Z = 79$). An atom with one more elementary charge on its nucleus is mercury; with one fewer, it is platinum. The K electrons, which play such a large role in the production of the x-ray spectrum, lie very close to the nucleus and are thus sensitive probes of its charge. The optical spectrum, on the other hand, involves transitions of the outermost electrons, which are heavily screened from the nucleus by the remaining electrons of the atom and thus are *not* sensitive probes of nuclear charge.

Accounting for the Moseley Plot

Moseley's experimental data, of which the Moseley plot of Fig. 40-17 is but a part, can be used directly to assign the elements to their proper places in the periodic table. This can be done even if no theoretical basis for Moseley's results can be established. However, there is such a basis.

According to Eqs. 39-32 and 39-33, the energy of the hydrogen atom is

$$E_n = -\frac{me^4}{8\varepsilon_0^2 h^2}\frac{1}{n^2} = -\frac{13.60 \text{ eV}}{n^2}, \qquad \text{for } n = 1, 2, 3, \ldots. \tag{40-24}$$

Consider now one of the two innermost electrons in the K shell of a multielectron atom. Because of the presence of the other K-shell electron, our electron "sees" an effective nuclear charge of approximately $(Z - 1)e$, where e is the elementary charge and Z is the atomic number of the element. The factor e^4 in Eq. 40-24 is the product of e^2—the square of hydrogen's nuclear charge—and $(-e)^2$—the square of an electron's charge. For a multielectron atom, we can approximate the effective energy of the atom by replacing the factor e^4 in Eq. 40-24 with $(Z - 1)^2 e^2 \times (-e)^2$, or $e^4(Z - 1)^2$. That gives us

$$E_n = -\frac{(13.60 \text{ eV})(Z - 1)^2}{n^2}. \tag{40-25}$$

We saw that the K_α x-ray photon (of energy hf) arises when an electron makes a transition from the L shell (with $n = 2$ and energy E_2) to the K shell (with $n = 1$ and energy E_1). Thus, using Eq. 40-25, we may write the energy change as

$$\begin{aligned} \Delta E &= E_2 - E_1 \\ &= \frac{-(13.60 \text{ eV})(Z - 1)^2}{2^2} - \frac{-(13.60 \text{ eV})(Z - 1)^2}{1^2} \\ &= (10.2 \text{ eV})(Z - 1)^2. \end{aligned}$$

Then the frequency f of the K_α line is

$$\begin{aligned} f &= \frac{\Delta E}{h} = \frac{(10.2 \text{ eV})(Z - 1)^2}{(4.14 \times 10^{-15} \text{ eV} \cdot \text{s})} \\ &= (2.46 \times 10^{15} \text{ Hz})(Z - 1)^2. \end{aligned} \tag{40-26}$$

Taking the square root of both sides yields

$$\sqrt{f} = CZ - C, \tag{40-27}$$

in which C is a constant ($= 4.96 \times 10^7 \text{ Hz}^{1/2}$). Equation 40-27 is the equation of a straight line. It shows that if we plot the square root of the frequency of the K_α x-ray spectral line against the atomic number Z, we should obtain a straight line. As Fig. 40-17 shows, that is exactly what Moseley found.

CHECKPOINT 3 The K_α x rays arising from a cobalt ($Z = 27$) target have a wavelength of about 179 pm. Is the wavelength of the K_α x rays arising from a nickel ($Z = 28$) target greater than or less than 179 pm?

Sample Problem 40-5

A cobalt target is bombarded with electrons, and the wavelengths of its characteristic x-ray spectrum are measured. There is also a second, fainter characteristic spectrum, which is due to an impurity in the cobalt. The wavelengths of the K_α lines are 178.9 pm (cobalt) and 143.5 pm (impurity), and the proton number for cobalt is $Z_{Co} = 27$. Determine the impurity using only these data.

Solution: The **Key Idea** here is that the wavelengths of the K_α lines for both the cobalt (Co) and the impurity (X) fall on a K_α Moseley plot, and Eq. 40-27 is the equation for that plot. Substituting c/λ for f in that equation, we obtain

$$\sqrt{\frac{c}{\lambda_{Co}}} = CZ_{Co} - C \quad \text{and} \quad \sqrt{\frac{c}{\lambda_X}} = CZ_X - C.$$

Dividing the second equation by the first neatly eliminates C, yielding

$$\sqrt{\frac{\lambda_{Co}}{\lambda_X}} = \frac{Z_X - 1}{Z_{Co} - 1}.$$

Substituting the given data yields

$$\sqrt{\frac{178.9 \text{ pm}}{143.5 \text{ pm}}} = \frac{Z_X - 1}{27 - 1}.$$

Solving for the unknown, we find that

$$Z_X = 30.0. \quad \text{(Answer)}$$

A glance at the periodic table identifies the impurity as zinc.

40-11 Lasers and Laser Light

In the early 1960s, quantum physics made one of its many contributions to technology: the **laser.** Laser light, like the light from an ordinary lightbulb, is emitted when atoms make a transition from one quantum state to a quantum state of lower energy. In a laser, however—but not in other light sources—the atoms act together to produce light with several special characteristics:

1. ***Laser light is highly monochromatic.*** Light from an ordinary incandescent lightbulb is spread over a continuous range of wavelengths and is certainly not monochromatic. The radiation from a fluorescent neon sign is monochromatic, true, to about 1 part in 10^6, but the sharpness of definition of laser light can be many times greater, as much as 1 part in 10^{15}.
2. ***Laser light is highly coherent.*** Individual long waves (*wave trains*) for laser light can be several hundred kilometers long. When two separated beams that have traveled such distances over separate paths are recombined, they "remember" their common origin and are able to form a pattern of interference fringes. The corresponding *coherence length* for wave trains emitted by a lightbulb is typically less than a meter.
3. ***Laser light is highly directional.*** A laser beam spreads very little; it departs from strict parallelism only because of diffraction at the exit aperture of the laser. For example, a laser pulse used to measure the distance to the Moon generates a spot on the Moon's surface with a diameter of only a few meters. Light from an ordinary bulb can be made into an approximately parallel beam by a lens, but the beam divergence is much greater than for laser light. Each point on a lightbulb's filament forms its own separate beam, and the angular divergence of the overall composite beam is set by the size of the filament.
4. ***Laser light can be sharply focused.*** If two light beams transport the same amount of energy, the beam that can be focused to the smaller spot will have the greater intensity at that spot. For laser light, the focused spot can be so small that an intensity of 10^{17} W/cm^2 is readily obtained. An oxyacetylene flame, by contrast, has an intensity of only about 10^3 W/cm^2.

Lasers Have Many Uses

The smallest lasers, used for voice and data transmission over optical fibers, have as their active medium a semiconducting crystal about the size of a pinhead. Small as they are, such lasers can generate about 200 mW of power. The largest lasers,

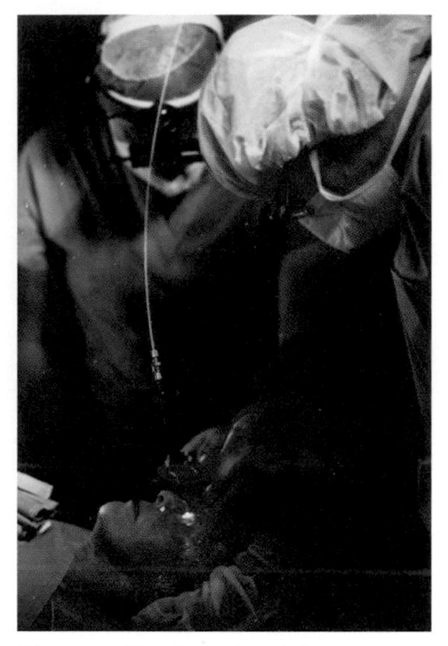

Fig. 40-18 A patient's loose retina is being welded into place by a laser directed into her eye.

used for nuclear fusion research and for astronomical and military applications, fill a large building. The largest such laser can generate brief pulses of laser light with a power level, during the pulse, of about 10^{14} W. This is a few hundred times greater than the total electrical power generating capacity of the United States. To avoid a brief national power blackout during a pulse, the energy required for each pulse is stored up at a steady rate during the relatively long interpulse interval.

Among the many uses of lasers are reading bar codes, manufacturing and reading compact discs and DVDs, performing surgery of many kinds (see the opening photo of this chapter and Fig. 40-18), surveying, cutting cloth in the garment industry (several hundred layers at a time), welding auto bodies, and generating holograms.

40-12 How Lasers Work

Because the word "laser" is an acronym for "light amplification by the stimulated emission of radiation," you should not be surprised that stimulated emission is the key to laser operation. Einstein introduced this concept in 1917. Although the world had to wait until 1960 to see an operating laser, the groundwork for its development was put in place decades earlier.

Consider an isolated atom that can exist either in its state of lowest energy (its ground state), whose energy is E_0, or in a state of higher energy (an excited state), whose energy is E_x. Here are three processes by which the atom can move from one of these states to the other:

1. ***Absorption.*** Figure 40-19*a* shows the atom initially in its ground state. If the atom is placed in an electromagnetic field that is alternating at frequency f, the atom can absorb an amount of energy hf from that field and move to the higher-energy state. From the principle of conservation of energy we have

$$hf = E_x - E_0. \tag{40-28}$$

We call this process **absorption.**

2. ***Spontaneous emission.*** In Fig. 40-19*b* the atom is in its excited state and no external radiation is present. After a time, the atom will move *of its own accord* to its ground state, emitting a photon of energy hf in the process. We call this process **spontaneous emission**—*spontaneous* because the event was not triggered by any outside influence. The light from the filament of an ordinary lightbulb is generated in this way.

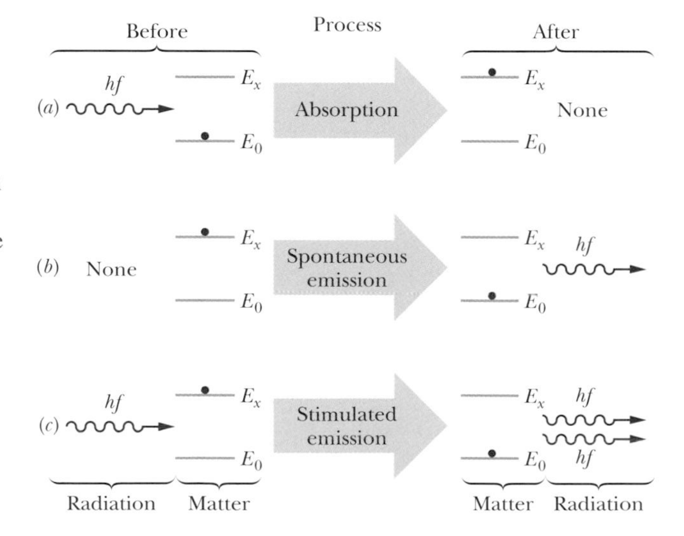

Fig. 40-19 The interaction of radiation and matter in the processes of (*a*) absorption, (*b*) spontaneous emission, and (*c*) stimulated emission. An atom (matter) is represented by the red dot; the atom is in either a lower quantum state with energy E_0 or a higher quantum state with energy E_x. In (*a*) the atom absorbs a photon of energy hf from a passing light wave. In (*b*) it emits a light wave by emitting a photon of energy hf. In (*c*) a passing light wave with photon energy hf causes the atom to emit a photon of the same energy, increasing the energy of the light wave.

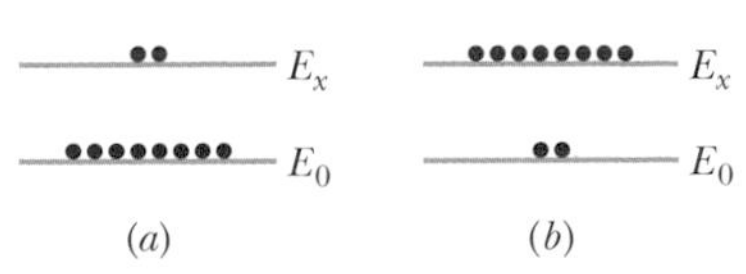

Fig. 40-20 (*a*) The equilibrium distribution of atoms between the ground state E_0 and excited state E_x accounted for by thermal agitation. (*b*) An inverted population, obtained by special methods. Such a population inversion is essential for laser action.

Normally, the mean life of excited atoms before spontaneous emission occurs is about 10^{-8} s. However, for some excited states, this mean life is perhaps as much as 10^5 times longer. We call such long-lived states **metastable;** they play an important role in laser operation.

3. ***Stimulated emission.*** In Fig. 40-19*c* the atom is again in its excited state, but this time radiation with a frequency given by Eq. 40-28 is present. A photon of energy hf can stimulate the atom to move to its ground state, during which process the atom emits an additional photon, whose energy is also hf. We call this process **stimulated emission**—*stimulated* because the event is triggered by the external photon. The emitted photon is in every way identical to the stimulating photon. Thus, the waves associated with the photons have the same energy, phase, polarization, and direction of travel.

Figure 40-19*c* describes stimulated emission for a single atom. Suppose now that a sample contains a large number of atoms in thermal equilibrium at temperature T. Before any radiation is directed at the sample, a number N_0 of these atoms are in their ground state with energy E_0 and a number N_x are in a state of higher energy E_x. Ludwig Boltzmann showed that N_x is given in terms of N_0 by

$$N_x = N_0 e^{-(E_x - E_0)/kT}, \tag{40-29}$$

in which k is Boltzmann's constant. This equation seems reasonable. The quantity kT is the mean kinetic energy of an atom at temperature T. The higher the temperature, the more atoms—on average—will have been "bumped up" by thermal agitation (that is, by atom–atom collisions) to the higher energy state E_x. Also, because $E_x > E_0$, Eq. 40-29 requires that $N_x < N_0$; that is, there will always be fewer atoms in the excited state than in the ground state. This is what we expect if the level populations N_0 and N_x are determined only by the action of thermal agitation. Figure 40-20*a* illustrates this situation.

If we now flood the atoms of Fig. 40-20*a* with photons of energy $E_x - E_0$, photons will disappear via absorption by ground-state atoms and photons will be generated largely via stimulated emission of excited-state atoms. Einstein showed that the probabilities per atom for these two processes are identical. Thus, because there are more atoms in the ground state, the *net* effect will be the absorption of photons.

To produce laser light, we must have more photons emitted than absorbed; that is, we must have a situation in which stimulated emission dominates. The direct way to bring this about is to start with more atoms in the excited state than in the ground state, as in Fig. 40-20*b*. However, because such a **population inversion** is not consistent with thermal equilibrium, we must think up clever ways to set up and maintain one.

The Helium–Neon Gas Laser

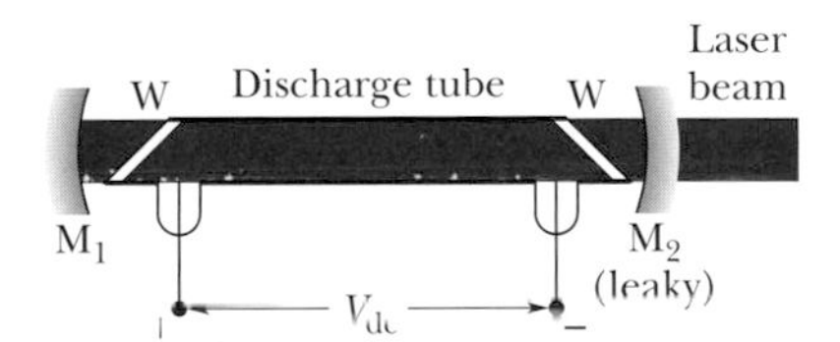

Fig. 40-21 The elements of a helium–neon gas laser. An applied potential V_{dc} sends electrons through a discharge tube containing a mixture of helium gas and neon gas. Electrons collide with helium atoms, which then collide with neon atoms, which emit light along the length of the tube. The light passes through transparent windows W and reflects back and forth through the tube from mirrors M_1 and M_2 to cause more neon atom emissions. Some of the light leaks through mirror M_2 to form the laser beam.

Figure 40-21 shows a type of laser commonly found in student laboratories. It was developed in 1961 by Ali Javan and his coworkers. The glass discharge tube is filled with a 20 : 80 mixture of helium and neon gases, neon being the medium in which laser action occurs.

Figure 40-22 shows simplified energy-level diagrams for the two types of atoms. An electric current passed through the helium–neon gas mixture serves—through collisions between helium atoms and electrons of the current—to raise many helium atoms to state E_3, which is metastable.

The energy of helium state E_3 (20.61 eV) is very close to the energy of neon state E_2 (20.66 eV). Thus, when a metastable (E_3) helium atom and a ground-state (E_0) neon atom collide, the excitation energy of the helium atom is often transferred to the neon atom, which then moves to state E_2. In this manner, neon level E_2 in Fig. 40-22 can become more heavily populated than neon level E_1.

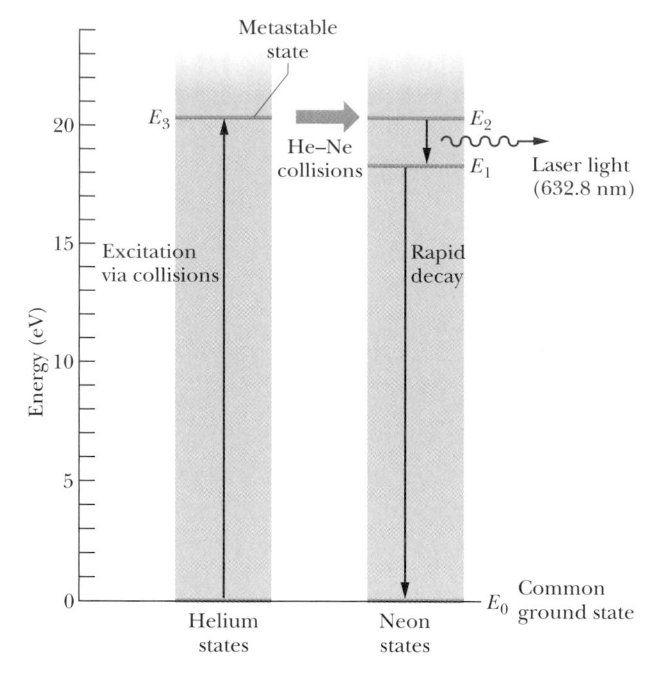

Fig. 40-22 Five essential energy levels for helium and neon atoms in a helium–neon gas laser. Laser action occurs between levels E_2 and E_1 of neon when more atoms are at the E_2 level than at the E_1 level.

This population inversion is relatively easy to set up because (1) initially there are essentially no neon atoms in state E_1, (2) the metastability of helium level E_3 ensures a ready supply of neon atoms in level E_2, and (3) neon atoms in level E_1 decay rapidly (through intermediate levels not shown) to the neon ground state E_0.

Suppose now that a single photon is spontaneously emitted as a neon atom transfers from state E_2 to state E_1. Such a photon can trigger a stimulated emission event, which, in turn, can trigger other stimulated emission events. Through such a chain reaction, a coherent beam of laser light, moving parallel to the tube axis, can build up rapidly. This light, of wavelength 632.8 nm (red), moves through the discharge tube many times by successive reflections from mirrors M_1 and M_2 shown in Fig. 40-21, accumulating additional stimulated emission photons with each passage. M_1 is totally reflecting, but M_2 is slightly "leaky" so that a small fraction of the laser light escapes to form a useful external beam.

CHECKPOINT 4 The wavelength of light from laser *A* (a helium–neon gas laser) is 632.8 nm; that from laser *B* (a carbon dioxide gas laser) is 10.6 μm; that from laser *C* (a gallium arsenide semiconductor laser) is 840 nm. Rank these lasers according to the energy interval between the two quantum states responsible for laser action, greatest first.

Sample Problem 40-6

In the helium–neon laser of Fig. 40-21, laser action occurs between two excited states of the neon atom. However, in many lasers, laser action (*lasing*) occurs between the ground state and an excited state, as suggested in Fig. 40-20.

(a) Consider such a laser that emits at wavelength $\lambda = 550$ nm. If a population inversion is not generated, what is the ratio of the population of atoms in state E_x to the population in the ground state E_0, with the atoms at room temperature?

Solution: One **Key Idea** here is that the naturally occurring population ratio N_x/N_0 of the two states is due to thermal agitation of the gas atoms, according to Eq. 40-29, which we can write as

$$N_x/N_0 = e^{-(E_x - E_0)/kT}. \tag{40-30}$$

To find N_x/N_0 with Eq. 40-30, we need to find the energy separation $E_x - E_0$ between the two states. Here we use another **Key Idea**: We can obtain $E_x - E_0$ from the given wavelength of 550 nm for the lasing between those two states. We find

$$E_x - E_0 = hf = \frac{hc}{\lambda}$$

$$= \frac{(6.63 \times 10^{-34}\ \text{J}\cdot\text{s})(3.00 \times 10^{8}\ \text{m/s})}{(550 \times 10^{-9}\ \text{m})(1.60 \times 10^{-19}\ \text{J/eV})}$$

$$= 2.26\ \text{eV}.$$

To solve Eq. 40-30, we also need the mean energy of thermal agitation kT for an atom at room temperature (assumed to be

300 K), which is

$$kT = (8.62 \times 10^{-5}\ \text{eV/K})(300\ \text{K}) = 0.0259\ \text{eV},$$

in which k is Boltzmann's constant.

Substituting the last two results into Eq. 40-30 gives us the population ratio at room temperature:

$$N_x/N_0 = e^{-(2.26\ \text{eV})/(0.0259\ \text{eV})} \approx 1.3 \times 10^{-38}. \quad \text{(Answer)}$$

This is an extremely small number. It is not unreasonable, however. Atoms with a mean thermal agitation energy of only 0.0259 eV will not often impart an energy of 2.26 eV to another atom in a collision.

(b) For the conditions of (a), at what temperature would the ratio N_x/N_0 be 1/2?

Solution: The two **Key Ideas** of (a) apply here, but this time we want the temperature T such that thermal agitation has bumped enough neon atoms up to the higher-energy state to give $N_x/N_0 = 1/2$. Substituting that ratio into Eq. 40-30, taking the natural logarithm of both sides, and solving for T yield

$$T = \frac{E_x - E_0}{k(\ln 2)} = \frac{2.26\ \text{eV}}{(8.62 \times 10^{-5}\ \text{eV/K})(\ln 2)} = 38\,000\ \text{K}. \quad \text{(Answer)}$$

This is much hotter than the surface of the Sun. It is clear that if we are to invert the populations of these two levels, some specific mechanism for bringing this about is needed—that is, we must "pump" the atoms. No temperature, however high, will naturally generate a population inversion by thermal agitation.

Review & Summary

Some Properties of Atoms The energies of atoms are quantized; that is, the atoms have only certain specific values of energy associated with different quantum states. Atoms can make transitions between different quantum states by emitting or absorbing a photon; the frequency f associated with that light is given by

$$hf = E_{\text{high}} - E_{\text{low}}, \quad (40\text{-}1)$$

where E_{high} is the higher energy and E_{low} is the lower energy of the pair of quantum states involved in the transition. Atoms also have quantized angular momenta and magnetic dipole moments.

Angular Momenta and Magnetic Dipole Moments An electron trapped in an atom has an *orbital angular momentum* $\vec{L}$ with a magnitude given by

$$L = \sqrt{\ell(\ell + 1)}\hbar, \quad (40\text{-}2)$$

where ℓ is the *orbital quantum number* (which can have the values given by Table 40-1) and where the constant "h-bar" is $\hbar = h/2\pi$. The projection L_z of $\vec{L}$ on an arbitrary z axis is quantized and measurable and can have the values

$$L_z = m_\ell\hbar, \quad (40\text{-}7)$$

where m_ℓ is the *orbital magnetic quantum number* (which can have the values given by Table 40-1).

A magnetic dipole is associated with the angular momentum $\vec{L}$ of an electron in an atom. This magnetic dipole has an **orbital magnetic dipole moment** $\vec{\mu}_{\text{orb}}$ that is directed opposite $\vec{L}$:

$$\vec{\mu}_{\text{orb}} = -\frac{e}{2m}\vec{L}, \quad (40\text{-}3)$$

where the minus sign indicates opposite directions. The projection $\mu_{\text{orb},z}$ of the orbital magnetic dipole moment on the z axis is quantized and measurable and can have the values

$$\mu_{\text{orb},z} = -m_\ell\mu_B, \quad (40\text{-}5)$$

where μ_B is the *Bohr magneton:*

$$\mu_B = \frac{eh}{4\pi m} = 9.274 \times 10^{-24}\ \text{J/T}. \quad (40\text{-}6)$$

An electron, whether trapped or free, has an intrinsic *spin angular momentum* (or just *spin*) $\vec{S}$ with a magnitude given by

$$S = \sqrt{s(s + 1)}\hbar, \quad (40\text{-}9)$$

where s is the *spin quantum number* of the electron, which is always $\frac{1}{2}$. The projection S_z of $\vec{S}$ on an arbitrary z axis is quantized and measurable and can have the values

$$S_z = m_s\hbar, \quad (40\text{-}12)$$

where m_s is the *spin magnetic quantum number* of the electron, which can be $+\frac{1}{2}$ or $-\frac{1}{2}$.

An electron has an intrinsic magnetic dipole that is associated with its spin angular momentum $\vec{S}$, whether the electron is confined to an atom or free. This magnetic dipole has a **spin magnetic dipole moment** $\vec{\mu}_s$ that is directed opposite $\vec{S}$:

$$\vec{\mu}_s = -\frac{e}{m}\vec{S}. \quad (40\text{-}10)$$

The projection $\mu_{s,z}$ of the spin magnetic dipole moment $\vec{\mu}_s$ on an arbitrary z axis is quantized and measurable and can have the values

$$\mu_{s,z} = -2m_s\mu_B. \quad (40\text{-}13)$$

Spin and Magnetic Resonance A proton has an intrinsic spin angular momentum $\vec{S}$ and an associated spin magnetic dipole moment $\vec{\mu}$ that is always in the *same* direction as $\vec{S}$. If a proton is located in magnetic field $\vec{B}$, the projection μ_z of $\vec{\mu}$ on a z axis (defined to be along the direction of $\vec{B}$) can have only two quantized orientations: parallel to $\vec{B}$ or antiparallel to $\vec{B}$. The energy difference between these orientations is $2\mu_z B$. The energy required of a photon to *spin-flip* the proton between the two orientations is

$$hf = 2\mu_z B. \quad (40\text{-}22)$$

In general, $\vec{B}$ is the vector sum of an external field $\vec{B}_{\text{ext}}$ set up by the magnetic-resonance equipment and an internal field $\vec{B}_{\text{int}}$ set up by the atoms and nuclei surrounding the proton. Detection of such spin-flips can lead to *nuclear magnetic resonance spectra* by which specific substances can be identified.

Pauli Exclusion Principle Electrons in atoms and other traps obey the **Pauli exclusion principle,** which requires that *no two electrons in the same atom or any other type of trap can have the same set of quantum numbers.*

Building the Periodic Table The elements are listed in the periodic table in order of increasing atomic number Z; the nuclear charge is Ze, and Z is both the number of protons in the nucleus and the number of electrons in the neutral atom.

States with the same value of n form a **shell,** and those with the same values of both n and ℓ form a **subshell.** In *closed* shells and subshells, which are those that contain the maximum number of electrons, the angular momenta and the magnetic moments of the individual electrons sum to zero.

X Rays and the Numbering of the Elements A **continuous spectrum** of x rays is emitted when high-energy electrons lose some of their energy in a collision with atomic nuclei. The **cutoff wavelength** $\lambda_{\min}$ is the wavelength emitted when such electrons lose *all* their initial energy in a single such encounter and is

$$\lambda_{\min} = \frac{hc}{K_0}, \tag{40-23}$$

in which K_0 is the initial kinetic energy of the electrons that strike the target.

The **characteristic x-ray spectrum** arises when high-energy electrons eject electrons from deep within the atom; when a resulting "hole" is filled by an electron from farther out in the atom, a photon of the characteristic x-ray spectrum is generated.

In 1913, British physicist H. G. J. Moseley measured the frequencies of the characteristic x rays from a number of elements. He noted that when the square root of the frequency is plotted against the position of the element in the periodic table, a straight line results, as in the **Moseley plot** of Fig. 40-17. This allowed Moseley to conclude that the property that determines the position of an element in the periodic table is not its atomic mass but its atomic number Z—that is, the number of protons in its nucleus.

Lasers and Laser Light Laser light arises by **stimulated emission.** That is, radiation of a frequency given by

$$hf = E_x - E_0 \tag{40-28}$$

can cause an atom to undergo a transition from an upper energy level (of energy E_x) to a lower energy level, with a photon of frequency f being emitted. The stimulating photon and the emitted photon are identical in every respect and combine to form laser light.

For the emission process to predominate, there must normally be a **population inversion;** that is, there must be more atoms in the upper energy level than in the lower one.

Questions

1 An atom of silver has closed $3d$ and $4d$ subshells. Which subshell has the greater number of electrons, or do they have the same number?

2 An electron in an atom of gold is in a state with $n = 4$. Which of these values of ℓ are possible for it: $-3, 0, 2, 3, 4, 5$?

3 How many (a) subshells and (b) electron states are in the $n = 2$ shell? How many (c) subshells and (d) electron states are in the $n = 5$ shell?

4 An atom of uranium has closed $6p$ and $7s$ subshells. Which subshell has the greater number of electrons?

5 An electron in a mercury atom is in the $3d$ subshell. Which of the following m_ℓ values are possible for it: $-3, -1, 0, 1, 2$?

6 From which atom of each of the following pairs is it easier to remove an electron: (a) krypton or bromine, (b) rubidium or cerium, (c) helium or hydrogen?

7 On which quantum numbers does the energy of an electron depend in (a) a hydrogen atom and (b) a vanadium atom?

8 Which (if any) of these statements about the Einstein–de Haas experiment or its results are true? (a) Atoms have angular momentum. (b) The angular momentum of atoms is quantized. (c) Atoms have magnetic moments. (d) The magnetic moments of atoms are quantized. (e) The angular momentum of an atom is strongly coupled to its magnetic moment. (f) The experiment relies on the conservation of angular momentum.

9 Label these statements as true or false: (a) One (and only one) of these subshells cannot exist: $2p$, $4f$, $3d$, $1p$. (b) The number of values of m_ℓ that are allowed depends only on ℓ and not on n. (c) There are four subshells with $n = 4$. (d) The smallest value of n for a given value of ℓ is $\ell + 1$. (e) All states with $\ell = 0$ also have $m_\ell = 0$. (f) There are n subshells for each value of n.

10 Consider the elements krypton and rubidium. (a) Which is more suitable for use in a Stern–Gerlach experiment of the kind described in connection with Fig. 40-8? (b) Which, if either, would not work at all?

11 The K_α x-ray line for any element arises because of a transition between the K shell ($n = 1$) and the L shell ($n = 2$). Figure 40-14 shows this line (for a molybdenum target) occurring at a single wavelength. With higher resolution, however, the line splits into several wavelength components because the L shell does not have a unique energy. (a) How many components does the K_α line have? (b) Similarly, how many components does the K_β line have?

12 The x-ray spectrum of Fig. 40-14 is for 35.0 keV electrons striking a molybdenum ($Z = 42$) target. If you substitute a silver ($Z = 47$) target for the molybdenum target, will (a) $\lambda_{\min}$, (b) the wavelength for the K_α line, and (c) the wavelength for the K_β line increase, decrease, or remain unchanged?

13 Figure 40-22 shows partial energy-level diagrams for the helium and neon atoms that are involved in the operation of a helium–neon laser. It is said that a helium atom in state E_3 can collide with a neon atom in its ground state and raise the neon atom to state E_2. The energy of helium state E_3

(20.61 eV) is close to, but not exactly equal to, the energy of neon state E_2 (20.66 eV). How can the energy transfer take place if these energies are not *exactly* equal?

14 Which (if any) of the following are essential for laser action to occur between two energy levels of an atom? (a) There are more atoms in the upper level than in the lower. (b) The upper level is metastable. (c) The lower level is metastable. (d) The lower level is the ground state of the atom. (e) The lasing medium is a gas.

Problems

SSM Solution is in the Student Solutions Manual.
WWW Solution is at http://www.wiley.com/college/halliday
• – ••• Number of dots indicates level of problem difficulty.

sec. 40-4 Angular Momenta and Magnetic Dipole Moments

•1 (a) How many ℓ values are associated with $n = 3$? (b) How many m_ℓ values are associated with $\ell = 1$?

•2 How many electron states are in these subshells: (a) $n = 4$, $\ell = 3$; (b) $n = 3$, $\ell = 1$; (c) $n = 4$, $\ell = 1$; (d) $n = 2$, $\ell = 0$?

•3 (a) What is the magnitude of the orbital angular momentum in a state with $\ell = 3$? (b) What is the magnitude of its largest projection on an imposed z axis?

•4 How many electron states are there in the following shells: (a) $n = 4$, (b) $n = 1$, (c) $n = 3$, (d) $n = 2$?

•5 An electron in a hydrogen atom is in a state with $\ell = 5$. What is the minimum possible value of the semiclassical angle between $\vec{L}$ and L_z?

•6 In the subshell $\ell = 3$, (a) what is the greatest (most positive) m_ℓ value, (b) how many states are available with the greatest m_ℓ value, and (c) what is the total number of states available in the subshell?

•7 An electron in a multielectron atom has $m_\ell = +4$. For this electron, what are (a) the value of ℓ, (b) the smallest possible value of n, and (c) the number of possible values of m_s?

•8 How many electron states are there in a shell defined by the quantum number $n = 5$?

••9 An electron is in a state with $\ell = 3$. (a) What multiple of $\hbar$ gives the magnitude of $\vec{L}$? (b) What multiple of μ_B gives the magnitude of $\vec{\mu}$? (c) What is the largest possible value of m_ℓ, (d) what multiple of $\hbar$ gives the corresponding value of L_z, and (e) what multiple of μ_B gives the corresponding value of $\mu_{orb,z}$. (f) What is the value of the semiclassical angle θ between the directions of L_z and $\vec{L}$? What is the value of angle θ for (g) the second largest possible value of m_ℓ and (h) the smallest (that is, most negative) possible value of m_ℓ? SSM WWW

••10 An electron is in a state with $n = 3$. What are (a) the number of possible values of ℓ, (b) the number of possible values of m_ℓ, (c) the number of possible values of m_s, (d) the number of states in the $n = 3$ shell, and (e) the number of subshells in the $n = 3$ shell?

••11 If orbital angular momentum $\vec{L}$ is measured along, say, a z axis to obtain a value for L_z, show that

$$(L_x^2 + L_y^2)^{1/2} = [\ell(\ell + 1) - m_\ell^2]^{1/2}\hbar$$

is the most that can be said about the other two components of the orbital angular momentum. SSM

sec. 40-5 The Stern–Gerlach Experiment

•12 Assume that in the Stern–Gerlach experiment as described for neutral silver atoms, the magnetic field $\vec{B}$ has a magnitude of 0.50 T. (a) What is the energy difference between the magnetic moment orientations of the silver atoms in the two subbeams? (b) What is the frequency of the radiation that would induce a transition between these two states? (c) What is the wavelength of this radiation, and (d) to what part of the electromagnetic spectrum does it belong?

•13 Calculate the (a) smaller and (b) larger value of the semiclassical angle between the electron spin angular momentum vector and the magnetic field in Sample Problem 40-1. Bear in mind that the orbital angular momentum of the valence electron in the silver atom is zero.

•14 Suppose that a hydrogen atom in its ground state moves 80 cm through and perpendicular to a vertical magnetic field that has a magnetic field gradient $dB/dz = 1.6 \times 10^2$ T/m. (a) What is the magnitude of force exerted by the field gradient on the atom due to the magnetic moment of the atom's electron, which we take to be 1 Bohr magneton? (b) What is the vertical displacement of the atom in the 80 cm of travel if its speed is 1.2×10^5 m/s?

•15 What is the acceleration of a silver atom as it passes through the deflecting magnet in the Stern–Gerlach experiment of Sample Problem 40-1? SSM

sec. 40-6 Magnetic Resonance

•16 A hydrogen atom in its ground state actually has two possible, closely spaced energy levels because the electron is in the magnetic field $\vec{B}$ of the proton (the nucleus). Accordingly, a potential energy is associated with the orientation of the electron's magnetic moment $\vec{\mu}$ relative to $\vec{B}$, and the electron is said to be either spin up (higher energy) or spin down (lower energy) in that field. If the electron is excited to the higher-energy level, it can de-excite by spin-flipping and emitting a photon. The wavelength associated with that photon is 21 cm. (Such a process occurs extensively in the Milky Way galaxy, and reception of the 21 cm radiation by radio telescopes reveals where hydrogen gas lies between stars.) What is the effective magnitude of $\vec{B}$ as experienced by the electron in the ground-state hydrogen atom?

•17 What is the wavelength associated with a photon that will induce a transition of an electron spin from parallel to antiparallel orientation in a magnetic field of magnitude 0.200 T? Assume that $\ell = 0$. SSM

•18 In an NMR experiment, the RF source oscillates at 34 MHz and magnetic resonance of the hydrogen atoms in the sample being investigated occurs when the external field $\vec{B}_{ext}$ has magnitude 0.78 T. Assume that $\vec{B}_{int}$ and $\vec{B}_{ext}$ are in the same direction and take the proton magnetic moment component μ_z to be 1.41×10^{-26} J/T. What is the magnitude of $\vec{B}_{\text{int}}$?

sec. 40-8 Multiple Electrons in Rectangular Traps

•19 Seven electrons are trapped in a one-dimensional infinite potential well of width L. What multiple of $h^2/8mL^2$ gives the energy of the ground state of this system? Assume that the electrons do not interact with one another, and do not neglect spin.

•20 A rectangular corral of widths $L_x = L$ and $L_y = 2L$ contains seven electrons. What multiple of $h^2/8mL^2$ gives the energy of the ground state of this system? Assume that the electrons do not interact with one another, and do not neglect spin.

••21 For the situation of Problem 19, what multiple of $h^2/8mL^2$ gives the energy of (a) the first excited state, (b) the second excited state, and (c) the third excited state of the system of seven electrons? (d) Construct an energy-level diagram for the lowest four energy levels of the system.

••22 For Problem 20, what multiple of $h^2/8mL^2$ gives the energy of (a) the first excited state, (b) the second excited state, and (c) the third excited state of the system of seven electrons? (d) Construct an energy-level diagram for the lowest four energy levels of the system.

••23 A cubical box of widths $L_x = L_y = L_z = L$ contains eight electrons. What multiple of $h^2/8mL^2$ gives the energy of the ground state of this system? Assume that the electrons do not interact with one another, and do not neglect spin. SSM

•••24 For the situation of Problem 23, what multiple of $h^2/8mL^2$ gives the energy of (a) the first excited state, (b) the second excited state, and (c) the third excited state of the system of eight electrons? (d) Construct an energy-level diagram for the lowest four energy levels of the system.

sec. 40-9 Building the Periodic Table

•25 Consider the elements selenium ($Z = 34$), bromine ($Z = 35$), and krypton ($Z = 36$). In their part of the periodic table, the subshells of the electronic states are filled in the sequence

$$1s\ 2s\ 2p\ 3s\ 3p\ 3d\ 4s\ 4p\ \cdots.$$

What are (a) the highest occupied subshell for selenium and (b) the number of electrons in it, (c) the highest occupied subshell for bromine and (d) the number of electrons in it, and (e) the highest occupied subshell for krypton and (f) the number of electrons in it?

•26 For a helium atom in its ground state, what are quantum numbers (n, ℓ, m_ℓ, and m_s) for the (a) spin-up electron and (b) spin-down electron?

•27 Two of the three electrons in a lithium atom have quantum numbers (n, ℓ, m_ℓ, m_s) of $(1, 0, 0, +\frac{1}{2})$ and $(1, 0, 0, -\frac{1}{2})$. What quantum numbers are possible for the third electron if the atom is (a) in the ground state and (b) in the first excited state? SSM WWW

•28 Suppose two electrons in an atom have quantum numbers $n = 2$ and $\ell = 1$. (a) How many states are possible for those two electrons? (Keep in mind that the electrons are indistinguishable.) (b) If the Pauli exclusion principle did not apply to the electrons, how many states would be possible?

•29 Show that the number of states with the same quantum number n is $2n^2$.

sec. 40-10 X Rays and the Ordering of the Elements

•30 Through what minimum potential difference must an electron in an x-ray tube be accelerated so that it can produce x rays with a wavelength of 0.100 nm?

••31 X rays are produced in an x-ray tube by electrons accelerated through an electric potential difference of 50.0 kV. Let K_0 be the kinetic energy of an electron at the end of the acceleration. The electron collides with a target nucleus (assume the nucleus remains stationary) and then has kinetic energy $K_1 = 0.500K_0$. (a) What wavelength is associated with the photon that is emitted? The electron collides with another target nucleus (assume it, too, remains stationary) and then has kinetic energy $K_2 = 0.500K_1$. (b) What wavelength is associated with the photon that is emitted?

••32 A 20 keV electron is brought to rest by colliding twice with target nuclei as in Fig. 40-15. (Assume the nuclei remain stationary.) The wavelength associated with the photon emitted in the second collision is 130 pm greater than that associated with the photon emitted in the first collision. (a) What is the kinetic energy of the electron after the first collision? What are (b) the wavelength λ_1 and (c) the energy E_1 associated with the first photon? What are (d) λ_2 and (e) E_2 associated with the second photon?

••33 In Fig. 40-14, the x rays shown are produced when 35.0 keV electrons strike a molybdenum ($Z = 42$) target. If the accelerating potential is maintained at this value but a silver ($Z = 47$) target is used instead, what values of (a) $\lambda_{\min}$, (b) the wavelength of the K_α line, and (c) the wavelength of the K_β line result? The K, L, and M atomic x-ray levels for silver (compare Fig. 40-16) are 25.51, 3.56, and 0.53 keV. SSM WWW

••34 The wavelength of the K_α line from iron is 193 pm. What is the energy difference between the two states of the iron atom that give rise to this transition?

••35 Show that a moving electron cannot spontaneously change into an x-ray photon in free space. A third body (atom or nucleus) must be present. Why is it needed? (*Hint:* Examine the conservation of energy and momentum.) SSM

••36 When electrons bombard a molybdenum target, they produce both continuous and characteristic x rays as shown in Fig. 40-14. In that figure the kinetic energy of the incident electrons is 35.0 keV. If the accelerating potential is increased to 50.0 keV, (a) what is the mean value of $\lambda_{\min}$, and (b) do the wavelengths of the K_α and K_β lines increase, decrease, or remain the same?

••37 Calculate the ratio of the wavelength of the K_α line for niobium (Nb) to that for gallium (Ga). Take needed data from the periodic table of Appendix G. SSM

••38 From Fig. 40-14, calculate approximately the energy difference $E_L - E_M$ for molybdenum. Compare it with the value that may be obtained from Fig. 40-16.

••39 A tungsten ($Z = 74$) target is bombarded by electrons in an x-ray tube. The K, L, and M energy levels for tungsten (compare Fig. 40-16) have the energies 69.5, 11.3, and 2.30 keV, respectively. (a) What is the minimum value of the accelerating potential that will permit the production of the characteristic K_α and K_β lines of tungsten? (b) For this same accelerating potential, what is λ_{min}? What are the (c) K_α and (d) K_β wavelengths?

••40 Here are the K_α wavelengths of a few elements:

Element	λ (pm)	Element	λ (pm)
Ti	275	Co	179
V	250	Ni	166
Cr	229	Cu	154
Mn	210	Zn	143
Fe	193	Ga	134

Make a Moseley plot (like that in Fig. 40-17) from these data and verify that its slope agrees with the value given for C in Section 40-10.

••41 The binding energies of K-shell and L-shell electrons in copper are 8.979 and 0.951 keV, respectively. If a K_α x ray from copper is incident on a sodium chloride crystal and gives a first-order Bragg reflection at an angle of 74.1° measured relative to parallel planes of sodium atoms, what is the spacing between these parallel planes?

••42 (a) From Eq. 40-26, what is the ratio of the photon energies due to K_α transitions in two atoms whose atomic numbers are Z and Z'? (b) What is this ratio for uranium and aluminum? (c) For uranium and lithium?

•••43 Determine the constant C in Eq. 40-27 to five significant figures by finding C in terms of the fundamental constants in Eq. 40-24 and then using data from Appendix B to evaluate those constants. Using this value of C in Eq. 40-27, determine the theoretical energy E_{theory} of the K_α photon for the low-mass elements listed in the following table. The table includes the value (eV) of the measured energy E_{exp} of the K_α photon for each listed element. The percentage deviation between E_{theory} and E_{exp} can be calculated as

$$\text{percentage deviation} = \frac{E_{theory} - E_{exp}}{E_{exp}} 100.$$

What is the percentage deviation for (a) Li, (b) Be, (c) B, (d) C, (e) N, (f) O, (g) F, (h) Ne, (i) Na, and (j) Mg?

Li	54.3	O	524.9
Be	108.5	F	676.8
B	183.3	Ne	848.6
C	277	Na	1041
N	392.4	Mg	1254

(There is actually more than one K_α ray because of the splitting of the L energy level, but that effect is negligible for the elements listed here.)

sec. 40-12 How Lasers Work

•44 For the conditions of Sample Problem 40-6a, how many moles of neon are needed to put 10 atoms in the excited state at energy E_x?

•45 A pulsed laser emits light at a wavelength of 694.4 nm. The pulse duration is 12 ps, and the energy per pulse is 0.150 J. (a) What is the length of the pulse? (b) How many photons are emitted in each pulse?

•46 A hypothetical atom has only two atomic energy levels, separated by 3.2 eV. Suppose that at a certain altitude in the atmosphere of a star there are $6.1 \times 10^{13}/\text{cm}^3$ of these atoms in the higher-energy state and $2.5 \times 10^{15}/\text{cm}^3$ in the lower-energy state. What is the temperature of the star's atmosphere at that altitude?

•47 A hypothetical atom has energy levels uniformly separated by 1.2 eV. At a temperature of 2000 K, what is the ratio of the number of atoms in the 13th excited state to the number in the 11th excited state? SSM

•48 A population inversion for two energy levels is often described by assigning a negative Kelvin temperature to the system. What negative temperature would describe a system in which the population of the upper energy level exceeds that of the lower level by 10% and the energy difference between the two levels is 2.26 eV?

•49 A helium–neon laser emits laser light at a wavelength of 632.8 nm and a power of 2.3 mW. At what rate are photons emitted by this device?

•50 A high-powered laser beam (λ = 600 nm) with a beam diameter of 12 cm is aimed at the Moon, 3.8×10^5 km distant. The beam spreads only because of diffraction. The angular location of the edge of the central diffraction disk (see Eq. 36-12) is given by

$$\sin\theta = \frac{1.22\lambda}{d},$$

where d is the diameter of the beam aperture. What is the diameter of the central diffraction disk on the Moon's surface?

•51 Assume that lasers are available whose wavelengths can be precisely "tuned" to anywhere in the visible range—that is, in the range 450 nm $< \lambda <$ 650 nm. If every television channel occupies a bandwidth of 10 MHz, how many channels can be accommodated within this wavelength range?

•52 The active volume of a laser constructed of the semiconductor GaAlAs is only 200 μm^3 (smaller than a grain of sand), and yet the laser can continuously deliver 5.0 mW of power at a wavelength of 0.80 μm. At what rate does it generate photons?

••53 The active medium in a particular laser that generates laser light at a wavelength of 694 nm is 6.00 cm long and 1.00 cm in diameter. (a) Treat the medium as an optical resonance cavity analogous to a closed organ pipe. How many standing-wave nodes are there along the laser axis? (b) By what amount Δf would the beam frequency have to shift to increase this number by one? (c) Show that Δf is just the inverse of the travel time of laser light for one round trip back and forth along the laser axis. (d) What is the corresponding fractional frequency shift $\Delta f/f$? The appropriate index of refraction of the lasing medium (a ruby crystal) is 1.75.

••54 The mirrors in the laser of Fig. 40-21, which are separated by 8.0 cm, form an optical cavity in which standing waves of laser light can be set up. Each standing wave has an integral number n of half wavelengths in the 8.0 cm length, where n is large and the waves differ slightly in wavelength. Near λ = 533 nm, how far apart in wavelength are the standing waves?

••55 A hypothetical atom has two energy levels, with a transition wavelength between them of 580 nm. In a particular sample at 300 K, 4.0×10^{20} such atoms are in the state of lower energy. (a) How many atoms are in the upper state, assuming conditions of thermal equilibrium? (b) Suppose, instead, that 3.0×10^{20} of these atoms are "pumped" into the upper state by an external process, with 1.0×10^{20} atoms remaining in the lower state. What is the maximum energy that could be released by the atoms in a single laser pulse if each atom jumps once between those two states (either via absorption or via stimulated emission).

••56 The beam from an argon laser (of wavelength 515 nm) has a diameter d of 3.00 mm and a continuous energy output rate of 5.00 W. The beam is focused onto a diffuse surface by a lens whose focal length f is 3.50 cm. A diffraction pattern such as that of Fig. 36-9 is formed, the radius of the central disk being given by

$$R = \frac{1.22 f\lambda}{d}$$

(see Eq. 36-12 and Sample Problem 36-4). The central disk can be shown to contain 84% of the incident power. (a) What is the radius of the central disk? (b) What is the average intensity (power per unit area) in the incident beam? (c) What is the average intensity in the central disk?

Additional Problems

57 Excited sodium atoms emit two closely spaced spectrum lines called the *sodium doublet* (Fig. 40-23) with wavelengths 588.995 nm and 589.592 nm. (a) What is the difference in energy between the two upper energy levels (n = 3, ℓ = 1)? (b) This energy difference occurs because the electron's spin magnetic moment can be oriented either parallel or antiparallel to the internal magnetic field associated with the electron's orbital motion. Use your result in (a) to find the magnitude of this internal magnetic field.

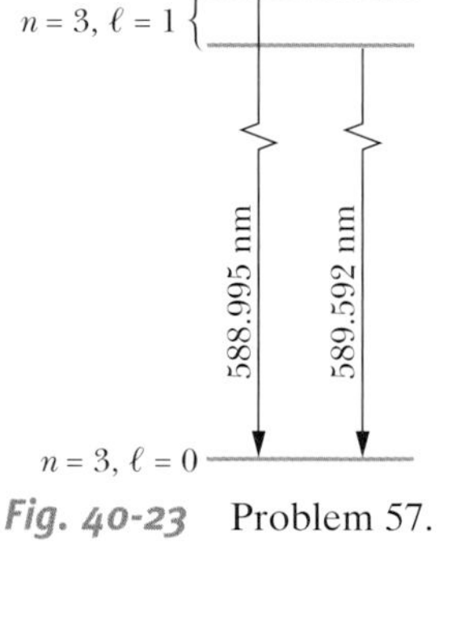

Fig. 40-23 Problem 57.

58 *Martian CO_2 laser.* Where sunlight shines on the atmosphere of Mars, carbon dioxide molecules at an altitude of about 75 km undergo natural laser action. The energy levels involved in the action are shown in Fig. 40-24; population inversion occurs between energy levels E_2 and E_1. (a) What wavelength of sunlight excites the molecules in the lasing action? (b) At what wavelength does lasing occur? (c) In what region of the electromagnetic spectrum do the excitation and lasing wavelengths lie?

$E_2 = 0.289$ eV
$E_1 = 0.165$ eV
$E_0 = 0$

Fig. 40-24 Problem 58.

59 Can an incoming intercontinental ballistic missile be destroyed by an intense laser beam? A beam of intensity 10^8 W/m^2 would probably burn into and destroy a nonspinning missile in 1 s. (a) If the laser had 5.0 MW power, 3.0 μm wavelength, and a 4.0 m beam diameter (a very powerful laser indeed), would it destroy a missile at a distance of 3000 km? (b) If the wavelength could be changed, what maximum value would work? Use the equation for the central diffraction maximum given by Eq. 36-12 ($\sin\theta = 1.22\lambda/d$).

60 *Comet stimulated emission.* When a comet approaches the Sun, the increased warmth evaporates water from the ice on the surface of the comet nucleus, producing a thin atmosphere of water vapor around the nucleus. Sunlight can then dissociate H_2O molecules in the vapor to H atoms and OH molecules. The sunlight can also excite the OH molecules to higher-energy levels, two of which are represented in Fig. 40-25.

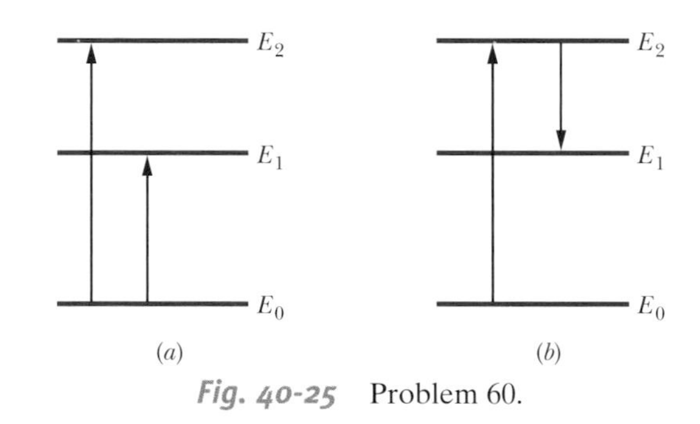

Fig. 40-25 Problem 60.

When the comet is still relatively far from the Sun, the sunlight causes equal excitation to the E_2 and E_1 levels (Fig. 40-25*a*). Hence, there is no population inversion between the two levels. However, as the comet approaches the Sun, the excitation to the E_1 level decreases and population inversion occurs. The reason has to do with one of the many wavelengths—said to be *Fraunhofer lines*—that are missing in sunlight because, as the light travels outward through the Sun's atmosphere, those particular wavelengths are absorbed by the atmosphere.

As a comet approaches the Sun, the Doppler effect due to the comet's speed relative to the Sun shifts the Fraunhofer lines in wavelength, apparently overlapping one of them with the wavelength required for excitation to the E_1 level in OH molecules. Population inversion then occurs in those molecules, and they radiate stimulated emission (Fig. 40-25*b*). For example, as comet Kouhoutek approached the Sun in December 1973 and January 1974, it radiated stimulated emission at about 1666 MHz during mid-January. (a) What was the energy difference $E_2 - E_1$ for that emission? (b) In what region of the electromagnetic spectrum was the emission?

61 Show that the cutoff wavelength (in picometers) in the continuous x-ray spectrum from any target is given by λ_{min} = $1240/V$, where V is the potential difference (in kilovolts) through which the electrons are accelerated before they strike the target.

62 A molybdenum (Z = 42) target is bombarded with 35.0

keV electrons and the x-ray spectrum of Fig. 40-14 results. The K_β and K_α wavelengths are 63.0 and 71.0 pm, respectively. What photon energy corresponds to the (a) K_β and (b) K_α radiation? The two radiations are to be filtered through one of the substances in the following table such that the substance absorbs the K_β line more strongly than the K_α line. A substance will absorb radiation x_1 more strongly than it absorbs radiation x_2 if a photon of x_1 has enough energy to eject a K electron from an atom of the substance but a photon of x_2 does not. The table gives the ionization energy of the K electron in molybdenum and four other substances. Which substance in the table will serve (c) best and (d) second best as the filter?

	Zr	Nb	Mo	Tc	Ru
Z	40	40	42	43	44
E_K (keV)	18.00	18.99	20.00	21.04	22.12

63 Lasers can be used to generate pulses of light whose durations are as short as 10 fs. (a) How many wavelengths of light ($\lambda = 500$ nm) are contained in such a pulse? (b) Supply the missing quantity X (in years):

$$\frac{10 \text{ fs}}{1 \text{ s}} = \frac{1 \text{ s}}{X}.$$

64 By measuring the go-and-return time for a laser pulse to travel from an Earth-bound observatory to a reflector on the Moon, it is possible to measure the separation between these bodies. (a) What is the predicted value of this time? (b) The separation can be measured to a precision of about 15 cm. To what uncertainty in travel time does this correspond? (c) If the laser beam forms a spot on the Moon 3 km in diameter, what is the angular divergence of the beam?

65 Knowing that the minimum x-ray wavelength produced by 40.0 keV electrons striking a target is 31.1 pm, determine the Planck constant h.

66 Show that if the 63 electrons in an atom of europium were assigned to shells according to the "logical" sequence of quantum numbers, this element would be chemically similar to sodium.

67 Suppose that the electron had no spin and that the Pauli exclusion principle still held. Which, if any, of the present noble gases would remain in that category?

68 (A correspondence principle problem.) Estimate (a) the quantum number ℓ for the orbital motion of Earth around the Sun and (b) the number of allowed orientations of the plane of Earth's orbit. (c) Find θ_{min}, the half-angle of the smallest cone that can be swept out by a perpendicular to Earth's orbit as Earth revolves around the Sun.

69 An electron in a multielectron atom is known to have the quantum number $\ell = 3$. What are its possible n, m_ℓ, and m_s quantum numbers?

70 Show that $\hbar = 1.06 \times 10^{-34}$ J·s $= 6.59 \times 10^{-16}$ eV·s.

41 Conduction of Electricity in Solids

Solid-state physics and solid-state electronics have radically changed modern life. For example, early computers relied on bulky vacuum tubes and took up the space of a large room. Today, far more powerful computers rely on tiny transistors in integrated circuits and take up the space of your lap (or even less). Seemingly, vacuum tubes are a thing of the past; indeed, they are no longer taught to electrical engineering majors. However, many of today's hard-rock guitar players, such as James Hetfield of Metallica shown here, insist on amplifiers using vacuum tubes and shun those using transistors.

Why do guitar rockers choose tube amplifiers instead of transistor amplifiers?

The answer is in this chapter.

41-1 What Is Physics?

In this chapter we focus on a goal of physics that has become enormously important in the last half-century. That goal is to answer the question: What are the mechanisms by which a material conducts, or does not conduct, electricity? The answers to that question are complex and still not well understood, largely because they involve the application of quantum physics—not to individual particles and atoms as we have seen in the last several chapters but to a tremendous number of particles and atoms grouped together and interacting. In spite of the difficulty, engineers and scientists have made great strides in the quantum physics of materials science, which is why we now have computers, calculators, cell phones, and many other types of *solid-state* electronic devices.

Our starting point in this chapter is to characterize the solids that can conduct electricity and those that cannot.

41-2 The Electrical Properties of Solids

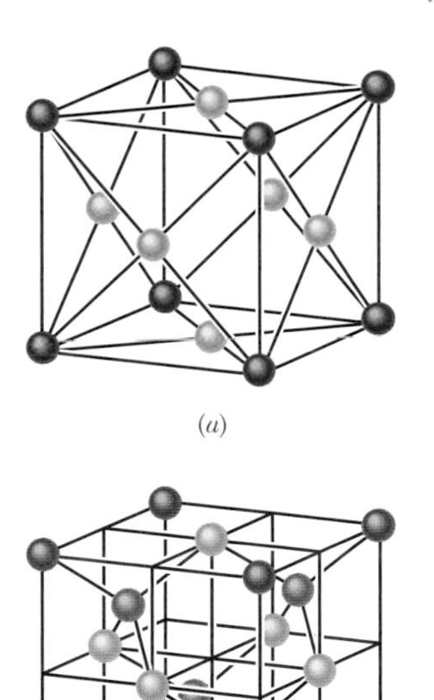

Fig. 41-1 (*a*) The unit cell for copper is a cube. There is one copper atom (darker) at each corner of the cube and one copper atom (lighter) at the center of each face of the cube. The arrangement is called *face-centered cubic.* (*b*) The unit cell for either silicon or the carbon atoms in diamond is also a cube, the atoms being arranged in what is called a *diamond lattice.* There is one atom (darkest) at each corner of the cube and one atom (lightest) at the center of each cube face; in addition, four atoms (medium color) lie within the cube. Every atom is bonded to its four nearest neighbors by a two-electron covalent bond (only the four atoms within the cube show all four nearest neighbors).

We shall examine only **crystalline solids**—that is, solids whose atoms are arranged in a repetitive three-dimensional structure called a **lattice.** We shall not consider such solids as wood, plastic, glass, and rubber, whose atoms are not arranged in such repetitive patterns. Figure 41-1 shows the basic repetitive units (the **unit cells**) of the lattice structures of copper, our prototype of a metal, and silicon and diamond (carbon), our prototypes of a semiconductor and an insulator, respectively.

We can classify solids electrically according to three basic properties:

1. Their **resistivity** ρ at room temperature, with the SI unit ohm-meter ($\Omega \cdot m$); resistivity is defined in Section 26-4.
2. Their **temperature coefficient of resistivity** α, defined as $\alpha = (1/\rho)(d\rho/dT)$ in Eq. 26-17 and having the SI unit inverse kelvin (K^{-1}). We can evaluate α for any solid by measuring ρ over a range of temperatures.
3. Their **number density of charge carriers** n. This quantity, the number of charge carriers per unit volume, can be found from measurements of the Hall effect, as discussed in Section 28-4, and from other measurements. It has the SI unit inverse cubic meter (m^{-3}).

From measurements of room-temperature resistivity alone, we discover that there are some materials—we call them **insulators**—that for all practical purposes do not conduct electricity at all. These are materials with very high resistivity. Diamond, an excellent example, has a resistivity greater than that of copper by the enormous factor of about 10^{24}.

We can then use measurements of ρ, α, and n to divide most noninsulators, at least at low temperatures, into two categories: **metals** and **semiconductors.**

Semiconductors have a considerably greater resistivity ρ than metals.

Semiconductors have a temperature coefficient of resistivity α that is both high and negative. That is, the resistivity of a semiconductor *decreases* with temperature, whereas that of a metal *increases.*

Semiconductors have a considerably lower number density of charge carriers n than metals.

Table 41-1 shows values of these quantities for copper, our prototype metal, and silicon, our prototype semiconductor.

Now, with measurements of ρ, α, and n in hand, we have an experimental basis for refining our central question about the conduction of electricity in solids: *What features make diamond an insulator, copper a metal, and silicon a semiconductor?* Again, quantum physics provides the answers.

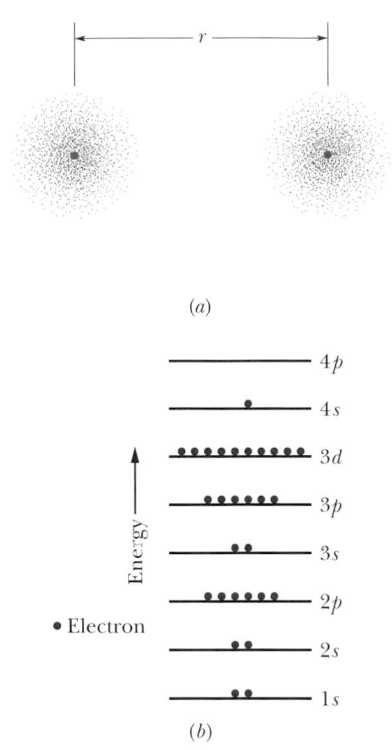

Fig. 41-2 (*a*) Two copper atoms separated by a large distance; their electron distributions are represented by dot plots. (*b*) Each copper atom has 29 electrons distributed among a set of subshells. In the neutral atom in its ground state, all subshells up through the 3*d* level are filled, the 4*s* subshell contains one electron (it can hold two), and higher subshells are empty. For simplicity, the subshells are shown as being evenly spaced in energy.

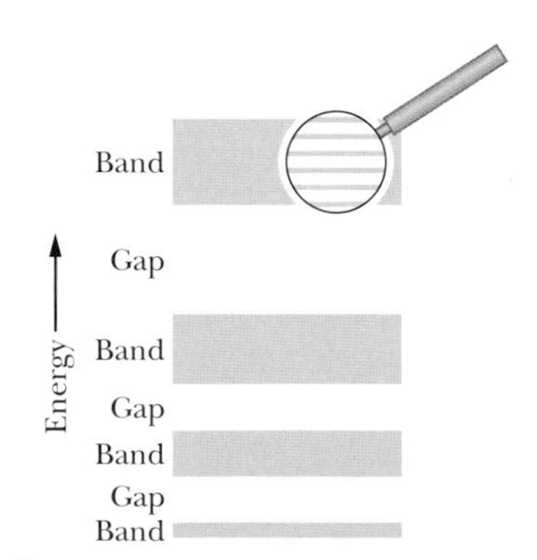

Fig. 41-3 The band–gap pattern of energy levels for an idealized crystalline solid. As the magnified view suggests, each band consists of a very large number of very closely spaced energy levels. (In many solids, adjacent bands may overlap; for clarity, we have not shown this condition.)

TABLE 41-1 Some Electrical Properties of Two Materials[a]

Property	Unit	Material	
		Copper	Silicon
Type of conductor		Metal	Semiconductor
Resistivity, ρ	$\Omega \cdot m$	2×10^{-8}	3×10^{3}
Temperature coefficient of resistivity, α	K^{-1}	$+4 \times 10^{-3}$	-70×10^{-3}
Number density of charge carriers, n	m^{-3}	9×10^{28}	1×10^{16}

[a]All values are for room temperature.

41-3 Energy Levels in a Crystalline Solid

The distance between adjacent copper atoms in solid copper is 260 pm. Figure 41-2*a* shows two isolated copper atoms separated by a distance r that is much greater than that. As Fig. 41-2*b* shows, each of these isolated neutral atoms stacks up its 29 electrons in an array of discrete subshells as follows:

$$1s^2\ 2s^2\ 2p^6\ 3s^2\ 3p^6\ 3d^{10}\ 4s^1.$$

Here we use the shorthand notation of Section 40-9 to identify the subshells. Recall, for example, that the subshell with principal quantum number $n = 3$ and orbital quantum number $\ell = 1$ is called the 3*p* subshell; it can hold up to $2(2\ell + 1) = 6$ electrons; the number it actually contains is indicated by a numerical superscript. We see above that the first six subshells in copper are filled, but the (outermost) 4*s* subshell, which can hold two electrons, holds only one.

If we bring the atoms of Fig. 41-2*a* closer together, they will—speaking loosely—begin to sense each other's presence. In the language of quantum physics, their wave functions will start to overlap, beginning with those of the outermost electrons.

When the wave functions of the two atoms overlap, we speak not of two independent atoms but of a single two-atom system; here the system contains $2 \times 29 = 58$ electrons. The Pauli exclusion principle also applies to this larger system and requires that each of these 58 electrons occupy a different quantum state. In fact, 58 quantum states are available because each energy level of the isolated atom splits into *two* levels for the two-atom system.

If we bring up more atoms, we gradually assemble a lattice of solid copper. If our lattice contains, say, N atoms, then each level of an isolated copper atom must split into N levels in the solid. Thus, the individual energy levels of the solid form energy **bands,** adjacent bands being separated by an energy **gap,** with the gap representing a range of energies that no electron can possess. A typical band ranges over only a few electron-volts. Since N may be of the order of 10^{24}, we see that the individual levels within a band are very close together indeed, and there are a vast number of levels.

Figure 41-3 suggests the band–gap structure of the energy levels in a generalized crystalline solid. Note that bands of lower energy are narrower than those of higher energy. This occurs because electrons that occupy the lower-energy bands spend most of their time deep within the atom's electron cloud. The wave functions of these core electrons do not overlap as much as the wave functions of the outer electrons do. Hence the splitting of the lower-energy levels (core electrons) is less than that of the higher-energy levels (outer electrons).

41-4 Insulators

A solid is said to be an electrical insulator if no current exists within it when we apply a potential difference across it. For a current to exist, the kinetic energy of the average electron must increase. In other words, some electrons in the solid must move to a higher-energy level. However, as Fig. 41-4*a* shows, in an insulator the highest band containing any electrons is fully occupied. Because the Pauli exclusion principle keeps electrons from moving to occupied levels, no electrons in the solid are allowed to move.

Thus, the electrons in the filled band of an insulator have no place to go; they are in gridlock. It is as if a child tries to climb a ladder that already has a child standing on each rung; since there are no unoccupied rungs, no one can move.

There are plenty of unoccupied levels (or *vacant levels*) in the band above the filled band in Fig. 41-4*a*. However, if an electron is to occupy one of those levels, it must acquire enough energy to jump across the substantial energy gap E_g that separates the two bands. In diamond, this gap is so wide (the energy needed to cross it is 5.5 eV, about 140 times the average thermal energy of a free particle at room temperature) that essentially no electron can jump across it. Diamond is thus an electrical insulator, and a very good one.

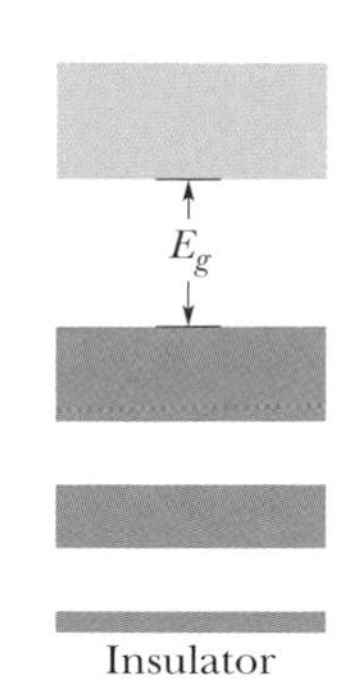

Fig. 41-4 (*a*) The band–gap pattern for an insulator; filled levels are shown in red and empty levels in blue.

Sample Problem 41-1

Approximately what is the probability that, at room temperature (300 K), an electron at the top of the highest filled band in diamond (an insulator) will jump the energy gap E_g in Fig. 41-4*a*? For diamond, E_g is 5.5 eV.

Solution: In Chapter 40 we used Eq. 40-29,

$$\frac{N_x}{N_0} = e^{-(E_x - E_0)/kT}, \tag{41-1}$$

to relate the population N_x of atoms at energy level E_x to the population N_0 at energy level E_0, where the atoms are part of a system at temperature T (measured in kelvins); k is the Boltzmann constant (8.62×10^{-5} eV/K).

A **Key Idea** here is that we can use Eq. 41-1 to *approximate* the probability P that an electron in an insulator will jump the energy gap E_g in Fig. 41-4*a*. To do so, we first set the energy difference $E_x - E_0$ to E_g. Then the probability P of the jump is approximately equal to the ratio N_x/N_0 of the number of electrons just above the energy gap to the number of electrons just below the gap.

For diamond, the exponent in Eq. 41-1 is

$$-\frac{E_g}{kT} = -\frac{5.5\text{ eV}}{(8.62 \times 10^{-5}\text{ eV/K})(300\text{ K})} = -213.$$

The required probability is then

$$P = \frac{N_x}{N_0} = e^{-(E_g/kT)} = e^{-213} \approx 3 \times 10^{-93}. \qquad \text{(Answer)}$$

This result tells us that approximately 3 electrons out of 10^{93} electrons would jump across the energy gap. Because any diamond stone has fewer than 10^{23} electrons, we see that the probability of the jump is vanishingly small. No wonder diamond is such a good insulator.

41-5 Metals

The feature that defines a metal is that, as Fig. 41-4*b* shows, the highest occupied energy level falls somewhere near the middle of an energy band. If we apply a potential difference across a metal, a current can exist because there are plenty of vacant levels at nearby higher energies into which electrons (the charge carriers in a metal) can jump. Thus, a metal can conduct electricity because electrons in its highest occupied band can easily move into higher energy levels.

In Section 26-6 we discussed the **free-electron model** of a metal, in which the **conduction electrons** are free to move throughout the volume of the sample like the molecules of a gas in a closed container. We used this model to derive an expression for the resistivity of a metal, assuming that the electrons follow the laws of Newtonian mechanics. Here we use that same model to explain the behavior of the conduction electrons in the partially filled band of Fig. 41-4*b*. However, we follow the laws of quantum physics by assuming the energies of these electrons to be quantized and the Pauli exclusion principle to hold.

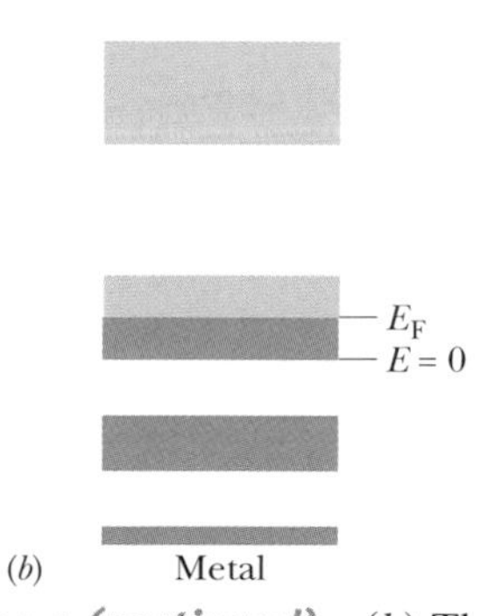

Fig. 41-4 (continued) (*b*) The band–gap pattern for a metal. The highest filled level, called the Fermi level, lies near the middle of a band. Since vacant levels are available within that band, electrons in the band can easily change levels, and conduction can take place.

We assume too that the electric potential energy of a conduction electron has the same constant value at all points within the lattice. If we choose this value of the potential energy to be zero, as we are free to do, then the mechanical energy E of the conduction electrons is entirely kinetic.

The level at the bottom of the partially filled band of Fig. 41-4*b* corresponds to $E = 0$. The highest occupied level in this band at absolute zero ($T = 0$ K) is called the **Fermi level,** and the energy corresponding to it is called the **Fermi energy** E_F; for copper, $E_F = 7.0$ eV.

The electron speed corresponding to the Fermi energy is called the **Fermi speed** v_F. For copper the Fermi speed is 1.6×10^6 m/s. This fact should be enough to shatter the popular misconception that all motion ceases at absolute zero; at that temperature—and solely because of the Pauli exclusion principle—the conduction electrons are stacked up in the partially filled band of Fig. 41-4*b* with energies that range from zero to the Fermi energy.

How Many Conduction Electrons Are There?

If we could bring individual atoms together to form a sample of a metal, we would find that the conduction electrons in the metal are the *valence electrons* of the atoms (the electrons in the outermost occupied shells of the atoms). A *monovalent* atom contributes one such electron to the conduction electrons in a metal; a *bivalent* atom contributes two such electrons. Thus, the total number of conduction electrons is

$$\begin{pmatrix}\text{number of conduction}\\ \text{electrons in sample}\end{pmatrix} = \begin{pmatrix}\text{number of atoms}\\ \text{in sample}\end{pmatrix}\begin{pmatrix}\text{number of valence}\\ \text{electrons per atom}\end{pmatrix}. \tag{41-2}$$

(In this chapter, we shall write several equations largely in words because the symbols we have previously used for the quantities in them now represent other quantities.) The *number density n* of conduction electrons in a sample is the number of conduction electrons per unit volume:

$$n = \frac{\text{number of conduction electrons in sample}}{\text{sample volume } V}. \tag{41-3}$$

We can relate the number of atoms in a sample to various other properties of the sample and to the material making up the sample with the following:

$$\begin{aligned}\begin{pmatrix}\text{number of atoms}\\ \text{in sample}\end{pmatrix} &= \frac{\text{sample mass } M_{\text{sam}}}{\text{atomic mass}} = \frac{\text{sample mass } M_{\text{sam}}}{(\text{molar mass } M)/N_A}\\ &= \frac{(\text{material's density})(\text{sample volume } V)}{(\text{molar mass } M)/N_A},\end{aligned} \tag{41-4}$$

where the molar mass M is the mass of one mole of the material in the sample and N_A is Avogadro's number (6.02×10^{23} mol^{-1}).

Sample Problem 41-2

How many conduction electrons are in a cube of magnesium with a volume of 2.00×10^{-6} m^3? Magnesium atoms are bivalent.

Solution: The **Key Ideas** here are these:

1. Because magnesium atoms are bivalent, each magnesium atom contributes two conduction electrons.
2. The cube's number of conduction electrons is related to its number of magnesium atoms by Eq. 41-2.
3. We can find the number of atoms with Eq. 41-4 and known data about the cube's volume and magnesium's properties.

We can write Eq. 41-4 as

$$\begin{pmatrix}\text{number}\\ \text{of atoms}\\ \text{in sample}\end{pmatrix} = \frac{(\text{material's density})(\text{sample volume } V)N_A}{\text{molar mass } M}.$$

Magnesium has a density of 1.738 g/cm^3 ($= 1.738 \times 10^3$ kg/m^3) and a molar mass of 24.312 g/mol ($= 24.312 \times 10^{-3}$ kg/mol) (see Appendix F). The numerator gives us

$$(1.738 \times 10^3 \text{ kg/m}^3)(2.00 \times 10^{-6} \text{ m}^3) \times (6.02 \times 10^{23} \text{ atoms/mol}) = 2.0926 \times 10^{21} \text{ kg/mol}.$$

Thus,

$$\begin{pmatrix}\text{number of atoms}\\ \text{in sample}\end{pmatrix} = \frac{2.0926 \times 10^{21} \text{ kg/mol}}{24.312 \times 10^{-3} \text{ kg/mol}} = 8.61 \times 10^{22}.$$

Using this result and the fact that magnesium atoms are bivalent, we find that Eq. 41-2 yields

$$\begin{pmatrix}\text{number of}\\ \text{conduction electrons}\\ \text{in sample}\end{pmatrix} = (8.61 \times 10^{22} \text{ atoms})\left(2\,\frac{\text{electrons}}{\text{atom}}\right) = 1.72 \times 10^{23} \text{ electrons.} \quad \text{(Answer)}$$

Conductivity Above Absolute Zero

Our practical interest in the conduction of electricity in metals is at temperatures above absolute zero. What happens to the electron distribution of Fig. 41-4*b* at such higher temperatures? As we shall see, surprisingly little.

Of the electrons in the partially filled band of Fig. 41-4*b*, only those that are close to the Fermi energy find unoccupied levels above them, and only those electrons are free to be boosted to these higher levels by thermal agitation. Even at $T = 1000$ K, a temperature at which copper would glow brightly in a dark room, the distribution of electrons among the available levels does not differ much from the distribution at $T = 0$ K.

Let us see why. The quantity kT, where k is the Boltzmann constant, is a convenient measure of the energy that may be given to a conduction electron by the random thermal motions of the lattice. At $T = 1000$ K, we have $kT = 0.086$ eV. No electron can hope to have its energy changed by more than a few times this relatively small amount by thermal agitation alone; so at best only those few conduction electrons whose energies are close to the Fermi energy are likely to jump to higher energy levels due to thermal agitation. Poetically stated, thermal agitation normally causes only ripples on the surface of the Fermi sea of electrons; the vast depths of that sea lie undisturbed.

How Many Quantum States Are There?

The ability of a metal to conduct electricity depends on how many quantum states are available to its electrons and what the energies of those states are. Thus, a question arises: What are the energies of the individual states in the partially filled band of Fig. 41-4*b*? This question is too difficult to answer because we cannot possibly list the energies of so many states individually. We ask instead: How many states in a unit volume of a sample have energies in the energy range E to $E + dE$? We write this number as $N(E)\,dE$, where $N(E)$ is called the **density of states** at energy E. The conventional unit for $N(E)\,dE$ is states per cubic meter (states/m^3, or simply m^{-3}), and the conventional unit for $N(E)$ is states per cubic meter per electron-volt (m^{-3} eV^{-1}).

We can find an expression for the density of states by counting the number of standing electron matter waves that can fit into a box the size of the metal sample we are considering. This is analogous to counting the number of standing waves of sound that can exist in a closed organ pipe. The differences are that our problem is three-dimensional (the organ pipe problem is one-dimensional) and the waves are matter waves (the organ-pipe waves are sound waves). The result of such counting can be shown to be

$$N(E) = \frac{8\sqrt{2}\pi m^{3/2}}{h^3} E^{1/2} \quad \text{(density of states, m}^{-3}\,\text{J}^{-1}\text{)}, \qquad (41\text{-}5)$$

where m ($= 9.109 \times 10^{-31}$ kg) is the electron mass, h ($= 6.626 \times 10^{-34}$ J $\cdot$ s) is the Planck constant, E is the energy in joules at which $N(E)$ is to be evaluated, and $N(E)$ is in states per cubic meter per joule (m^{-3} J^{-1}). To modify this equation so that the value of E is in electron-volts and the value of $N(E)$ is in states per cubic meter per electron-volt (m^{-3} eV^{-1}), multiply the right side of the equation by $e^{3/2}$, where e is the fundamental charge, 1.602×10^{-19} C. Figure 41-5 is a plot of such a modified version of Eq. 41-5. Note that nothing in Eq. 41-5 or Fig. 41-5 involves the shape, temperature, or composition of the sample.

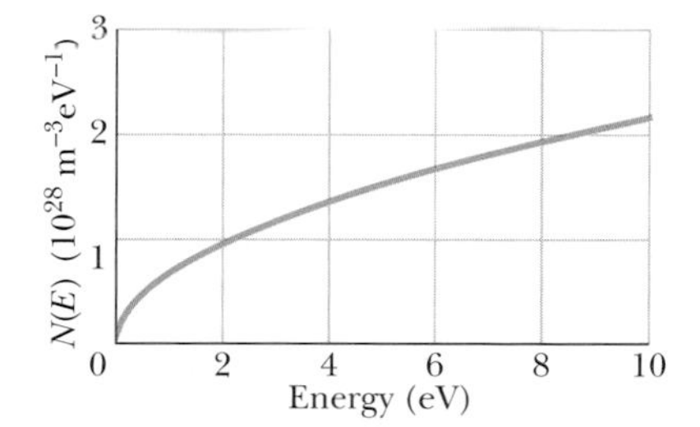

Fig. 41-5 The density of states $N(E)$—that is, the number of electron energy levels per unit energy interval per unit volume—plotted as a function of electron energy. The density of states function simply counts the available states; it says nothing about whether these states are occupied by electrons.

CHECKPOINT 1 (a) Is the spacing between adjacent energy levels at $E = 4$ eV in copper larger than, the same as, or smaller than the spacing at $E = 6$ eV? (b) Is the spacing between adjacent energy levels at $E = 4$ eV in copper larger than, the same as, or smaller than the spacing for an identical volume of aluminum at that same energy?

Sample Problem 41-3

(a) Using the data of Fig. 41-5, determine the number of states per electron-volt at 7 eV in a metal sample with a volume V of $2 \times 10^{-9}\ m^3$.

Solution: The **Key Idea** is that we can obtain the number of states per electron-volt at a given energy by using the density of states $N(E)$ at that energy and the sample's volume V. At an energy of 7 eV, this means that

$$\begin{pmatrix}\text{number of states}\\ \text{per eV at 7 eV}\end{pmatrix} = \begin{pmatrix}\text{density of states}\\ N(E) \text{ at 7 eV}\end{pmatrix}\begin{pmatrix}\text{volume } V\\ \text{of sample}\end{pmatrix}.$$

From Fig. 41-5, we see that at an energy E of 7 eV, the density of states is about $1.8 \times 10^{28}\ m^{-3}\ eV^{-1}$. Thus,

$$\begin{pmatrix}\text{number of states}\\ \text{per eV at 7 eV}\end{pmatrix} = (1.8 \times 10^{28}\ m^{-3}\ eV^{-1})(2 \times 10^{-9}\ m^3)$$
$$= 3.6 \times 10^{19}\ eV^{-1}$$
$$\approx 4 \times 10^{19}\ eV^{-1}. \qquad \text{(Answer)}$$

(b) Next, determine the number of states N in the sample within a *small* energy range ΔE of 0.003 eV centered at 7 eV.

Solution: From Eq. 41-5 and Fig. 41-5, we know that the density of states is a function of energy E. However, for an energy range ΔE that is small relative to E, we can approximate the density of states (and thus the number of states per electron-volt) to be constant. Thus, at an energy of 7 eV, we find the number of states N in the energy range ΔE of 0.003 eV as

$$\begin{pmatrix}\text{number of states } N\\ \text{in range } \Delta E \text{ at 7 eV}\end{pmatrix} = \begin{pmatrix}\text{number of states}\\ \text{per eV at 7 eV}\end{pmatrix}\begin{pmatrix}\text{energy}\\ \text{range } \Delta E\end{pmatrix}$$

or

$$N = (3.6 \times 10^{19}\ eV^{-1})(0.003\ eV)$$
$$= 1.1 \times 10^{17} \approx 1 \times 10^{17}. \qquad \text{(Answer)}$$

The Occupancy Probability P(E)

The ability of a metal to conduct electricity depends on the probability that available vacant levels will actually be occupied. Thus, another question arises: If an energy level is available at energy E, what is the probability $P(E)$ that it is actually occupied by an electron? At $T = 0$ K, we know that for all levels with energies below the Fermi energy, $P(E) = 1$, corresponding to a certainty that the level is occupied. We also know that, at $T = 0$ K, for all levels with energies above the Fermi energy, $P(E) = 0$, corresponding to a certainty that the level is *not* occupied. Figure 41-6*a* illustrates this situation.

To find $P(E)$ at temperatures above absolute zero, we must use a set of quantum counting rules called **Fermi–Dirac statistics,** named for the physicists who introduced them. Using these rules, it is possible to show that the **occupancy probability** $P(E)$ is

$$P(E) = \frac{1}{e^{(E - E_F)/kT} + 1} \quad \text{(occupancy probability)}, \qquad (41\text{-}6)$$

in which E_F is the Fermi energy. Note that $P(E)$ depends not on the energy E of the level but only on the difference $E - E_F$, which may be positive or negative.

To see whether Eq. 41-6 describes Fig. 41-6*a*, we substitute $T = 0$ K in it. Then,

For $E < E_F$, the exponential term in Eq. 41-6 is $e^{-\infty}$, or zero; so $P(E) = 1$, in agreement with Fig. 41-6*a*.

For $E > E_F$, the exponential term is $e^{+\infty}$; so $P(E) = 0$, again in agreement with Fig. 41-6*a*.

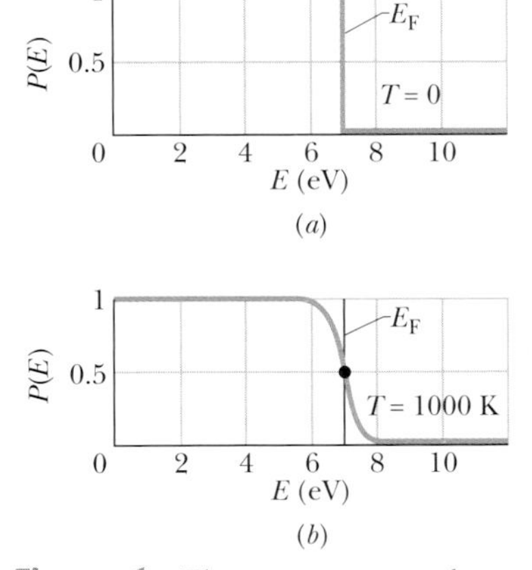

Fig. 41-6 The occupancy probability $P(E)$ is the probability that an energy level will be occupied by an electron. (*a*) At $T = 0$ K, $P(E)$ is unity for levels with energies E up to the Fermi energy E_F and zero for levels with higher energies. (*b*) At $T = 1000$ K, a few electrons whose energies were slightly less than the Fermi energy at $T = 0$ K move up to states with energies slightly greater than the Fermi energy. The dot on the curve shows that, for $E = E_F$, $P(E) = 0.5$.

Figure 41-6*b* is a plot of $P(E)$ for $T = 1000$ K. It shows that, as stated above, changes in the distribution of electrons among the available states involve only states whose energies are near the Fermi energy E_F. Note that if $E = E_F$ (no matter what the temperature T), the exponential term in Eq. 41-6 is $e^0 = 1$ and $P(E) = 0.5$. This leads us to a more useful definition of the Fermi energy:

> The Fermi energy of a given material is the energy of a quantum state that has the probability 0.5 of being occupied by an electron.

Figures 41-6*a* and *b* are plotted for copper, which has a Fermi energy of 7.0 eV. Thus, for copper both at $T = 0$ K and at $T = 1000$ K, a state at energy $E = 7.0$ eV has a probability of 0.5 of being occupied.

Sample Problem 41-4

(a) What is the probability that a quantum state whose energy is 0.10 eV above the Fermi energy will be occupied? Assume a sample temperature of 800 K.

Solution: The **Key Idea** here is that the occupancy probability of any state in a metal can be found from Fermi–Dirac statistics according to Eq. 41-6. To apply that equation, let us first calculate its dimensionless exponent:

$$\frac{E - E_F}{kT} = \frac{0.10 \text{ eV}}{(8.62 \times 10^{-5} \text{ eV/K})(800 \text{ K})} = 1.45.$$

Inserting this exponent into Eq. 41-6 yields

$$P(E) = \frac{1}{e^{1.45} + 1} = 0.19 \text{ or } 19\%. \qquad \text{(Answer)}$$

(b) What is the probability of occupancy for a state that is 0.10 eV *below* the Fermi energy?

Solution: The **Key Idea** of part (a) applies here also except that now the state has an energy *below* the Fermi energy. Thus, the exponent in Eq. 41-6 has the same magnitude we found in part (a) but is negative; so Eq. 41-6 now yields

$$P(E) = \frac{1}{e^{-1.45} + 1} = 0.81 \text{ or } 81\%. \qquad \text{(Answer)}$$

For states below the Fermi energy, we are often more interested in the probability that the state is *not* occupied. This probability is just $1 - P(E)$, or 19%. Note that it is the same as the probability of occupancy in (a).

How Many Occupied States Are There?

Equation 41-5 and Fig. 41-5 tell us how the available states are distributed in energy. The occupancy probability of Eq. 41-6 gives us the probability that any given state will actually be occupied by an electron. To find $N_o(E)$, the **density of occupied states**, we must multiply each available state by the corresponding value of the occupancy probability; that is,

$$\begin{pmatrix}\text{density of occupied states}\\ N_o(E) \text{ at energy } E\end{pmatrix} = \begin{pmatrix}\text{density of states}\\ N(E) \text{ at energy } E\end{pmatrix}\begin{pmatrix}\text{occupancy probability}\\ P(E) \text{ at energy } E\end{pmatrix}$$

or

$$N_o(E) = N(E)\,P(E) \qquad \text{(density of occupied states).} \qquad (41\text{-}7)$$

Figure 41-7*a* is a plot of Eq. 41-7 for copper at $T = 0$ K. It is found by multiplying, at each energy, the value of the density of states function (Fig. 41-5) by the value of the occupancy probability for absolute zero (Fig. 41-6*a*). Figure 41-7*b*, calculated similarly, shows the density of occupied states for copper at $T = 1000$ K.

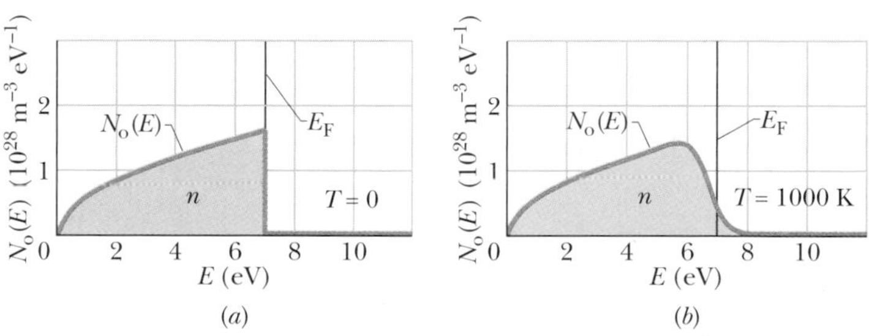

Fig. 41-7 (*a*) The density of occupied states $N_o(E)$ for copper at absolute zero. The area under the curve is the number density of electrons n. Note that all states with energies up to the Fermi energy $E_F = 7$ eV are occupied, and all those with energies above the Fermi energy are vacant. (*b*) The same for copper at $T = 1000$ K. Note that only electrons with energies near the Fermi energy have been affected and redistributed.

Sample Problem 41-5

If the metal in Sample Problem 41-3 is copper, which has a Fermi energy of 7.0 eV, how many occupied states per eV lie in a narrow energy range around 7.0 eV?

Solution: The **Key Idea** of Sample Problem 41-3a applies here also, except that now we use the density of *occupied* states $N_o(E)$ as given by Eq. 41-7 ($N_o(E) = N(E)\,P(E)$). A second **Key Idea** is that because we want to evaluate quantities for a narrow energy range around 7.0 eV (the Fermi energy for copper), the occupancy probability $P(E)$ is 0.50. From Fig. 41-5, we see that the density of states at 7 eV is about 1.8×10^{28} m^{-3} eV^{-1}. Thus, Eq. 41-7 tells us that the density of occupied states is

$$N_o(E) = N(E)\,P(E) = (1.8 \times 10^{28} \text{ m}^{-3} \text{ eV}^{-1})(0.50)$$
$$= 0.9 \times 10^{28} \text{ m}^{-3} \text{ eV}^{-1}.$$

Next, we rewrite the equation in Sample Problem 41-3a in terms of occupied states:

$$\begin{pmatrix}\text{number of } \textit{occupied}\\ \text{states per eV at 7 eV}\end{pmatrix} = \begin{pmatrix}\text{density of } \textit{occupied}\\ \text{states } N_o(E) \text{ at 7 eV}\end{pmatrix} \times \begin{pmatrix}\text{volume } V\\ \text{of sample}\end{pmatrix}.$$

Substituting our result for $N_o(E)$ and the previously given volume 2×10^{-9} m^3 for V then gives us

$$\begin{pmatrix}\text{number of occupied}\\ \text{states per eV}\\ \text{at 7 eV}\end{pmatrix} = (0.9 \times 10^{28} \text{ m}^{-3} \text{ eV}^{-1})(2 \times 10^{-9} \text{ m}^3)$$
$$= 1.8 \times 10^{19} \text{ eV}^{-1}$$
$$\approx 2 \times 10^{19} \text{ eV}^{-1}. \qquad \text{(Answer)}$$

Calculating the Fermi Energy

Suppose we add up (via integration) the number of occupied states per unit volume in Fig. 41-7*a* (for $T = 0$ K) at all energies between $E = 0$ and $E = E_F$. The result must equal n, the number of conduction electrons per unit volume for the metal. In equation form, we have

$$n = \int_0^{E_F} N_o(E)\, dE. \tag{41-8}$$

(Graphically, the integral here represents the area under the distribution curve of Fig. 41-7*a*.) Because $P(E) = 1$ for all energies below the Fermi energy when $T = 0$ K, Eq. 41-7 tells us we can replace $N_o(E)$ in Eq. 41-8 with $N(E)$ and then use Eq. 41-8 to find the Fermi energy E_F. If we substitute Eq. 41-5 into Eq. 41-8, we find that

$$n = \frac{8\sqrt{2}\pi m^{3/2}}{h^3}\int_0^{E_F} E^{1/2}\, dE = \frac{8\sqrt{2}\pi m^{3/2}}{h^3}\frac{2E_F^{3/2}}{3},$$

in which m is the electron mass. Solving for E_F now leads to

$$E_F = \left(\frac{3}{16\sqrt{2}\pi}\right)^{2/3}\frac{h^2}{m}n^{2/3} = \frac{0.121h^2}{m}n^{2/3}. \tag{41-9}$$

Thus, when we know n, the number of conduction electrons per unit volume for a metal, we can find the Fermi energy for that metal.

41-6 Semiconductors

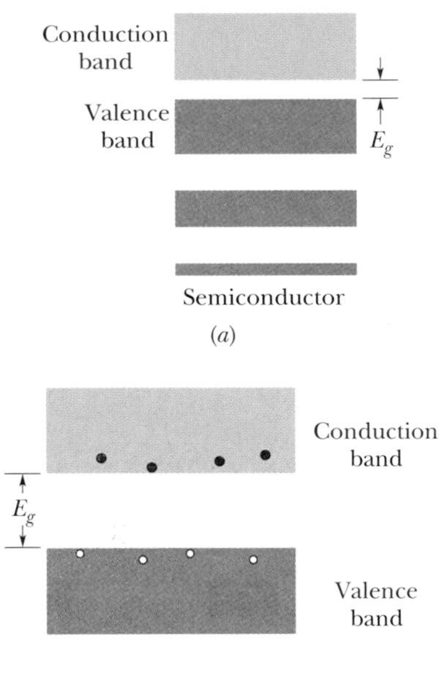

Fig. 41-8 (*a*) The band–gap pattern for a semiconductor. It resembles that of an insulator (see Fig. 41-4*a*) except that here the energy gap E_g is much smaller; thus electrons, because of their thermal agitation, have some reasonable probability of being able to jump the gap. (*b*) Thermal agitation has caused a few electrons to jump the gap from the valence band to the conduction band, leaving an equal number of holes in the valence band.

If you compare Fig. 41-8*a* with Fig. 41-4*a*, you can see that the band structure of a semiconductor is like that of an insulator. The main difference is that the semiconductor has a much smaller energy gap E_g between the top of the highest filled band (called the **valence band**) and the bottom of the vacant band just above it (called the **conduction band**). Thus, there is no doubt that silicon ($E_g = 1.1$ eV) is a semiconductor and diamond ($E_g = 5.5$ eV) is an insulator. In silicon—but not in diamond—there is a real possibility that thermal agitation at room temperature will cause electrons to jump the gap from valence to conduction band.

In Table 41-1 we compared three basic electrical properties of copper, our prototype metallic conductor, and silicon, our prototype semiconductor. Let us look again at that table, one row at a time, to see how a semiconductor differs from a metal.

Number Density of Charge Carriers n

The bottom row of Table 41-1 shows that copper has far more charge carriers per unit volume than silicon, by a factor of about 10^{13}. For copper, each atom contributes one electron, its single valence electron, to the conduction process. Charge carriers in silicon arise only because, at thermal equilibrium, thermal agitation causes a certain (very small) number of valence-band electrons to jump the energy gap into the conduction band, leaving an equal number of unoccupied energy states, called **holes,** in the valence band. Figure 41-8*b* shows the situation.

Both the electrons in the conduction band and the holes in the valence band serve as charge carriers. The holes do so by permitting a certain freedom of movement to the electrons remaining in the valence band, electrons that, in the absence of holes, would be gridlocked. If an electric field $\vec{E}$ is set up in a semiconductor, the electrons in the valence band, being negatively charged, tend to drift in the direction opposite $\vec{E}$. This causes the positions of the holes to drift in the direction of $\vec{E}$. In effect, the holes behave like moving particles of charge $+e$.

It may help to think of a row of cars parked bumper to bumper, with the lead car at one car's length from a barrier and the empty one-car-length distance being an available parking space. If the leading car moves forward to the barrier, it opens up a parking space behind it. The second car can then move up to fill that space, allowing the third car to move up, and so on. The motions of the many cars toward the barrier are most simply analyzed by focusing attention on the drift of the single "hole" (parking space) away from the barrier.

In semiconductors, conduction by holes is just as important as conduction by electrons. In thinking about hole conduction, we can assume that all unoccupied states in the valence band are occupied by particles of charge $+e$ and that all electrons in the valence band have been removed, so that these positive charge carriers can move freely throughout the band.

Resistivity ρ

Recall from Chapter 26 that the resistivity ρ of a material is $m/e^2n\tau$, where m is the electron mass, e is the fundamental charge, n is the number of charge carriers per unit volume, and τ is the mean time between collisions of the charge carriers. Table 41-1 shows that, at room temperature, the resistivity of silicon is higher than that of copper by a factor of about 10^{11}. This vast difference can be accounted for by the vast difference in n. Other factors enter, but their effect on the resistivity is swamped by the enormous difference in n.

Temperature Coefficient of Resistivity α

Recall that α (see Eq. 26-17) is the fractional change in resistivity per unit change in temperature:

$$\alpha = \frac{1}{\rho}\frac{d\rho}{dT}. \qquad (41\text{-}10)$$

The resistivity of copper *increases* with temperature (that is, $d\rho/dT > 0$) because collisions of copper's charge carriers occur more frequently at higher temperatures. Thus, α is *positive* for copper.

The collision frequency also increases with temperature for silicon. However, the resistivity of silicon actually *decreases* with temperature ($d\rho/dT < 0$) because the number of charge carriers n (electrons in the conduction band and holes in the valence band) increases so rapidly with temperature. (More electrons jump the gap from the valence band to the conduction band.) Thus, the fractional change α is *negative* for silicon.

CHECKPOINT 2 The research laboratory of a large corporation developed three new solid materials whose electrical properties are shown here. Anticipating patent applications, the laboratory identified these materials with code names. Classify each material as a metal, an insulator, a semiconductor, or none of the above:

Material (Code Name)	n (m^{-3})	ρ ($\Omega\cdot$m)	α (K^{-1})
Cleveland	10^{29}	10^{-8}	$+10^{-3}$
Boca Raton	10^{28}	10^{-9}	-10^{-3}
Seattle	10^{15}	10^{3}	-10^{-2}

41-7 Doped Semiconductors

The usefulness of semiconductors in technology can be greatly improved by introducing a small number of suitable replacement atoms (called impurities) into the semiconductor lattice—a process called **doping.** Typically, only about 1 silicon atom in 10^7 is replaced by a dopant atom in the doped semiconductor. Essentially all modern semiconducting devices are based on doped material. Such materials are of two types, called ***n*-type** and ***p*-type**; we discuss each in turn.

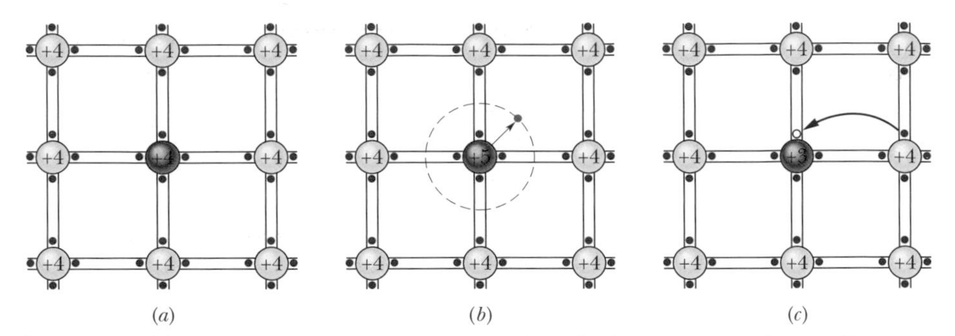

Fig. 41-9 (a) A flattened-out representation of the lattice structure of pure silicon. Each silicon ion is coupled to its four nearest neighbors by a two-electron covalent bond (represented by a pair of red dots between two parallel black lines). The electrons belong to the bond—not to the individual atoms—and form the valence band of the sample. (b) One silicon atom is replaced by a phosphorus atom (valence = 5). The "extra" electron is only loosely bound to its ion core and may easily be elevated to the conduction band, where it is free to wander through the volume of the lattice. (c) One silicon atom is replaced by an aluminum atom (valence = 3). There is now a hole in one of the covalent bonds and thus in the valence band of the sample. The hole can easily migrate through the lattice as electrons from neighboring bonds move in to fill it. Here the hole migrates rightward.

n-Type Semiconductors

The electrons in an isolated silicon atom are arranged in subshells according to the scheme

$$1s^2\ 2s^2\ 2p^6\ 3s^2\ 3p^2,$$

in which, as usual, the superscripts (which add to 14, the atomic number of silicon) represent the numbers of electrons in the specified subshells.

Figure 41-9a is a flattened-out representation of a portion of the lattice of pure silicon in which the portion has been projected onto a plane; compare the figure with Fig. 41-1b, which represents the unit cell of the lattice in three dimensions. Each silicon atom contributes its pair of $3s$ electrons and its pair of $3p$ electrons to form a rigid two-electron covalent bond with each of its four nearest neighbors. (A covalent bond is a link between two atoms in which the atoms share a pair of electrons.) The four atoms that lie within the unit cell in Fig. 41-1b show these four bonds.

The electrons that form the silicon–silicon bonds constitute the valence band of the silicon sample. If an electron is torn from one of these bonds so that it becomes free to wander throughout the lattice, we say that the electron has been raised from the valence band to the conduction band. The minimum energy required to do this is the gap energy E_g.

Because four of its electrons are involved in bonds, each silicon "atom" is actually an ion consisting of an inert neon-like electron cloud (containing 10 electrons) surrounding a nucleus whose charge is $+14e$, where 14 is the atomic number of silicon. The net charge of each of these ions is thus $+4e$, and the ions are said to have a *valence number* of 4.

In Fig. 41-9b the central silicon ion has been replaced by an atom of phosphorus (valence = 5). Four of the valence electrons of the phosphorus form bonds with the four surrounding silicon ions. The fifth ("extra") electron is only loosely bound to the phosphorus ion core. On an energy-band diagram, we usually say that such an electron occupies a localized energy state that lies within the energy gap, at an average energy interval E_d below the bottom of the conduction band; this is indicated in Fig. 41-10a. Because $E_d \ll E_g$, the energy required to excite electrons from *these* levels into the conduction band is much less than that required to excite silicon valence electrons into the conduction band.

The phosphorus atom is called a **donor** atom because it readily *donates* an electron to the conduction band. In fact, at room temperature virtually *all* the electrons contributed by the donor atoms are in the conduction band. By adding donor atoms, it is possible to increase greatly the number of electrons in the conduction band, by a factor very much larger than Fig. 41-10*a* suggests.

Semiconductors doped with donor atoms are called ***n*-type semiconductors;** the *n* stands for *negative,* to imply that the negative charge carriers introduced into the conduction band greatly outnumber the positive charge carriers, which are the holes in the valence band. In *n*-type semiconductors, the electrons are called the **majority carriers** and the holes are called the **minority carriers.**

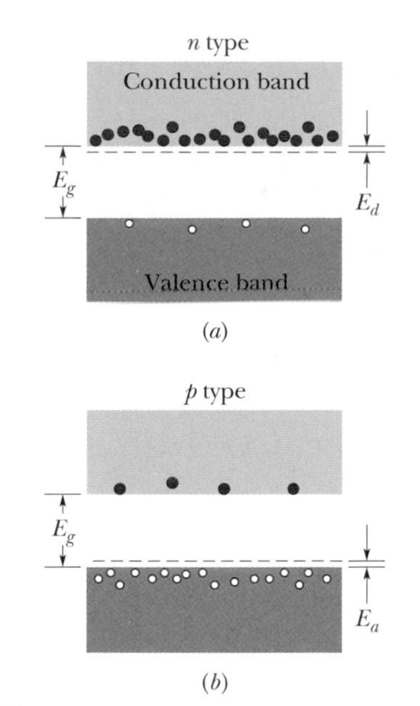

Fig. 41-10 (*a*) In a doped *n*-type semiconductor, the energy levels of donor electrons lie a small interval E_d below the bottom of the conduction band. Because donor electrons can be easily excited to the conduction band, there are now many more electrons in that band. The valence band contains the same small number of holes as before the dopant was added. (*b*) In a doped *p*-type semiconductor, the acceptor levels lie a small energy interval E_a above the top of the valence band. There are now many more holes in the valence band. The conduction band contains the same small number of electrons as before the dopant was added. The ratio of majority carriers to minority carriers in both (*a*) and (*b*) is very much greater than is suggested by these diagrams.

p-Type Semiconductors

Now consider Fig. 41-9*c*, in which one of the silicon atoms (valence = 4) has been replaced by an atom of aluminum (valence = 3). The aluminum atom can bond covalently with only three silicon atoms, and so there is now a "missing" electron (a hole) in one aluminum–silicon bond. With a small expenditure of energy, an electron can be torn from a neighboring silicon–silicon bond to fill this hole, thereby creating a hole in *that* bond. Similarly, an electron from some other bond can be moved to fill the newly created hole. In this way, the hole can migrate through the lattice.

The aluminum atom is called an **acceptor** atom because it readily *accepts* an electron from a neighboring bond—that is, from the valence band of silicon. As Fig. 41-10*b* suggests, this electron occupies a localized energy state that lies within the energy gap, at an average energy interval E_a above the top of the valence band. Because this energy interval E_a is small, valence electrons are easily bumped up to the acceptor level, leaving holes in the valence band. Thus, by adding acceptor atoms, it is possible to greatly increase the number of holes in the valence band, by a factor much larger than Fig. 41-10*b* suggests. In silicon at room temperature, virtually *all* the acceptor levels are occupied by electrons.

Semiconductors doped with acceptor atoms are called ***p*-type semiconductors;** the *p* stands for *positive* to imply that the holes introduced into the valence band, which behave like positive charge carriers, greatly outnumber the electrons in the conduction band. In *p*-type semiconductors, holes are the majority carriers and electrons are the minority carriers.

Table 41-2 summarizes the properties of a typical *n*-type and a typical *p*-type semiconductor. Note particularly that the donor and acceptor ion cores, although they are charged, are not charge *carriers* because at normal temperatures they remain fixed in their lattice sites.

TABLE 41-2
Properties of Two Doped Semiconductors

	Type of Semiconductor	
Property	*n*	*p*
Matrix material	Silicon	Silicon
Matrix nuclear charge	$+14e$	$+14e$
Matrix energy gap	1.2 eV	1.2 eV
Dopant	Phosphorus	Aluminum
Type of dopant	Donor	Acceptor
Majority carriers	Electrons	Holes
Minority carriers	Holes	Electrons
Dopant energy gap	$E_d = 0.045$ eV	$E_a = 0.067$ eV
Dopant valence	5	3
Dopant nuclear charge	$+15e$	$+13e$
Dopant net ion charge	$+e$	$-e$

Sample Problem 41-6

The number density n_0 of conduction electrons in pure silicon at room temperature is about $10^{16}\ \mathrm{m}^{-3}$. Assume that, by doping the silicon lattice with phosphorus, we want to increase this number by a factor of a million (10^6). What fraction of silicon atoms must we replace with phosphorus atoms? (Recall that at room temperature, thermal agitation is so effective that essentially every phosphorus atom donates its "extra" electron to the conduction band.)

Solution: One **Key Idea** here is that, because each phosphorus atom contributes one conduction electron and because we want the total number density of conduction electrons to be $10^6 n_0$, the number density of phosphorus atoms n_P must be given by

$$10^6 n_0 = n_0 + n_P.$$

Then

$$n_P = 10^6 n_0 - n_0 \approx 10^6 n_0 = (10^6)(10^{16}\ \mathrm{m}^{-3}) = 10^{22}\ \mathrm{m}^{-3}.$$

This tells us that we must add 10^{22} atoms of phosphorus per cubic meter of silicon.

A second **Key Idea** is that we can find the number density n_{Si} of silicon atoms in pure silicon (before the doping) from Eq. 41-4, which we can write as

$$\begin{pmatrix}\text{number of atoms}\\ \text{in sample}\end{pmatrix} = \frac{(\text{silicon density})(\text{sample volume } V)}{(\text{silicon molar mass } M_{Si})/N_A}.$$

Dividing both sides by the sample volume V to get the number density of silicon atoms n_{Si} on the left, we then have

$$n_{Si} = \frac{(\text{silicon density})N_A}{M_{Si}}.$$

Appendix F tells us that the density of silicon is $2.33\ \mathrm{g/cm^3}$ ($= 2330\ \mathrm{kg/m^3}$) and the molar mass of silicon is 28.1 g/mol ($= 0.0281$ kg/mol). Thus, we have

$$n_{Si} = \frac{(2330\ \mathrm{kg/m^3})(6.02 \times 10^{23}\ \mathrm{atoms/mol})}{0.0281\ \mathrm{kg/mol}} = 5 \times 10^{28}\ \mathrm{atoms/m^3} = 5 \times 10^{28}\ \mathrm{m}^{-3}.$$

The fraction we seek is approximately

$$\frac{n_P}{n_{Si}} = \frac{10^{22}\ \mathrm{m}^{-3}}{5 \times 10^{28}\ \mathrm{m}^{-3}} = \frac{1}{5 \times 10^6}. \quad \text{(Answer)}$$

If we replace only *one silicon atom in five million* with a phosphorus atom, the number of electrons in the conduction band will be increased by a factor of a million.

How can such a tiny admixture of phosphorus have what seems to be such a big effect? The answer is that, although the effect is very significant, it is not "big." The number density of conduction electrons was $10^{16}\ \mathrm{m}^{-3}$ before doping and $10^{22}\ \mathrm{m}^{-3}$ after doping. For copper, however, the conduction-electron number density (given in Table 41-1) is about $10^{29}\ \mathrm{m}^{-3}$. Thus, even after doping, the number density of conduction electrons in silicon remains much less than that of a typical metal, such as copper, by a factor of about 10^7.

41-8 The p-n Junction

A ***p-n* junction** (Fig. 41-11*a*) is a single semiconductor crystal that has been selectively doped so that one region is *n*-type material and the adjacent region is *p*-type material. Such junctions are at the heart of essentially all semiconductor devices.

We assume, for simplicity, that the junction has been formed mechanically, by jamming together a bar of *n*-type semiconductor and a bar of *p*-type semiconductor. Thus, the transition from one region to the other is perfectly sharp, occurring at a single **junction plane.**

Let us discuss the motions of electrons and holes just after the *n*-type bar and the *p*-type bar, both electrically neutral, have been jammed together to form the junction. We first examine the majority carriers, which are electrons in the *n*-type material and holes in the *p*-type material.

Motions of the Majority Carriers

If you burst a helium-filled balloon, helium atoms will diffuse (spread) outward into the surrounding air. This happens because there are very few helium atoms in normal air. In more formal language, there is a helium *density gradient* at the balloon–air interface (the number density of helium atoms varies across the interface); the helium atoms move so as to reduce the gradient.

In the same way, electrons on the *n* side of Fig. 41-11*a* that are close to the junction plane tend to diffuse across it (from right to left in the figure) and into the *p* side, where there are very few free electrons. Similarly, holes on the *p* side that are close to the junction plane tend to diffuse across that plane (from left to right) and into the *n* side, where there are very few holes. The motions of both the electrons and the holes contribute to a **diffusion current** I_{diff}, conventionally directed from left to right as indicated in Fig. 41-11*d*.

Recall that the *n*-side is studded throughout with positively charged donor ions, fixed firmly in their lattice sites. Normally, the excess positive charge of each of these ions is compensated electrically by one of the conduction-band electrons. When an *n*-side electron diffuses across the junction plane, however, the diffusion "uncovers" one of these donor ions, thus introducing a fixed positive charge near the junction plane on the *n* side. When the diffusing electron arrives on the *p* side, it quickly combines with an acceptor ion (which lacks one electron), thus introducing a fixed negative charge near the junction plane on the *p* side.

In this way electrons diffusing through the junction plane from right to left in Fig. 41-11*a* result in a buildup of **space charge** on each side of the junction plane, as indicated in Fig. 41-11*b*. Holes diffusing through the junction plane from left to right have exactly the same effect. (Take the time now to convince yourself of that.) The motions of both majority carriers—electrons and holes—contribute to the buildup of these two space charge regions, one positive and one negative. These two regions form a **depletion zone,** so named because it is relatively free of *mobile* charge carriers; its width is shown as d_0 in Fig. 41-11*b*.

The buildup of space charge generates an associated **contact potential difference** V_0 across the depletion zone, as Fig. 41-11*c* shows. This potential difference limits further diffusion of electrons and holes across the junction plane. Negative charges tend to avoid regions of low potential. Thus, an electron approaching the junction plane from the right in Fig. 41-11*b* is moving toward a region of low potential and would tend to turn back into the *n* side. Similarly, a positive charge (a hole) approaching the junction plane from the left is moving toward a region of high potential and would tend to turn back into the *p* side.

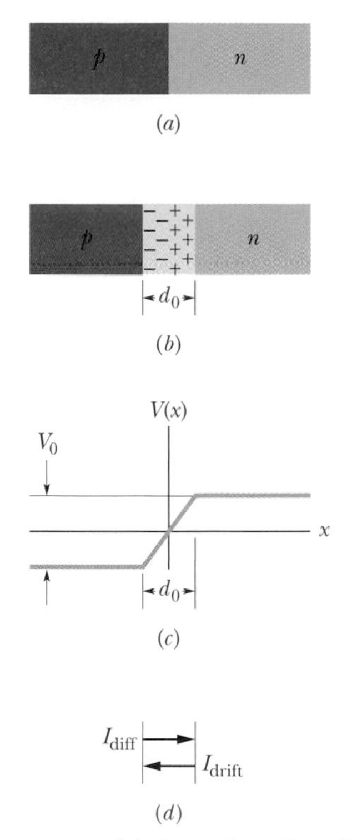

Fig. 41-11 (*a*) A *p-n* junction. (*b*) Motions of the majority charge carriers across the junction plane uncover a space charge associated with uncompensated donor ions (to the right of the plane) and acceptor ions (to the left). (*c*) Associated with the space charge is a contact potential difference V_0 across d_0. (*d*) The diffusion of majority carriers (both electrons and holes) across the junction plane produces a diffusion current I_{diff}. (In a real *p-n* junction, the boundaries of the depletion zone would not be sharp, as shown here, and the contact potential curve (*c*) would be smooth, with no sharp corners.)

Motions of the Minority Carriers

As Fig. 41-10*a* shows, although the majority carriers in *n*-type material are electrons, there are a few holes. Likewise in *p*-type material (Fig. 41-10*b*), although the majority carriers are holes, there are also a few electrons. These few holes and electrons are the minority carriers in the corresponding materials.

Although the potential difference V_0 in Fig. 41-11*c* acts as a barrier for the majority carriers, it is a downhill trip for the minority carriers, be they electrons on the *p* side or holes on the *n* side. Positive charges (holes) tend to seek regions of low potential; negative charges (electrons) tend to seek regions of high potential. Thus, both types of minority carriers are *swept across* the junction plane by the contact potential difference and together constitute a **drift current** I_{drift} across the junction plane from right to left, as Fig. 41-11*d* indicates.

Thus, an isolated *p-n* junction is in an equilibrium state in which a contact potential difference V_0 exists between its ends. At equilibrium, the average diffusion current I_{diff} that moves through the junction plane from the *p* side to the *n* side is just balanced by an average drift current I_{drift} that moves in the opposite direction. These two currents cancel because the net current through the junction plane must be zero; otherwise charge would be transferred without limit from one end of the junction to the other.

CHECKPOINT 3 Which of the following five currents across the junction plane of Fig. 41-11*a* must be zero?

(a) the net current due to holes, both majority and minority carriers included
(b) the net current due to electrons, both majority and minority carriers included
(c) the net current due to both holes and electrons, both majority and minority carriers included
(d) the net current due to majority carriers, both holes and electrons included
(e) the net current due to minority carriers, both holes and electrons included

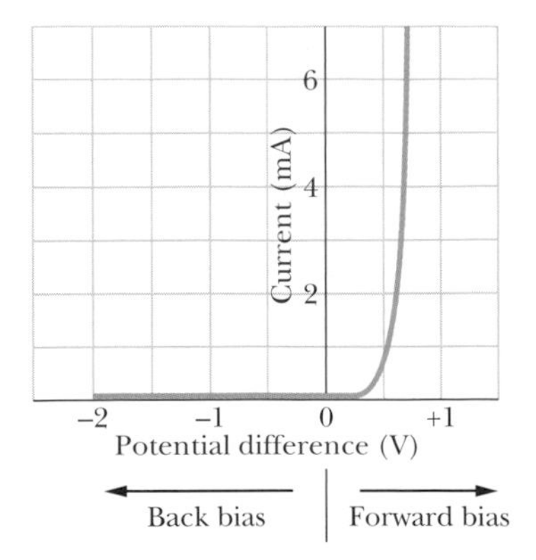

Fig. 41-12 A current–voltage plot for a *p-n* junction, showing that the junction is highly conducting when forward-biased and essentially nonconducting when back-biased.

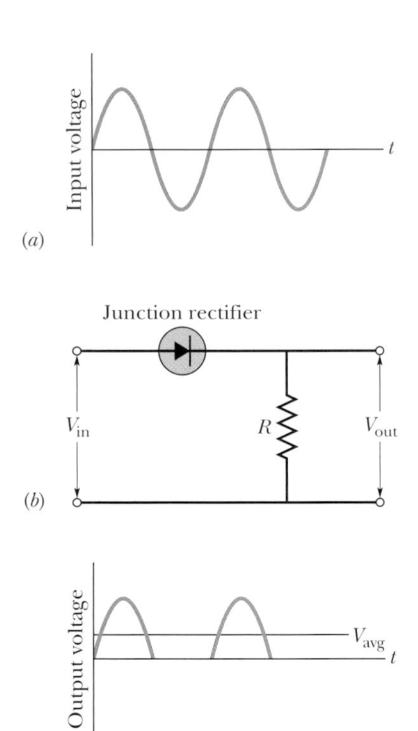

Fig. 41-13 A *p-n* junction connected as a junction rectifier. The action of the circuit in (*b*) is to pass the positive half of the input wave form (*a*) but to suppress the negative half. The average potential of the input wave form is zero; that of the output wave form (*c*) has a positive value V_{avg}.

41-9 The Junction Rectifier

Look now at Fig. 41-12. It shows that, if we place a potential difference across a *p-n* junction in one direction (here labeled + and "Forward bias"), there will be a current through the junction. However, if we reverse the direction of the potential difference, there will be approximately zero current through the junction.

One application of this property is the **junction rectifier,** whose symbol is shown in Fig. 41-13*b*; the arrowhead corresponds to the *p*-type end of the device and points in the allowed direction of conventional current. A sine wave input potential to the device (Fig. 41-13*a*) is transformed to a half-wave output potential (Fig. 41-13*c*) by the junction rectifier; that is, the rectifier acts as essentially a closed switch (zero resistance) for one polarity of the input potential and as essentially an open switch (infinite resistance) for the other.

The average value of the input voltage in Fig. 41-13*a* is zero, but that of the output voltage in Fig. 41-13*c* is not. Thus, a junction rectifier can be used as part of an apparatus to convert an alternating potential difference into a constant potential difference, as for an electronic power supply.

Figure 41-14 shows why a *p-n* junction operates as a junction rectifier. In Fig. 41-14*a*, a battery is connected across the junction with its positive terminal connected at the *p* side. In this **forward-bias connection,** the *p* side becomes more positive and the *n* side becomes more negative, thus *decreasing* the height of the potential barrier V_0 of Fig. 41-11*c*. More of the majority carriers can now surmount this smaller barrier; hence, the diffusion current I_{diff} increases markedly.

The minority carriers that form the drift current, however, sense no barrier; so the drift current I_{drift} is not affected by the external battery. The nice current balance that existed at zero bias (see Fig. 41-11*d*) is thus upset, and, as shown in Fig. 41-14*a*, a large net forward current I_F appears in the circuit.

Another effect of forward bias is to narrow the depletion zone, as a comparison of Figs. 41-11*b* and Fig. 41-14*a* shows. The depletion zone narrows because the reduced potential barrier associated with forward bias must be associated with a smaller space charge. Because the ions producing the space charge are fixed in their lattice sites, a reduction in their number can come about only through a reduction in the width of the depletion zone.

Because the depletion zone normally contains very few charge carriers, it is normally a region of high resistivity. However, when its width is substantially reduced by a forward bias, its resistance is also reduced substantially, as is consistent with the large forward current.

Figure 41-14*b* shows the **back-bias** connection, in which the negative terminal of the battery is connected at the *p*-type end of the *p-n* junction. Now the applied emf *increases* the contact potential difference, the diffusion current *decreases* substantially while the drift current remains unchanged, and a relatively *small* back current I_B results. The depletion zone *widens,* its *high* resistance being consistent with the *small* back current I_B.

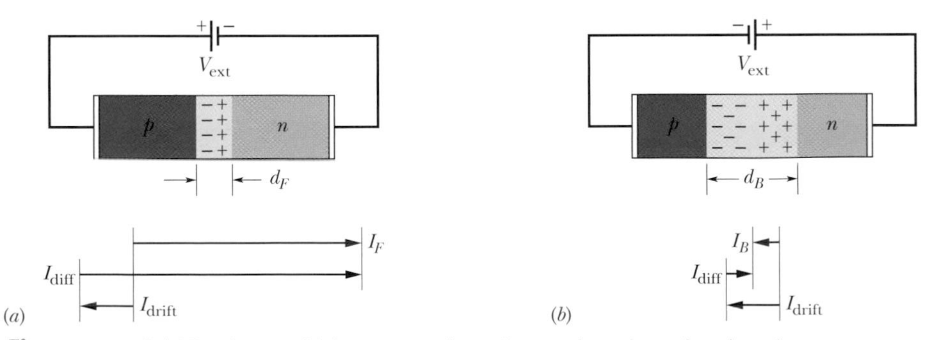

Fig. 41-14 (*a*) The forward-bias connection of a *p-n* junction, showing the narrowed depletion zone and the large forward current I_F. (*b*) The back-bias connection, showing the widened depletion zone and the small back current I_B.

41-10 The Light-Emitting Diode (LED)

Nowadays, we can hardly avoid the brightly colored "electronic" numbers that glow at us from cash registers and gasoline pumps, microwave ovens and alarm clocks, and we cannot seem to do without the invisible infrared beams that control elevator doors and operate television sets via remote control. In nearly all cases this light is emitted from a *p-n* junction operating as a **light-emitting diode** (LED). How can a *p-n* junction generate light?

Consider first a simple semiconductor. When an electron from the bottom of the conduction band falls into a hole at the top of the valence band, an energy E_g equal to the gap width is released. In silicon, germanium, and many other semiconductors, this energy is largely transformed into thermal energy of the vibrating lattice, and as a result, no light is emitted.

In some semiconductors, however, including gallium arsenide, the energy can be emitted as a photon of energy hf at wavelength

$$\lambda = \frac{c}{f} = \frac{c}{E_g/h} = \frac{hc}{E_g}. \tag{41-11}$$

To emit enough light to be useful as an LED, the material must have a suitably large number of electron–hole transitions. This condition is *not* satisfied by a pure semiconductor because, at room temperature, there are simply not enough electron–hole pairs. As Fig. 41-10 suggests, doping will not help. In doped *n*-type material the number of conduction electrons is greatly increased, but there are not enough holes for them to combine with; in doped *p*-type material there are plenty of holes but not enough electrons to combine with them. Thus, neither a pure semiconductor nor a doped semiconductor can provide enough electron–hole transitions to serve as a practical LED.

What we need is a semiconductor material with a very large number of electrons in the conduction band *and* a correspondingly large number of holes in the valence band. A device with this property can be fabricated by placing a strong forward bias on a heavily doped *p-n* junction, as in Fig. 41-15. In such an arrangement the current I through the device serves to inject electrons into the *n*-type material and to inject holes into the *p*-type material. If the doping is heavy enough and the current is great enough, the depletion zone can become very narrow, perhaps only a few micrometers wide. The result is a great number density of electrons in the *n*-type material facing a correspondingly great number density of holes in the *p*-type material, across the narrow depletion zone. With such great number densities so near each other, many electron–hole combinations occur, causing light to be emitted from that zone. Figure 41-16 shows the construction of an actual LED.

Commercial LEDs designed for the visible region are commonly based on gallium suitably doped with arsenic and phosphorus atoms. An arrangement in which 60% of the nongallium sites are occupied by arsenic ions and 40% by phosphorus ions results in a gap width E_g of about 1.8 eV, corresponding to red light. Other doping and transition-level arrangements make it possible to construct LEDs that emit light in essentially any desired region of the visible and near-visible spectra.

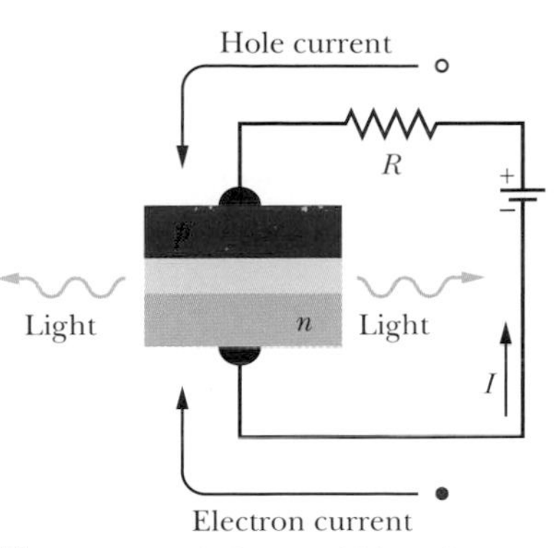

Fig. 41-15 A forward-biased *p-n* junction, showing electrons being injected into the *n*-type material and holes into the *p*-type material. (Holes move in the conventional direction of the current I, equivalent to electrons moving in the opposite direction.) Light is emitted from the narrow depletion zone each time an electron and a hole combine across that zone.

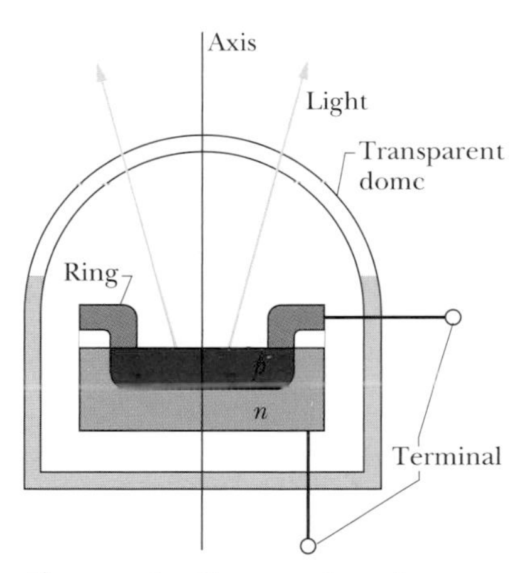

Fig. 41-16 Cross section of an LED (the device has rotational symmetry about the central axis). The *p*-type material, which is thin enough to transmit light, is in the form of a circular disk. A connection is made to the *p*-type material through a circular metal ring that touches the disk at its periphery. The depletion zone between the *n*-type material and the *p*-type material is not shown.

The Photo-Diode

Passing a current through a suitably arranged *p-n* junction can generate light. The reverse is also true; that is, shining light on a suitably arranged *p-n* junction can produce a current in a circuit that includes the junction. This is the basis for the **photo-diode.**

When you click your television remote control, an LED in the device sends out a coded sequence of pulses of infrared light. The receiving device in your television set is an elaboration of the simple (two-terminal) photo-diode that not

Fig. 41-17 A junction laser developed at the AT&T Bell Laboratories. The cube at the right is a grain of salt.

only detects the infrared signals but also amplifies them and transforms them into electrical signals that change the channel or adjust the volume, among other tasks.

The Junction Laser

In the arrangement of Fig. 41-15, there are many electrons in the conduction band of the n-type material and many holes in the valence band of the p-type material. Thus, there is a **population inversion** for the electrons; that is, there are more electrons in higher energy levels than in lower energy levels. As we discussed in Section 40-12, this is normally a necessary—but not a sufficient—condition for laser action.

When a single electron moves from the conduction band to the valence band, it can release its energy as a photon. This photon can stimulate a second electron to fall into the valence band, producing a second photon by stimulated emission. In this way, if the current through the junction is great enough, a chain reaction of stimulated emission events can occur and laser light can be generated. To bring this about, opposite faces of the p-n junction crystal must be flat and parallel, so that light can be reflected back and forth within the crystal. (Recall that in the helium–neon laser of Fig. 40-21, a pair of mirrors served this purpose.) Thus, a p-n junction can act as a **junction laser,** its light output being highly coherent and much more sharply defined in wavelength than light from an LED.

Junction lasers are built into compact disc (CD) players, where, by detecting reflections from the rotating disc, they are used to translate microscopic pits in the disc into sound. They are also much used in optical communication systems based on optical fibers. Figure 41-17 suggests their tiny scale. Junction lasers are usually designed to operate in the infrared region of the electromagnetic spectrum because optical fibers have two "windows" in that region (at $\lambda = 1.31$ and $1.55\ \mu$m) for which the energy absorption per unit length of the fiber is a minimum.

Sample Problem 41-7

An LED is constructed from a p-n junction based on a certain Ga-As-P semiconducting material whose energy gap is 1.9 eV. What is the wavelength of the emitted light?

Solution: The **Key Idea** here is to assume that the transitions are from the bottom of the conduction band to the top of the valence band; then Eq. 41-11 holds. From this equation

$$\lambda = \frac{hc}{E_g} = \frac{(6.63 \times 10^{-34}\ \text{J}\cdot\text{s})(3.00 \times 10^{8}\ \text{m/s})}{(1.9\ \text{eV})(1.60 \times 10^{-19}\ \text{J/eV})}$$
$$= 6.5 \times 10^{-7}\ \text{m} = 650\ \text{nm}. \quad \text{(Answer)}$$

Light of this wavelength is red.

CHECKPOINT 4 For the LED in this sample problem, is 650 nm (a) the only wavelength that can be emitted, (b) the maximum emitted wavelength, (c) the minimum emitted wavelength, or (d) the average emitted wavelength? (Consider the quantum jump assumed in the solution.)

41-11 The Transistor

A **transistor** is a three-terminal semiconducting device that can be used to amplify input signals. Figure 41-18 shows a generalized **f**ield-**e**ffect **t**ransistor (FET); in it, the flow of electrons from terminal S (the *source*) leftward through the shaded

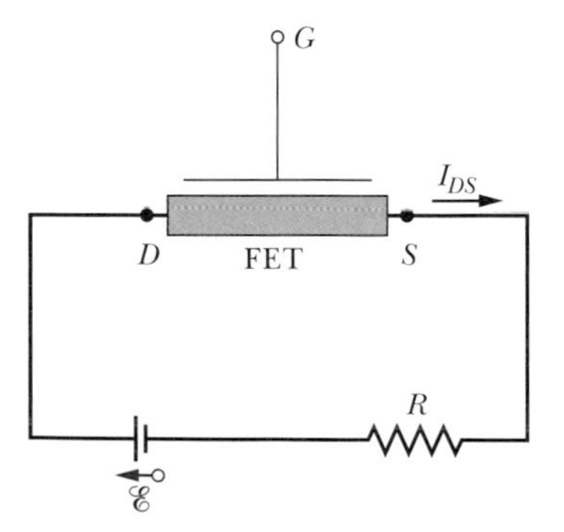

Fig. 41-18 A circuit containing a generalized field-effect transistor through which electrons flow from the source terminal S to the drain terminal D. (The conventional current I_{DS} is in the opposite direction.) The magnitude of I_{DS} is controlled by the electric field set up within the FET by a potential applied to G, the gate terminal.

region to terminal D (the *drain*) can be controlled by an electric field (hence field effect) set up within the device by a suitable electric potential applied to terminal G (the *gate*). Transistors are available in many types; we shall discuss only a particular FET called a MOSFET, or **m**etal-**o**xide-**s**emiconductor-**f**ield-**e**ffect **t**ransistor. The MOSFET has been described as the workhorse of the modern electronics industry.

For many applications the MOSFET is operated in only two states: with the drain-to-source current I_{DS} ON (gate open) or with it OFF (gate closed). The first of these can represent a 1 and the other a 0 in the binary arithmetic on which digital logic is based, and therefore MOSFETs can be used in digital logic circuits. Switching between the ON and OFF states can occur at high speed, so that binary logic data can be moved through MOSFET-based circuits very rapidly. MOSFETs about 500 nm in length—about the same as the wavelength of yellow light—are routinely fabricated for use in electronic devices of all kinds.

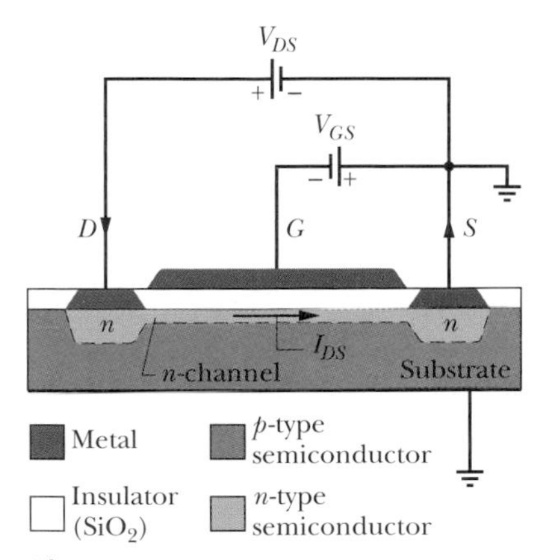

Fig. 41-19 A particular type of field-effect transistor known as a MOSFET. The magnitude of the drain-to-source conventional current I_{DS} through the n channel is controlled by the potential difference V_{GS} applied between the source S and the gate G. A depletion zone that exists between the n-type material and the p-type substrate is not shown.

Figure 41-19 shows the basic structure of a MOSFET. A single crystal of silicon or other semiconductor is lightly doped to form p-type material that serves as the *substrate*. Embedded in this substrate, by heavily "overdoping" with n-type dopants, are two "islands" of n-type material, forming the drain D and the source S. The drain and source are connected by a thin channel of n-type material, called the ***n* channel.** A thin insulating layer of silicon dioxide (hence the O in MOSFET) is deposited on the crystal and penetrated by two metallic terminals (hence the M) at D and S, so that electrical contact can be made with the drain and the source. A thin metallic layer—the gate G— is deposited facing the n channel. Note that the gate makes no electrical contact with the transistor proper, being separated from it by the insulating oxide layer.

Consider first that the source and p-type substrate are grounded (at zero potential) and the gate is "floating"; that is, the gate is not connected to an external source of emf. Let a potential V_{DS} be applied between the drain and the source, such that the drain is positive. Electrons will then flow through the n channel from source to drain, and the conventional current I_{DS}, as shown in Fig. 41-19, will be from drain to source through the n channel.

Now let a potential V_{GS} be applied to the gate, making it negative with respect to the source. The negative gate sets up within the device an electric field (hence the "field effect") that tends to repel electrons from the n channel down into the substrate. This electron movement widens the (naturally occurring) depletion zone between the n channel and the substrate, at the expense of the n channel. The reduced width of the n channel, coupled with a reduction in the number of charge carriers in that channel, increases the resistance of that channel and thus decreases the current I_{DS}. With the proper value of V_{GS}, this current can be shut off completely; hence, by controlling V_{GS}, the MOSFET can be switched between its ON and OFF modes.

Charge carriers do not flow through the substrate because it (1) is lightly doped, (2) is not a good conductor, and (3) is separated from the n channel and the two n-type islands by an insulating depletion zone, not specifically shown in Fig. 41-19. Such a depletion zone always exists at a boundary between n-type material and p-type material, as Fig. 41-11b shows.

Integrated Circuits

Computers and other electronic devices employ thousands (if not millions) of transistors and other electronic components, such as capacitors and resistors. These are not assembled as separate units but are crafted into a single semiconducting **chip,** forming an **integrated circuit.** Figure 41-20 shows a Pentium microprocessor chip, manufactured by Intel Corporation. It contains almost 7 million transistors, along with many other electronic components.

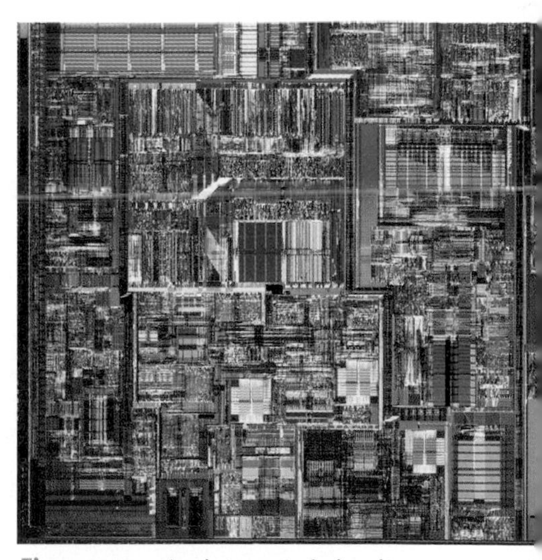

Fig. 41-20 An integrated circuit for the Intel Pentium chip, used mainly in computers.

Amplifiers for Guitar Rock

The mechanical oscillations that a player sets up on a string of an electric guitar induce electrical oscillations in a pickup coil positioned just under the string. Those electrical oscillations must be amplified so that they can drive a speaker system, to produce sound for an audience to hear. When the electric guitar became popular in rock in the early 1960s, the amplifiers used tubes because transistor amplifiers were not yet dependable. As rock moved into psychedelia and then into heavy metal, guitarists cranked up their amplifiers to the maximum in order to shake their audiences. Such high amplification by a tube amplifier introduces significant distortion in the final sound, and that distortion quickly became identified with the sound of rock.

Transistor amplifiers do not produce the same type of distortion when driven hard—they are said to produce a "clean" sound. Thus, rock guitarists shun them even today because they do not produce the "proper" sound of rock. Jimi Hendrix, who was the first person to understand the electric guitar and its amplifier as a combined musical instrument, once put it, "I really like my old Marshall tube amps, because when . . . the volume is turned up all the way, there's nothing [that] can beat them . . . ."

Review & Summary

Metals, Semiconductors, and Insulators Three electrical properties that can be used to distinguish among crystalline solids are **resistivity** ρ, **temperature coefficient of resistivity** α, and **number density of charge carriers** n. Solids can be broadly divided into three categories: **insulators** (very high ρ), **metals** (low ρ, positive and low α, large n), and **semiconductors** (high ρ, negative and high α, small n).

Energy Levels and Gaps in a Crystalline Solid An isolated atom can exist in only a discrete set of energy levels. As atoms come together to form a solid, the levels of the individual atoms merge to form the discrete energy **bands** of the solid. These energy bands are separated by energy **gaps,** each of which corresponds to a range of energies that no electron may possess.

Any energy band is made up of an enormous number of very closely spaced levels. The Pauli exclusion principle asserts that only one electron may occupy each of these levels.

Insulators In an insulator, the highest band containing electrons is completely filled and is separated from the vacant band above it by an energy gap so large that electrons can essentially never become thermally agitated enough to jump across the gap.

Metals In a metal, the highest band that contains any electrons is only partially filled. The energy of the highest filled level at a temperature of 0 K is called the **Fermi energy** E_F for the metal; for copper, $E_F = 7.0$ eV.

The electrons in the partially filled band are the **conduction electrons** and their number is

$$\begin{pmatrix}\text{number of conduction}\\ \text{electrons in sample}\end{pmatrix} = \begin{pmatrix}\text{number of atoms}\\ \text{in sample}\end{pmatrix} \times \begin{pmatrix}\text{number of valence}\\ \text{electrons per atom}\end{pmatrix}. \quad (41\text{-}2)$$

The number of atoms in a sample is given by

$$\begin{aligned}\begin{pmatrix}\text{number of atoms}\\ \text{in sample}\end{pmatrix} &= \frac{\text{sample mass } M_{\text{sam}}}{\text{atomic mass}}\\ &= \frac{\text{sample mass } M_{\text{sam}}}{(\text{molar mass } M)/N_A}\\ &= \frac{\begin{pmatrix}\text{material's}\\ \text{density}\end{pmatrix}\begin{pmatrix}\text{sample}\\ \text{volume } V\end{pmatrix}}{(\text{molar mass } M)/N_A}.\end{aligned} \quad (41\text{-}4)$$

The number density n of the conduction electrons is

$$n = \frac{\text{number of conduction electrons in sample}}{\text{sample volume } V}. \quad (41\text{-}3)$$

The **density of states** function $N(E)$ is the number of available energy levels per unit volume of the sample and per unit energy interval and is given by

$$N(E) = \frac{8\sqrt{2}\pi m^{3/2}}{h^3}E^{1/2} \quad \text{(density of states, m}^{-3}\text{ J}^{-1}\text{)}, \quad (41\text{-}5)$$

where m ($= 9.109 \times 10^{-31}$ kg) is the electron mass, h ($= 6.626 \times 10^{-34}$ J·s) is the Planck constant, and E is the energy in joules at which $N(E)$ is to be evaluated. To modify the equation so that the value of E is in eV and the value of $N(E)$ is in m^{-3} eV^{-1}, multiply the right side by $e^{3/2}$ (where $e = 1.602 \times 10^{-19}$ C).

The **occupancy probability** $P(E)$, the probability that a given available state will be occupied by an electron, is

$$P(E) = \frac{1}{e^{(E - E_F)/kT} + 1} \quad \text{(occupancy probability)}. \quad (41\text{-}6)$$

The **density of occupied states** $N_o(E)$ is given by the product of the two quantities in Eqs. (41-5) and (41-6):

$$N_o(E) = N(E)\,P(E) \quad \text{(density of occupied states)}. \quad (41\text{-}7)$$

The Fermi energy for a metal can be found by integrating $N_o(E)$

for $T = 0$ from $E = 0$ to $E = E_F$. The result is

$$E_F = \left(\frac{3}{16\sqrt{2}\pi}\right)^{2/3} \frac{h^2}{m} n^{2/3} = \frac{0.121h^2}{m} n^{2/3}. \quad (41\text{-}9)$$

Semiconductors The band structure of a semiconductor is like that of an insulator except that the gap width E_g is much smaller in the semiconductor. For silicon (a semiconductor) at room temperature, thermal agitation raises a few electrons to the **conduction band,** leaving an equal number of **holes** in the **valence band.** Both electrons and holes serve as charge carriers.

The number of electrons in the conduction band of silicon can be increased greatly by doping with small amounts of phosphorus, thus forming ***n*-type material.** The number of holes in the valence band can be greatly increased by doping with aluminum, thus forming ***p*-type material.**

The *p-n* Junction A ***p-n* junction** is a single semiconducting crystal with one end doped to form *p*-type material and the other end doped to form *n*-type material, the two types meeting at a **junction plane.** At thermal equilibrium, the following occurs at that plane:

The **majority carriers** (electrons on the *n* side and holes on the *p* side) diffuse across the junction plane, producing a **diffusion current** I_{diff}.

The **minority carriers** (holes on the *n* side and electrons on the *p* side) are swept across the junction plane, forming a **drift current** I_{drift}. These two currents are equal in magnitude, making the net current zero.

A **depletion zone,** consisting largely of charged donor and acceptor ions, forms across the junction plane.

A **contact potential difference** V_0 develops across the depletion zone.

Applications of the *p-n* Junction When a potential difference is applied across a *p-n* junction, the device conducts electricity more readily for one polarity of the applied potential difference than for the other. Thus, a *p-n* junction can serve as a **junction rectifier.**

When a *p-n* junction is forward biased, it can emit light, hence can serve as a **light-emitting diode** (LED). The wavelength of the emitted light is given by

$$\lambda = \frac{c}{f} = \frac{hc}{E_g}. \quad (41\text{-}11)$$

A strongly forward-biased *p-n* junction with parallel end faces can operate as a **junction laser,** emitting light of a sharply defined wavelength.

MOSFETs In a MOSFET, a type of three-terminal transistor, a potential applied to the **gate** terminal G controls the internal flow of electrons from the **source** terminal S to the **drain** terminal D. Commonly, a MOSFET is operated only in its ON (conducting) or OFF (not conducting) condition. Installed by the thousands and millions on silicon wafers (**chips**) to form **integrated circuits,** MOSFETs form the basis for computer hardware.

Questions

1 Figure 41-1*a* shows 14 atoms that represent the unit cell of copper. However, because each of these atoms is shared with one or more adjoining unit cells, only a fraction of each atom belongs to the unit cell shown. What is the number of atoms per unit cell for copper? (To answer, count up the fractional atoms belonging to a single unit cell.)

2 Figure 41-1*b* shows 18 atoms that represent the unit cell of silicon. Fourteen of these atoms, however, are shared with one or more adjoining unit cells. What is the number of atoms per unit cell for silicon? (See Question 1.)

3 Is the drift speed v_d of the conduction electrons in a current-carrying copper wire about equal to, much greater than, or much less than the Fermi speed v_F for copper?

4 In a silicon lattice, where should you look if you want to find (a) a conduction electron, (b) a valence electron, and (c) an electron associated with the 2*p* subshell of the isolated silicon atom?

5 On which of the following does the interval between adjacent energy levels in the highest occupied band of a metal depend: (a) the material of which the sample is made, (b) the size of the sample, (c) the position of the level in the band, (d) the temperature of the sample, (e) the Fermi energy of the metal?

6 An isolated atom of germanium has 32 electrons, arranged in subshells according to this scheme:

$$1s^2\ 2s^2\ 2p^6\ 3s^2\ 3p^6\ 3d^{10}\ 4s^2\ 4p^2.$$

This element has the same crystal structure as silicon and, like silicon, is a semiconductor. Which of these electrons form the valence band of crystalline germanium?

7 The energy gaps E_g for the semiconductors silicon and germanium are, respectively, 1.12 and 0.67 eV. Which of the following statements, if any, are true? (a) Both substances have the same number density of charge carriers at room temperature. (b) At room temperature, germanium has a greater number density of charge carriers than silicon. (c) Both substances have a greater number density of conduction electrons than holes. (d) For each substance, the number density of electrons equals that of holes.

8 Which of the following statements, if any, are true? (a) At low enough temperatures, silicon behaves like an insulator. (b) At high enough temperatures, silicon becomes a good conductor. (c) At high enough temperatures, silicon behaves like a metal.

9 A sample of silicon is doped with phosphorus. Which of the following statements, if any, are true? (a) The number of holes in the sample is slightly increased. (b) The sample's resistivity is increased. (c) The sample becomes positively charged. (d) The sample becomes negatively charged. (e) The gap between the valence band and the conduction band decreases slightly.

10 (a) Of the elements arsenic, indium, tin, gallium, antimony, and boron, which would produce n-type material if used as a dopant in silicon? (b) Which would produce p-type material? (c) Which would be unsuitable as a dopant? (*Hint*: Consult the periodic table in Appendix G.)

11 Germanium ($Z = 32$) has the same crystal structure and the same bonding pattern as silicon. Is the net charge on a germanium ion within its lattice $+e$, $+2e$, $+4e$, $+28e$, or $+32e$?

12 A certain isolated p-n junction develops a contact potential difference V_0 of 0.78 V across its depletion zone. A voltmeter is connected across the terminals of the junction, the positive terminal of the meter being connected to the p side of the junction. Will the meter read (a) +0.78 V, (b) −0.78 V, (c) zero, or (d) something else? (*Hint:* Contact potentials appear at the connections between the p-n junction and the voltmeter leads.)

13 In the biased p-n junctions shown in Fig. 41-14, there is an electric field $\vec{E}$ in each of the two depletion zones, associated with the potential difference that exists across that zone. (a) Is $\vec{E}$ directed from left to right or from right to left? (b) Is its magnitude greater for forward bias or for back bias?

14 To fabricate an n-type semiconductor, would you use silicon doped with arsenic or germanium doped with indium? (*Hint:* Consult the periodic table in Appendix G.)

15 An LED based on a gallium–arsenic–phosphorus semiconducting crystal emits red light. If you look at a white surface through such a crystal, will you see red, blue, nothing because the crystal is opaque, or white?

16 Which of the following obey Ohm's law: (a) a bar of pure silicon, (b) a bar of n-type silicon, (c) a bar of p-type silicon, (d) a p-n junction?

Problems

SSM Solution is in the Student Solutions Manual.
WWW Solution is at http://www.wiley.com/college/halliday
• – ••• Number of dots indicates level of problem difficulty.

sec. 41-5 Metals

•1 Copper, a monovalent metal, has molar mass 63.54 g/mol and density 8.96 g/cm^3. What is the number density n of conduction electrons in copper?

•2 Use Eq. 41-9 to verify 7.0 eV as copper's Fermi energy.

•3 (a) Show that Eq. 41-5 can be written as $N(E) = CE^{1/2}$. (b) Evaluate C in terms of meters and electron-volts. (c) Calculate $N(E)$ for $E = 5.00$ eV.

•4 What is the number density of conduction electrons in gold, which is a monovalent metal? Use the molar mass and density provided in Appendix F.

•5 What is the probability that a state 0.0620 eV above the Fermi energy will be occupied at (a) $T = 0$ K and (b) $T = 320$ K? SSM

•6 Calculate the density of states $N(E)$ for a metal at energy $E = 8.0$ eV and show that your result is consistent with the curve of Fig. 41-5.

•7 Show that Eq. 41-9 can be written as $E_F = An^{2/3}$, where the constant A has the value $3.65 \times 10^{-19}\ \text{m}^2 \cdot \text{eV}$.

•8 A state 63 meV above the Fermi level has a probability of occupancy of 0.090. What is the probability of occupancy for a state 63 meV *below* the Fermi level?

••9 In Eq. 41-6 let $E - E_F = \Delta E = 1.00$ eV. (a) At what temperature does the result of using this equation differ by 1.0% from the result of using the classical Boltzmann equation $P(E) = e^{-\Delta E/kT}$ (which is Eq. 41-1 with two changes in notation)? (b) At what temperature do the results from these two equations differ by 10%? SSM WWW

••10 What is the Fermi energy of gold, which is a monovalent metal with molar mass 197 g/mol and density 19.3 g/cm^3?

••11 The Fermi energy for copper is 7.00 eV. For copper at 1000 K, (a) find the energy of the energy level whose probability of being occupied by an electron is 0.900. For this energy, evaluate (b) the density of states $N(E)$ and (c) the density of occupied states $N_o(E)$.

••12 Assume that the total volume of a metal sample is the sum of the volume occupied by the metal ions making up the lattice and the (separate) volume occupied by the conduction electrons. The density and molar mass of sodium (a metal) are 971 kg/m^3 and 23.0 g/mol, respectively; assume the radius of the Na^+ ion is 98.0 pm. (a) What percent of the volume of a sample of metallic sodium is occupied by its conduction electrons? (b) Carry out the same calculation for copper, which has density, molar mass, and ionic radius of 8960 kg/m^3, 63.5 g/mol, and 135 pm, respectively. (c) For which of these metals do you think the conduction electrons behave more like a free-electron gas?

••13 Calculate $N_o(E)$, the density of occupied states, for copper at $T = 1000$ K for an energy E of (a) 4.00 eV, (b) 6.75 eV, (c) 7.00 eV, (d) 7.25 eV, and (e) 9.00 eV. Compare your results with the graph of Fig. 41-7b. The Fermi energy for copper is 7.00 eV.

••14 Show that the probability $P(E)$ that an energy level having energy E is not occupied is

$$P(E) = \frac{1}{e^{-\Delta E/kT} + 1},$$

where $\Delta E = E - E_F$.

••15 The Fermi energy for silver is 5.5 eV. At $T = 0$°C, what are the probabilities that states with the following energies are occupied: (a) 4.4 eV, (b) 5.4 eV, (c) 5.5 eV, (d) 5.6 eV, and (e) 6.4 eV? (f) At what temperature is the probability 0.16 that a state with energy $E = 5.6$ eV is occupied? SSM

••16 What is the probability that an electron will jump across the energy gap E_g in a diamond that has a mass equal to the mass of Earth? Use the result of Sample Problem 41-1 and the molar mass of carbon in Appendix F; assume that in diamond there is one valence electron per carbon atom.

••17 The Fermi energy of aluminum is 11.6 eV; its density and molar mass are 2.70 g/cm^3 and 27.0 g/mol, respectively.

From these data, determine the number of conduction electrons per atom.

••18 Calculate the number density (number per unit volume) for (a) molecules of oxygen gas at 0.0°C and 1.0 atm pressure and (b) conduction electrons in copper. (c) What is the ratio of the latter to the former? What is the average distance between (d) the oxygen molecules and (e) the conduction electrons, assuming this distance is the edge length of a cube with a volume equal to the available volume per particle (molecule or electron)?

••19 Silver is a monovalent metal. Calculate (a) the number density of conduction electrons, (b) the Fermi energy, (c) the Fermi speed, and (d) the de Broglie wavelength corresponding to this electron speed. See Appendix F for the needed data on silver. SSM WWW

••20 At $T = 300$ K, how far above the Fermi energy is a state for which the probability of occupation by a conduction electron is 0.10?

••21 Show that, at $T = 0$ K, the average energy E_{avg} of the conduction electrons in a metal is equal to $\frac{3}{5}E_F$. (*Hint:* By definition of average, $E_{avg} = (1/n) \int E\, N_o(E)\, dE$, where n is the number density of charge carriers.) SSM

••22 Zinc is a bivalent metal. Calculate (a) the number density of conduction electrons, (b) the Fermi energy, (c) the Fermi speed, and (d) the de Broglie wavelength corresponding to this electron speed. See Appendix F for the needed data on zinc.

••23 Use the result of Problem 21 to calculate the total translational kinetic energy of the conduction electrons in 1.00 cm^3 of copper at $T = 0$ K.

••24 At 1000 K, the fraction of the conduction electrons in a metal that have energies greater than the Fermi energy is equal to the area under the curve of Fig. 41-7*b* beyond E_F divided by the area under the entire curve. It is difficult to find these areas by direct integration. However, an approximation to this fraction at any temperature T is

$$frac = \frac{3kT}{2E_F}.$$

Note that $frac = 0$ for $T = 0$ K, just as we would expect. What is this fraction for copper at (a) 300 K and (b) 1000 K? For copper, $E_F = 7.0$ eV. (c) Check your answers by numerical integration using Eq. 41-7.

••25 At what temperature do 1.30% of the conduction electrons in lithium (a metal) have energies greater than the Fermi energy E_F, which is 4.70 eV? (See Problem 24.)

••26 (a) Using the result of Problem 21 and 7.00 eV for copper's Fermi energy, determine how much energy would be released by the conduction electrons in a copper coin with mass 3.10 g if we could suddenly turn off the Pauli exclusion principle. (b) For how long would this amount of energy light a 100 W lamp? (*Note:* There is no way to turn off the Pauli principle!)

sec. 41-6 Semiconductors

•27 (a) What maximum light wavelength will excite an electron in the valence band of diamond to the conduction band? The energy gap is 5.50 eV. (b) In what part of the electromagnetic spectrum does this wavelength lie? SSM

••28 The compound gallium arsenide is a commonly used semiconductor, having an energy gap E_g of 1.43 eV. Its crystal structure is like that of silicon, except that half the silicon atoms are replaced by gallium atoms and half by arsenic atoms. Draw a flattened-out sketch of the gallium arsenide lattice, following the pattern of Fig. 41-9*a*. What is the net charge of the (a) gallium and (b) arsenic ion core? (c) How many electrons per bond are there? (*Hint:* Consult the periodic table in Appendix G.)

••29 The occupancy probability function (Eq. 41-6) can be applied to semiconductors as well as to metals. In semiconductors the Fermi energy is close to the midpoint of the gap between the valence band and the conduction band. For germanium, the gap width is 0.67 eV. What is the probability that (a) a state at the bottom of the conduction band is occupied and (b) a state at the top of the valence band is not occupied? Assume that $T = 290$ K. (*Note:* In a pure semiconductor, the Fermi energy lies symmetrically between the population of conduction electrons and the population of holes and thus is at the center of the gap. There need not be an available state at the location of the Fermi energy.)

••30 In a simplified model of an undoped semiconductor, the actual distribution of energy states may be replaced by one in which there are N_v states in the valence band, all these states having the same energy E_v, and N_c states in the conduction band, all these states having the same energy E_c. The number of electrons in the conduction band equals the number of holes in the valence band. (a) Show that this last condition implies that

$$\frac{N_c}{\exp(\Delta E_c/kT) + 1} = \frac{N_v}{\exp(\Delta E_v/kT) + 1},$$

in which

$$\Delta E_c = E_c - E_F \quad \text{and} \quad \Delta E_v = -(E_v - E_F).$$

(b) If the Fermi level is in the gap between the two bands and its distance from each band is large relative to kT, then the exponentials dominate in the denominators. Under these conditions, show that

$$E_F = \frac{(E_c + E_v)}{2} + \frac{kT \ln(N_v/N_c)}{2}$$

and that, if $N_v \approx N_c$, the Fermi level for the undoped semiconductor is close to the gap's center.

sec. 41-7 Doped Semiconductors

••31 What mass of phosphorus is needed to dope 1.0 g of silicon to the extent described in Sample Problem 41-6? SSM WWW

••32 Pure silicon at room temperature has an electron number density in the conduction band of about 5×10^{15} m^{-3} and an equal density of holes in the valence band. Suppose that one of every 10^7 silicon atoms is replaced by a phosphorus atom. (a) Which type will the doped semiconductor be, n or p? (b) What charge carrier number density will the phosphorus add? (c) What is the ratio of the charge carrier number density (electrons in the conduction band and holes in the valence band) in the doped silicon to that in pure silicon?

••33 Doping changes the Fermi energy of a semiconductor. Consider silicon, with a gap of 1.11 eV between the top of the valence band and the bottom of the conduction band. At 300 K the Fermi level of the pure material is nearly at the midpoint of the gap. Suppose that silicon is doped with donor atoms, each of which has a state 0.15 eV below the bottom of the silicon conduction band, and suppose further that doping raises the Fermi level to 0.11 eV below the bottom of that band (Fig. 41-21). For (a) pure and (b) doped silicon, calculate the probability that a state at the bottom of the silicon conduction band is occupied. (c) Calculate the probability that a donor state in the doped material is occupied.

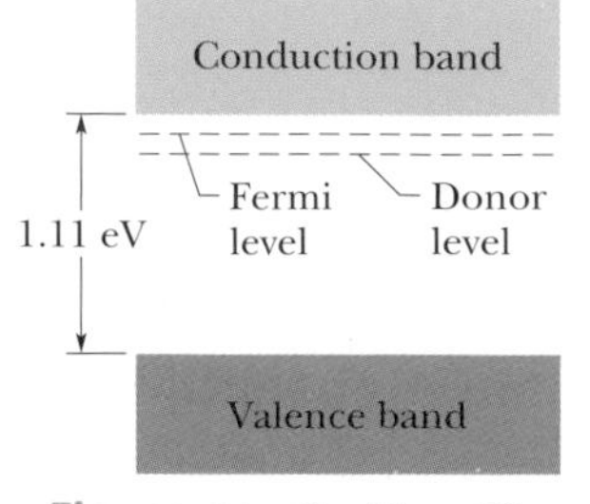

Fig. 41-21 Problem 33.

••34 A silicon sample is doped with atoms having donor states 0.110 eV below the bottom of the conduction band. (The energy gap in silicon is 1.11 eV.) If each of these donor states is occupied with a probability of 5.00×10^{-5} at $T = 300$ K, (a) is the Fermi level above or below the top of the silicon valence band and (b) how far above or below? (c) What then is the probability that a state at the bottom of the silicon conduction band is occupied?

sec. 41-9 The Junction Rectifier

•35 When a photon enters the depletion zone of a *p-n* junction, the photon can scatter from the valence electrons there, transferring part of its energy to each electron, which then jumps to the conduction band. Thus, the photon creates electron–hole pairs. For this reason, the junctions are often used as light detectors, especially in the x-ray and gamma-ray regions of the electromagnetic spectrum. Suppose a single 662 keV gamma-ray photon transfers its energy to electrons in multiple scattering events inside a semiconductor with an energy gap of 1.1 eV, until all the energy is transferred. Assuming that each electron jumps the gap from the top of the valence band to the bottom of the conduction band, find the number of electron–hole pairs created by the process. SSM

•36 For an ideal *p-n* junction rectifier with a sharp boundary between its two semiconducting sides, the current I is related to the potential difference V across the rectifier by

$$I = I_0(e^{eV/kT} - 1),$$

where I_0, which depends on the materials but not on I or V, is called the *reverse saturation current*. The potential difference V is positive if the rectifier is forward-biased and negative if it is back-biased. (a) Verify that this expression predicts the behavior of a junction rectifier by graphing I versus V from -0.12 V to $+0.12$ V. Take $T = 300$ K and $I_0 = 5.0$ nA. (b) For the same temperature, calculate the ratio of the current for a 0.50 V forward bias to the current for a 0.50 V back bias.

sec. 41-10 The Light-Emitting Diode (LED)

•37 In a particular crystal, the highest occupied band is full. The crystal is transparent to light of wavelengths longer than 295 nm but opaque at shorter wavelengths. Calculate, in electron-volts, the gap between the highest occupied band and the next higher (empty) band for this material. SSM

•38 A potassium chloride crystal has an energy band gap of 7.6 eV above the topmost occupied band, which is full. Is this crystal opaque or transparent to light of wavelength 140 nm?

sec. 41-11 The Transistor

•39 A certain computer chip that is about the size of a postage stamp (2.54 cm × 2.22 cm) contains about 3.5 million transistors. If the transistors are square, what must be their *maximum* dimension? (*Note:* Devices other than transistors are also on the chip, and there must be room for the interconnections among the circuit elements. Transistors smaller than 0.7 μm are now commonly and inexpensively fabricated.)

•40 A silicon-based MOSFET has a square gate 0.50 μm on edge. The insulating silicon oxide layer that separates the gate from the *p*-type substrate is 0.20 μm thick and has a dielectric constant of 4.5. (a) What is the equivalent gate–substrate capacitance (treating the gate as one plate and the substrate as the other plate)? (b) Approximately how many elementary charges e appear in the gate when there is a gate–source potential difference of 1.0 V?

Additional Problems

41 (a) Show that the density of states at the Fermi energy is given by

$$N(E_F) = \frac{(4)(3^{1/3})(\pi^{2/3})mn^{1/3}}{h^2}$$
$$= (4.11 \times 10^{18}\ \text{m}^{-2}\ \text{eV}^{-1})n^{1/3},$$

in which n is the number density of conduction electrons. (b) Calculate $N(E_F)$ for copper, which is a monovalent metal with molar mass 63.54 g/mol and density 8.96 g/cm^3. (c) Verify your calculation with the curve of Fig. 41-5, recalling that $E_F = 7.0$ eV for copper.

42 Silver melts at 961°C. At the melting point, what fraction of the conduction electrons are in states with energies greater than the Fermi energy of 5.5 eV? (See Problem 24.)

43 (a) Find the angle θ between adjacent nearest-neighbor bonds in the silicon lattice. Recall that each silicon atom is bonded to four of its nearest neighbors. The four neighbors form a regular tetrahedron—a pyramid whose sides and base are equilateral triangles. (b) Find the bond length, given that the atoms at the corners of the tetrahedron are 388 pm apart.

44 Show that $P(E)$, the occupancy probability in Eq. 41-6, is symmetrical about the value of the Fermi energy; that is, show that

$$P(E_F + \Delta E) + P(E_F - \Delta E) = 1.$$

45 (a) Show that the slope dP/dE of Eq. 41-6 at $E = E_F$ is $-1/4kT$. (b) Show that the tangent line to the curve of Fig. 41-6*b* at $E = E_F$ intercepts the horizontal axis at $E = E_F + 2kT$.

46 Calculate $d\rho/dT$ at room temperature for (a) copper and (b) silicon, using data from Table 41-1.

47 At what pressure, in atmospheres, would the number of molecules per unit volume in an ideal gas be equal to the number density of the conduction electrons in copper, with both gas and copper at temperature $T = 300$ K?

48 The Fermi energy of copper is 7.0 eV. Verify that the corresponding Fermi speed is 1600 km/s.

49 Verify the numerical factor 0.121 in Eq. 41-9.

Nuclear Physics

Most people are well aware of the danger of nuclear radiation from radioactive building materials, nuclear plants, and (especially) nuclear weapons. However, most people are not aware that radiation is a concern in long airplane trips, not so much for the passengers as for the crew. Indeed, some flights are of more concern than others, and the total flight time per year of crew members on those flights is restricted by airline companies.

What poses a radiation danger to air crews?

The answer is in this chapter.

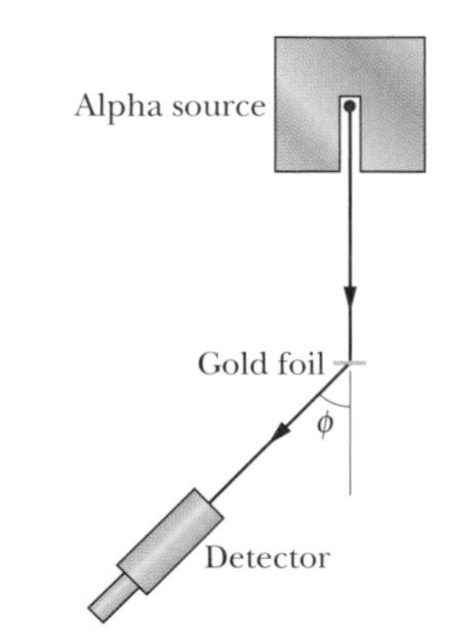

Fig. 42-1 An arrangement (top view) used in Rutherford's laboratory in 1911–1913 to study the scattering of α particles by thin metal foils. The detector can be rotated to various values of the scattering angle ϕ. The alpha source was radon gas, a decay product of radium. With this simple "tabletop" apparatus, the atomic nucleus was discovered.

42-1 What Is Physics?

Thus far in our exploration of the quantum amusement park, we have examined electrons trapped within various potential wells, including atoms, but we have largely neglected what lies at the center of an atom—the nucleus. For the last 90 years, a principal goal of physics has been to work out the quantum physics of nuclei, and, for almost as long, a principal goal of some types of engineering has been to apply that quantum physics with applications ranging from radiation therapy in the war on cancer to detectors of radon gas in basements.

Before we get to such applications and the quantum physics of nuclei, let's first discuss how physicists discovered that an atom has a nucleus. As obvious as that fact is today, it initially came as an incredible surprise.

42-2 Discovering the Nucleus

In the first years of the 20th century, not much was known about the structure of atoms beyond the fact that they contain electrons. The electron had been discovered (by J. J. Thomson) in 1897, and its mass was unknown in those early days. Thus, it was not possible even to say how many negatively charged electrons a given atom contained. Scientists reasoned that because atoms were electrically neutral, they must also contain some positive charge, but nobody knew what form this compensating positive charge took.

In 1911 Ernest Rutherford proposed that the positive charge of the atom is densely concentrated at the center of the atom, forming its **nucleus,** and that, furthermore, the nucleus is responsible for most of the mass of the atom. Rutherford's proposal was no mere conjecture but was based firmly on the results of an experiment suggested by him and carried out by his collaborators, Hans Geiger (of Geiger counter fame) and Ernest Marsden, a 20-year-old student who had not yet earned his bachelor's degree.

In Rutherford's day it was known that certain elements, called **radioactive,** transform into other elements spontaneously, emitting particles in the process. One such element is radon, which emits alpha (α) particles that have an energy of about 5.5 MeV. We now know that these particles are helium nuclei.

Rutherford's idea was to direct energetic alpha particles at a thin target foil and measure the extent to which they were deflected as they passed through the foil. Alpha particles, which are about 7300 times more massive than electrons, have a charge of $+2e$.

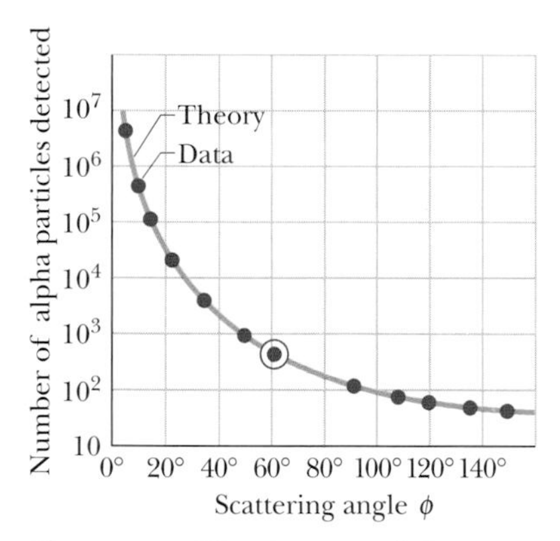

Fig. 42-2 The dots are alpha-particle scattering data for a gold foil, obtained by Geiger and Marsden using the apparatus of Fig. 42-1. The solid curve is the theoretical prediction, based on the assumption that the atom has a small, massive, positively charged nucleus. Note that the vertical scale is logarithmic, covering six orders of magnitude. The data have been adjusted to fit the theoretical curve at the experimental point that is enclosed in a circle.

Figure 42-1 shows the experimental arrangement of Geiger and Marsden. Their alpha source was a thin-walled glass tube of radon gas. The experiment involves counting the number of alpha particles that are deflected through various scattering angles ϕ.

Figure 42-2 shows their results. Note especially that the vertical scale is logarithmic. We see that most of the particles are scattered through rather small angles but—and this was the big surprise—a very small fraction of them are scattered through very large angles, approaching 180°. In Rutherford's words: "It was quite the most incredible event that ever happened to me in my life. It was almost as incredible as if you had fired a 15-inch shell at a piece of tissue paper and it [the shell] came back and hit you."

Why was Rutherford so surprised? At the time of these experiments, most physicists believed in the so-called plum pudding model of the atom, which had been advanced by J. J. Thomson. In this view the positive charge of the atom was thought to be spread out through the entire volume of the atom. The electrons (the "plums") were thought to vibrate about fixed points within this sphere of positive charge (the "pudding").

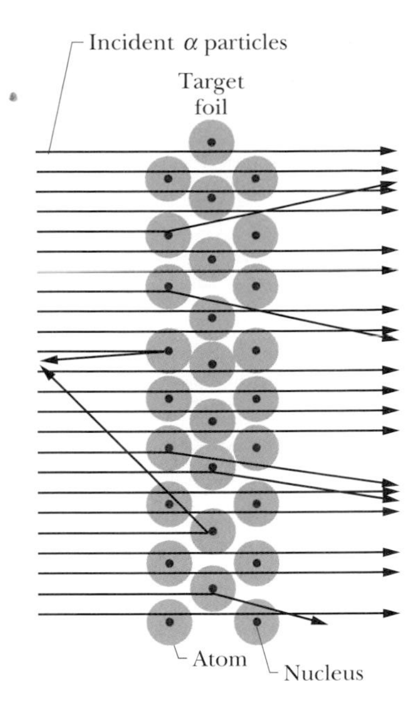

Fig. 42-3 The angle through which an incident alpha particle is scattered depends on how close the particle's path lies to an atomic nucleus. Large deflections result only from very close encounters.

The maximum deflecting force that could act on an alpha particle as it passed through such a large positive sphere of charge would be far too small to deflect the alpha particle by even as much as 1°. (The expected deflection has been compared to what you would observe if you fired a bullet through a sack of snowballs.) The electrons in the atom would also have very little effect on the massive, energetic alpha particle. They would, in fact, be themselves strongly deflected, much as a swarm of gnats would be brushed aside by a stone thrown through them.

Rutherford saw that, to deflect the alpha particle backward, there must be a large force; this force could be provided if the positive charge, instead of being spread throughout the atom, were concentrated tightly at its center. Then the incoming alpha particle could get very close to the positive charge without penetrating it; such a close encounter would result in a large deflecting force.

Figure 42-3 shows possible paths taken by typical alpha particles as they pass through the atoms of the target foil. As we see, most are either undeflected or only slightly deflected, but a few (those whose incoming paths pass, by chance, very close to a nucleus) are deflected through large angles. From an analysis of the data, Rutherford concluded that the radius of the nucleus must be smaller than the radius of an atom by a factor of about 10^4. In other words, the atom is mostly empty space.

Sample Problem 42-1

A 5.30 MeV alpha particle happens, by chance, to be headed directly toward the nucleus of an atom of gold, which contains 79 protons. How close does the alpha particle get to the center of the nucleus before coming momentarily to rest and reversing its motion? Neglect the recoil of the relatively massive nucleus.

Solution: The **Key Idea** here is that throughout this process, the total mechanical energy E of the system of alpha particle and gold nucleus is conserved. In particular, the system's initial mechanical energy E_i, before the particle and nucleus interact, is equal to its mechanical energy E_f when the alpha particle momentarily stops. The initial energy E_i is just the kinetic energy K_α of the incoming alpha particle. The final energy E_f is just the electric potential energy U of the system (the kinetic energy is then zero). We can find U with Eq. 24-4 ($U = q_1 q_2 / 4\pi\varepsilon_0 r$).

Let d be the center-to-center distance between the alpha particle and the gold nucleus when the alpha particle is at its stopping point. Then we can write the conservation of energy $E_i = E_f$ as

$$K_\alpha = \frac{1}{4\pi\varepsilon_0}\frac{q_\alpha q_{\text{Au}}}{d},$$

in which q_α ($= 2e$) is the charge of the alpha particle (2 protons) and q_{Au} ($= 79e$) is the charge of the gold nucleus (79 protons).

Substituting for the charges and solving for d yield

$$d = \frac{(2e)(79e)}{4\pi\varepsilon_0 K_\alpha}$$
$$= \frac{(2 \times 79)(1.60 \times 10^{-19}\ \text{C})^2}{(4\pi)(8.85 \times 10^{-12}\ \text{F/m})(5.30\ \text{MeV})(1.60 \times 10^{-13}\ \text{J/MeV})}$$
$$= 4.29 \times 10^{-14}\ \text{m}. \qquad \text{(Answer)}$$

This is a small distance by atomic standards but not by nuclear standards. It is, in fact, considerably larger than the sum of the radii of the gold nucleus and the alpha particle. Thus, this alpha particle reverses its motion without ever actually "touching" the gold nucleus.

42-3 Some Nuclear Properties

Table 42-1 shows some properties of a few atomic nuclei. When we are interested primarily in their properties as specific nuclear species (rather than as parts of atoms), we call these particles **nuclides.**

TABLE 42-1 Some Properties of Selected Nuclides

Nuclide	Z	N	A	Stability[a]	Mass[b] (u)	Spin[c]	Binding Energy (MeV/nucleon)
^{1}H	1	0	1	99.985%	1.007 825	$\frac{1}{2}$	—
^{7}Li	3	4	7	92.5%	7.016 004	$\frac{3}{2}$	5.60
^{31}P	15	16	31	100%	30.973 762	$\frac{1}{2}$	8.48
^{84}Kr	36	48	84	57.0%	83.911 507	0	8.72
^{120}Sn	50	70	120	32.4%	119.902 197	0	8.51
^{157}Gd	64	93	157	15.7%	156.923 957	$\frac{3}{2}$	8.21
^{197}Au	79	118	197	100%	196.966 552	$\frac{3}{2}$	7.91
^{227}Ac	89	138	227	21.8 y	227.027 747	$\frac{3}{2}$	7.65
^{239}Pu	94	145	239	24 100 y	239.052 157	$\frac{1}{2}$	7.56

[a]For stable nuclides, the **isotopic abundance** is given; this is the fraction of atoms of this type found in a typical sample of the element. For radioactive nuclides, the half-life is given.

[b]Following standard practice, the reported mass is that of the neutral atom, not that of the bare nucleus.

[c]Spin angular momentum in units of $\hbar$.

Some Nuclear Terminology

Nuclei are made up of protons and neutrons. The number of protons in a nucleus (called the **atomic number** or **proton number** of the nucleus) is represented by the symbol Z; the number of neutrons (the **neutron number**) is represented by the symbol N. The total number of neutrons and protons in a nucleus is called its **mass number** A; thus

$$A = Z + N. \tag{42-1}$$

Neutrons and protons, when considered collectively, are called **nucleons.**

We represent nuclides with symbols such as those displayed in the first column of Table 42-1. Consider ^{197}Au, for example. The superscript 197 is the mass number A. The chemical symbol Au tells us that this element is gold, whose atomic number is 79. From Eq. 42-1, the neutron number of this nuclide is $197 - 79$, or 118.

Nuclides with the same atomic number Z but different neutron numbers N are called **isotopes** of one another. The element gold has 32 isotopes, ranging from ^{173}Au to ^{204}Au. Only one of them (^{197}Au) is stable; the remaining 31 are radioactive. Such **radionuclides** undergo **decay** (or **disintegration**) by emitting a particle and thereby transforming to a different nuclide.

Organizing the Nuclides

The neutral atoms of all isotopes of an element (all with the same Z) have the same number of electrons and the same chemical properties, and they fit into the same box in the periodic table of the elements. The *nuclear* properties of the isotopes of a given element, however, are very different from one isotope to another. Thus, the periodic table is of limited use to the nuclear physicist, the nuclear chemist, or the nuclear engineer.

We organize the nuclides on a **nuclidic chart** like that in Fig. 42-4, in which a nuclide is represented by plotting its proton number against its neutron number. The stable nuclides in this figure are represented by the green, the radionuclides by the beige. As you can see, the radionuclides tend to lie on either side of— and at the upper end of—a well-defined band of stable nuclides. Note too that light stable nuclides tend to lie close to the line $N = Z$, which means that they have about the same numbers of neutrons and protons. Heavier nuclides, how-

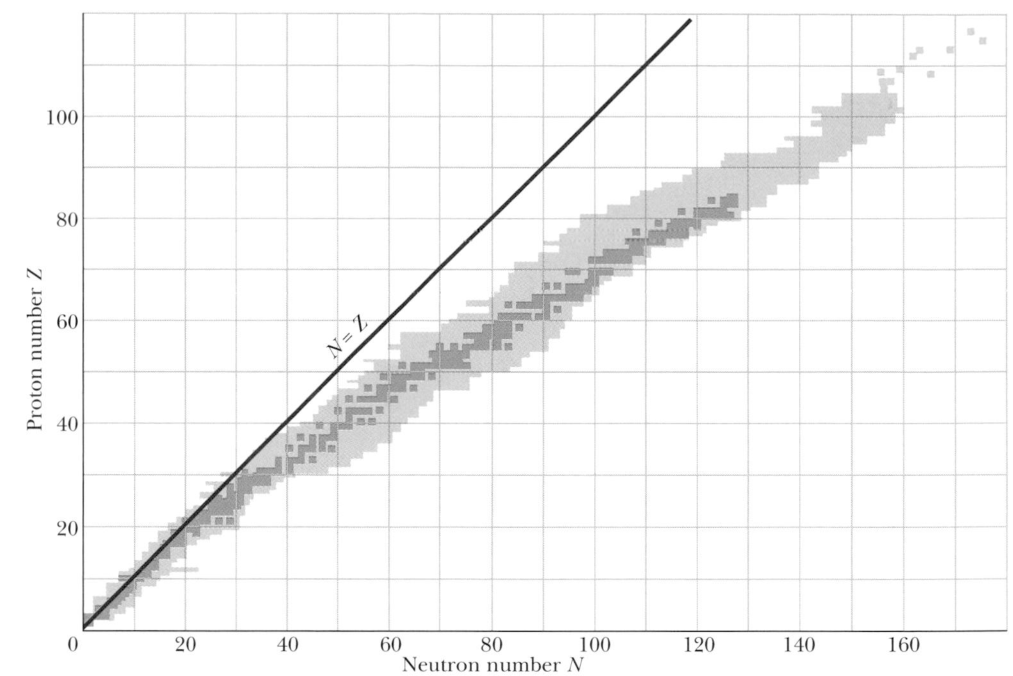

Fig. 42-4 A plot of the known nuclides. The green shading identifies the band of stable nuclides, the beige shading the radionuclides. Low-mass, stable nuclides have essentially equal numbers of neutrons and protons, but more massive nuclides have an increasing excess of neutrons. The figure shows that there are no stable nuclides with $Z > 83$ (bismuth).

ever, tend to have many more neutrons than protons. As an example, we saw that ^{197}Au has 118 neutrons and only 79 protons, a *neutron excess* of 39.

Nuclidic charts are available as wall charts, in which each small box on the chart is filled with data about the nuclide it represents. Figure 42-5 shows a section of such a chart, centered on ^{197}Au. Relative abundances (usually, as found on Earth) are shown for stable nuclides, and half-lives (a measure of decay rate) are shown for radionuclides. The sloping line points out a line of **isobars**—nuclides of the same mass number, $A = 198$ in this case.

As of early 2004, nuclides with atomic numbers as high as $Z = 116$ ($A = 292$) had been found in laboratory experiments (no elements with Z greater than

Proton number Z \ Neutron number N	115	116	117	118	119	120	121
82	^{197}Pb 43 min	^{198}Pb 2.4 h	^{199}Pb 1.5 h	^{200}Pb 21.5 h	^{201}Pb 9.33 h	^{202}Pb 53000 y	^{203}Pb 2.16 d
81	^{196}Tl 1.84 h	^{197}Tl 2.83 h	^{198}Tl 5.3 h	^{199}Tl 7.4 h	^{200}Tl 26.1 h	^{201}Tl 72.9 h	^{202}Tl 12.2 d
80	^{195}Hg 9.5 h	^{196}Hg 0.15%	^{197}Hg 64.1 h	^{198}Hg 10.0%	^{199}Hg 16.9%	^{200}Hg 23.1%	^{201}Hg 13.2%
79	^{194}Au 39.4 h	^{195}Au 186 d	^{196}Au 6.18 d	^{197}Au 100%	^{198}Au 2.69 d	^{199}Au 3.14 d	^{200}Au 48.4 min
78	^{193}Pt 60 y	^{194}Pt 32.9%	^{195}Pt 33.8%	^{196}Pt 25.3%	^{197}Pt 18.3 h	^{198}Pt 7.2%	^{199}Pt 30.8 min
77	^{192}Ir 73.8 d	^{193}Ir 62.7%	^{194}Ir 19.2 h	^{195}Ir 2.8 h	^{196}Ir 52 s	^{197}Ir 5.8 min	^{198}Ir ≈8 s
76	^{191}Os 15.4 d	^{192}Os 41.0%	^{193}Os 30.5 h	^{194}Os 6.0 y	^{195}Os 6.5 min	^{196}Os 35 min	–

A = 198

Fig. 42-5 An enlarged and detailed section of the nuclidic chart of Fig. 42-4, centered on ^{197}Au. Green squares represent stable nuclides, for which relative isotopic abundances are given. Beige squares represent radionuclides, for which half-lives are given. Isobaric lines of constant mass number A slope as shown by the example line for $A = 198$.

92 occur naturally). Although large nuclides generally should be highly unstable and last only a very brief time, certain supermassive nuclides are relatively stable, with fairly long lifetimes. These stable supermassive nuclides and other predicted ones form an *island of stability* at high values of Z and N on a nuclidic chart like Fig. 42-4.

CHECKPOINT 1 Based on Fig. 42-4, which of the following nuclides do you conclude are not likely to be detected: ^{52}Fe ($Z = 26$), ^{90}As ($Z = 33$), ^{158}Nd ($Z = 60$), ^{175}Lu ($Z = 71$), ^{208}Pb ($Z = 82$)?

Nuclear Radii

A convenient unit for measuring distances on the scale of nuclei is the *femtometer* ($= 10^{-15}$ m). This unit is often called the *fermi;* the two names share the same abbreviation. Thus,

$$1 \text{ femtometer} = 1 \text{ fermi} = 1 \text{ fm} = 10^{-15} \text{ m}. \tag{42-2}$$

We can learn about the size and structure of nuclei by bombarding them with a beam of high-energy electrons and observing how the nuclei deflect the incident electrons. The electrons must be energetic enough (at least 200 MeV) to have de Broglie wavelengths that are smaller than the nuclear structures they are to probe.

The nucleus, like the atom, is not a solid object with a well-defined surface. Furthermore, although most nuclides are spherical, some are notably ellipsoidal. Nevertheless, electron-scattering experiments (as well as experiments of other kinds) allow us to assign to each nuclide an effective radius given by

$$r = r_0 A^{1/3}, \tag{42-3}$$

in which A is the mass number and $r_0 \approx 1.2$ fm. We see that the volume of a nucleus, which is proportional to r^3, is directly proportional to the mass number A and is independent of the separate values of Z and N.

Equation 42-3 does not apply to *halo nuclides,* which are neutron-rich nuclides that were first produced in laboratories in the 1980s. These nuclides are larger than predicted by Eq. 42-3, because some of the neutrons form a *halo* around a spherical core of the protons and the rest of the neutrons. Lithium isotopes give an example. When a neutron is added to ^8Li to form ^9Li, neither of which are halo nuclides, the effective radius increases by about 4%. However, when two neutrons are added to ^9Li to form the neutron-rich isotope ^{11}Li (the largest of the lithium isotopes), they do not join that existing nucleus but instead form a halo around it, increasing the effective radius by about 30%. Apparently this halo configuration involves less energy than a core containing all 11 nucleons. (In this chapter we shall generally assume that Eq. 42-3 applies.)

Atomic Masses

Atomic masses are now measured to great precision, but usually nuclear masses are not directly measurable because stripping off all the electrons from an atom is difficult. As we briefly discussed in Section 37-12, atomic masses are often reported in *atomic mass units*, a system in which the atomic mass of neutral ^{12}C is defined to be exactly 12 u.

Precise atomic masses are available in tables on the Web and are usually provided in homework problems. However, sometimes we need only an approximation of the mass of either a nucleus alone or a neutral atom. The mass number A of a nuclide gives such an approximate mass in atomic mass units. For example, the approximate mass of both the nucleus and the neutral atom for ^{197}Au is 197 u, which is close to the actual atomic mass of 196.966 573 u.

As we saw in Section 37-12,

$$1 \text{ u} = 1.660\,538\,73 \times 10^{-27} \text{ kg}. \tag{42-4}$$

We also saw that if the total mass of the participants in a nuclear reaction changes by an amount Δm, there is an energy release or absorption given by Eq. 37-50 ($Q = -\Delta m\, c^2$). As we shall now see, nuclear energies are often reported in multiples of 1 MeV. Thus, a convenient conversion between mass units and energy units is provided by Eq. 37-46:

$$c^2 = 931.494\,013 \text{ MeV/u}. \tag{42-5}$$

Scientists and engineers working with atomic masses often prefer to report the mass of an atom by means of the atom's *mass excess* Δ, defined as

$$\Delta = M - A \qquad \text{(mass excess)}, \tag{42-6}$$

where M is the actual mass of the atom in atomic mass units and A is the mass number for that atom's nucleus.

Nuclear Binding Energies

The mass M of a nucleus is *less* than the total mass Σm of its individual protons and neutrons. That means that the mass energy Mc^2 of a nucleus is *less* than the total mass energy $\Sigma(mc^2)$ of its individual protons and neutrons. The difference between these two energies is called the **binding energy** of the nucleus:

$$\Delta E_{\text{be}} = \Sigma(mc^2) - Mc^2 \qquad \text{(binding energy)}. \tag{42-7}$$

Caution: Binding energy is not an energy that resides in the nucleus. Rather, it is a *difference* in mass energy between a nucleus and its individual nucleons. If we were able to separate a nucleus into its nucleons, we would have to transfer a total energy equal to ΔE_{be} to those particles during the separating process. Although we cannot actually tear apart a nucleus in this way, the nuclear binding energy is still a convenient measure of how well a nucleus is held together.

A better measure is the **binding energy per nucleon** ΔE_{ben}, which is the ratio of the binding energy ΔE_{be} of a nucleus to the number A of nucleons in that nucleus:

$$\Delta E_{\text{ben}} = \frac{\Delta E_{\text{be}}}{A} \qquad \text{(binding energy per nucleon)}. \tag{42-8}$$

We can think of the binding energy per nucleon as the average energy needed to separate a nucleus into its individual nucleons.

Figure 42-6 is a plot of the binding energy per nucleon ΔE_{ben} versus mass number A for a large number of nuclei. Those high on the plot are very tightly bound; that is, we would have to supply a great amount of energy per nucleon to break apart one of those nuclei. The nuclei that are lower on the plot, at the left and right sides, are less tightly bound, and less energy per nucleon would be required to break them apart.

These simple statements about Fig. 42-6 have profound consequences. The nucleons in a nucleus on the right side of the plot would be more tightly bound if that nucleus were to split into two nuclei that lie near the top of the plot. Such a process, called **fission,** occurs naturally with large (high mass number A) nuclei such as uranium, which can undergo fission spontaneously (that is, without an external cause or source of energy). The process can also occur in nuclear weapons, in which many uranium or plutonium nuclei are made to fission all at once, to create an explosion.

The nucleons in any pair of nuclei on the left side of the plot would be more tightly bound if the pair were to combine to form a single nucleus that lies near

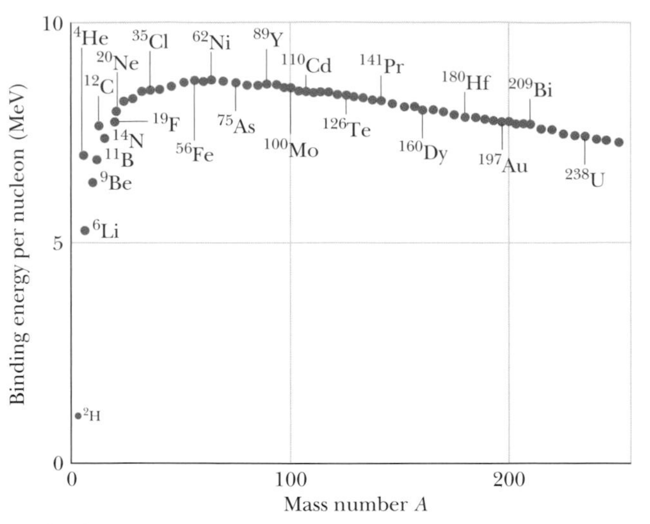

Fig. 42-6 The binding energy per nucleon for some representative nuclides. The nickel nuclide ^{62}Ni has the highest binding energy per nucleon (about 8.794 60 MeV/nucleon) of any known stable nuclide. Note that the alpha particle (^{4}He) has a higher binding energy per nucleon than its neighbors in the periodic table and thus is also particularly stable.

the top of the plot. Such a process, called **fusion,** occurs naturally in stars. Were this not true, the Sun would not shine and thus life could not exist on Earth.

Nuclear Energy Levels

The energy of nuclei, like that of atoms, is quantized. That is, nuclei can exist only in discrete quantum states, each with a well-defined energy. Figure 42-7 shows some of these energy levels for ^{28}Al, a typical low-mass nuclide. Note that the energy scale is in millions of electron-volts, rather than the electron-volts used for atoms. When a nucleus makes a transition from one level to a level of lower energy, the emitted photon is typically in the gamma-ray region of the electromagnetic spectrum.

Nuclear Spin and Magnetism

Many nuclides have an intrinsic *nuclear angular momentum,* or spin, and an associated intrinsic *nuclear magnetic moment.* Although nuclear angular momenta are roughly of the same magnitude as the angular momenta of atomic electrons, nuclear magnetic moments are much smaller than typical atomic magnetic moments.

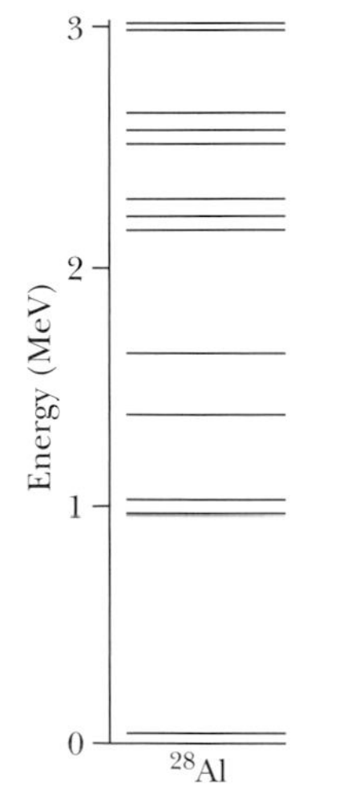

Fig. 42-7 Energy levels for the nuclide ^{28}Al, deduced from nuclear reaction experiments.

The Nuclear Force

The force that controls the motions of atomic electrons is the familiar electromagnetic force. To bind the nucleus together, however, there must be a strong attractive nuclear force of a totally different kind, strong enough to overcome the repulsive force between the (positively charged) nuclear protons and to bind both protons and neutrons into the tiny nuclear volume. The nuclear force must also be of short range because its influence does not extend very far beyond the nuclear "surface."

The present view is that the nuclear force that binds neutrons and protons in the nucleus is not a fundamental force of nature but is a secondary, or "spillover," effect of the **strong force** that binds quarks together to form neutrons and protons. In much the same way, the attractive force between certain neutral molecules is a spillover effect of the Coulomb electric force that acts within each molecule to bind it together.

Sample Problem 42-2

We can think of all nuclides as made up of a neutron-proton mixture that we can call *nuclear matter*. What is the density of nuclear matter?

Solution: One **Key Idea** here is that we can find the (average) density ρ of a nucleus by dividing its total mass by its volume. Let m represent the mass of a nucleon (either a proton or a neutron, because those particles have about the same mass). Then the mass of a nucleus containing A nucleons is Am. Next, we assume the nucleus is spherical with radius r. Then its volume is $\frac{4}{3}\pi r^3$, and we can write the density of the nucleus as

$$\rho = \frac{Am}{\frac{4}{3}\pi r^3}.$$

A second **Key Idea** is that the radius r is given by Eq. 42-3 ($r = r_0 A^{1/3}$), where r_0 is 1.2 fm ($= 1.2 \times 10^{-15}$ m). Substituting for r then leads to

$$\rho = \frac{Am}{\frac{4}{3}\pi r_0^3 A} = \frac{m}{\frac{4}{3}\pi r_0^3}.$$

Note that A has canceled out; thus, this equation for density ρ applies to any nucleus that can be treated as spherical with a radius given by Eq. 42-3. Using 1.67×10^{-27} kg for the mass m of a nucleon, we then have

$$\rho = \frac{1.67 \times 10^{-27}\ \text{kg}}{\frac{4}{3}\pi(1.2 \times 10^{-15}\ \text{m})^3} \approx 2 \times 10^{17}\ \text{kg/m}^3. \quad \text{(Answer)}$$

This is about 2×10^{14} times the density of water.

Sample Problem 42-3

What is the binding energy per nucleon for ^{120}Sn?

Solution: We need two **Key Ideas** here:

1. We can find the binding energy per nucleon ΔE_{ben} if we first find the binding energy ΔE_{be} and then divide by the number of nucleons A in the nucleus, according to Eq. 42-8 ($\Delta E_{ben} = \Delta E_{be}/A$).
2. We can find ΔE_{be} by finding the difference between the mass energy Mc^2 of the nucleus and the total mass energy $\Sigma(mc^2)$ of the individual nucleons that make up the nucleus, according to Eq. 42-7 ($\Delta E_{be} = \Sigma(mc^2) - Mc^2$).

From Table 42-1, we see that a ^{120}Sn nucleus consists of 50 protons ($Z = 50$) and 70 neutrons ($N = A - Z = 120 - 50 = 70$). Thus, we need to imagine a ^{120}Sn nucleus being separated into its 50 protons and 70 neutrons,

$$(^{120}\text{Sn nucleus}) \rightarrow 50\begin{pmatrix}\text{separate}\\ \text{protons}\end{pmatrix} + 70\begin{pmatrix}\text{separate}\\ \text{neutrons}\end{pmatrix}, \quad (42\text{-}9)$$

and then compute the resulting change in mass energy.

For that computation, we need the masses of a ^{120}Sn nucleus, a proton, and a neutron. However, because the mass of a neutral atom (nucleus *plus* electrons) is much easier to measure than the mass of a bare nucleus, calculations of binding energies are traditionally done with atomic masses. Thus, let's modify Eq. 42-9 so that it has a neutral ^{120}Sn atom on the left side. To do that, we include 50 electrons on the left side (to match the 50 protons in the ^{120}Sn nucleus). We must also add 50 electrons on the right side to balance Eq. 42-9. Those 50 electrons can be combined with the 50 protons, to form 50 neutral hydrogen atoms. We then have

$$(^{120}\text{Sn atom}) \rightarrow 50\begin{pmatrix}\text{separate}\\ \text{H atoms}\end{pmatrix} + 70\begin{pmatrix}\text{separate}\\ \text{neutrons}\end{pmatrix}. \quad (42\text{-}10)$$

In Table 42-1, the mass M_{Sn} of a ^{120}Sn atom is 119.902 197 u and the mass m_H of a hydrogen atom is 1.007 825 u; the mass m_n of a neutron is 1.008 664 u. Thus, Eq. 42-7 yields

$$\begin{aligned}\Delta E_{be} &= \Sigma(mc^2) - Mc^2\\ &= 50(m_H c^2) + 70(m_n c^2) - M_{Sn}c^2\\ &= 50(1.007\,825\ \text{u})c^2 + 70(1.008\,664\ \text{u})c^2\\ &\quad - (119.902\,197\ \text{u})c^2\\ &= (1.095\,553\ \text{u})c^2\\ &= (1.095\,553\ \text{u})(931.494\,013\ \text{MeV/u}) = 1020.5\ \text{MeV},\end{aligned}$$

where Eq. 42-5 ($c^2 = 931.494\,013$ MeV/u) provides an easy unit conversion. Note that using atomic masses instead of nuclear masses does not affect the result because the mass of the 50 electrons in the ^{120}Sn atom subtracts out from the mass of the electrons in the 50 hydrogen atoms.

Now Eq. 42-8 gives us the binding energy per nucleon as

$$\begin{aligned}\Delta E_{ben} &= \frac{\Delta E_{be}}{A} = \frac{1020.5\ \text{MeV}}{120}\\ &= 8.50\ \text{MeV/nucleon}. \quad \text{(Answer)}\end{aligned}$$

42-4 Radioactive Decay

As Fig. 42-4 shows, most of the nuclides that have been identified are radioactive. A radioactive nuclide spontaneously emits a particle, transforming itself in the process into a different nuclide, occupying a different square on the nuclidic chart.

Radioactive decay provided the first evidence that the laws that govern the subatomic world are statistical. Consider, for example, a 1 mg sample of uranium metal. It contains 2.5×10^{18} atoms of the very long-lived radionuclide ^{238}U. The nuclei of these particular atoms have existed without decaying since they were created—well before the formation of our solar system. During any given second, only about 12 of the nuclei in our sample will happen to decay by emitting an alpha particle, transforming themselves into nuclei of ^{234}Th. However,

> There is absolutely no way to predict whether any given nucleus in a radioactive sample will be among the small number of nuclei that decay during the next second. All have the same chance.

Although we cannot predict which nuclei in a sample will decay, we can say that if a sample contains N radioactive nuclei, then the rate ($= -dN/dt$) at which nuclei will decay is proportional to N:

$$-\frac{dN}{dt} = \lambda N, \tag{42-11}$$

in which λ, the **disintegration constant** (or **decay constant**) has a characteristic value for every radionuclide. Its SI unit is the inverse second (s^{-1}).

To find N as a function of time t, we first rearrange Eq. 42-11 as

$$\frac{dN}{N} = -\lambda\, dt, \tag{42-12}$$

and then integrate both sides, obtaining

$$\int_{N_0}^{N} \frac{dN}{N} = -\lambda \int_{t_0}^{t} dt,$$

or

$$\ln N - \ln N_0 = -\lambda(t - t_0). \tag{42-13}$$

Here N_0 is the number of radioactive nuclei in the sample at some arbitrary initial time t_0. Setting $t_0 = 0$ and rearranging Eq. 42-13 give us

$$\ln \frac{N}{N_0} = -\lambda t. \tag{42-14}$$

Taking the exponential of both sides (the exponential function is the antifunction of the natural logarithm) leads to

$$\frac{N}{N_0} = e^{-\lambda t}$$

or

$$N = N_0 e^{-\lambda t} \quad \text{(radioactive decay)}, \tag{42-15}$$

in which N_0 is the number of radioactive nuclei in the sample at $t = 0$ and N is the number remaining at any subsequent time t. Note that lightbulbs (for one example) follow no such exponential decay law. If we life-test 1000 bulbs, we expect that they will all "decay" (that is, burn out) at more or less the same time. The decay of radionuclides follows quite a different law.

We are often more interested in the decay rate R ($= -dN/dt$) than in N itself. Differentiating Eq. 42-15, we find

$$R = -\frac{dN}{dt} = \lambda N_0 e^{-\lambda t}$$

or

$$R = R_0 e^{-\lambda t} \quad \text{(radioactive decay)}, \tag{42-16}$$

an alternative form of the law of radioactive decay (Eq. 42-15). Here R_0 is the decay rate at time $t = 0$ and R is the rate at any subsequent time t. We can now rewrite Eq. 42-11 in terms of the decay rate R of the sample as

$$R = \lambda N, \tag{42-17}$$

where R and the number of radioactive nuclei N that have not yet undergone decay must be evaluated at the same instant.

The total decay rate R of a sample of one or more radionuclides is called the **activity** of that sample. The SI unit for activity is the **becquerel,** named for Henri Becquerel, the discoverer of radioactivity:

$$1 \text{ becquerel} = 1 \text{ Bq} = 1 \text{ decay per second.}$$

An older unit, the **curie,** is still in common use:

$$1 \text{ curie} = 1 \text{ Ci} = 3.7 \times 10^{10} \text{ Bq.}$$

Here is an example using these units: "The activity of spent reactor fuel rod #5658 on January 15, 2004, was 3.5×10^{15} Bq ($= 9.5 \times 10^4$ Ci)." Thus, on that day 3.5×10^{15} radioactive nuclei in the rod decayed each second. The identities of the radionuclides in the fuel rod, their disintegration constants λ, and the types of radiation they emit have no bearing on this measure of activity.

Often a radioactive sample will be placed near a detector that, for reasons of geometry or detector inefficiency, does not record all the disintegrations that occur in the sample. The reading of the detector under these circumstances is proportional to (and smaller than) the true activity of the sample. Such proportional activity measurements are reported not in becquerel units but simply in counts per unit time.

There are two common time measures of how long any given type of radionuclides lasts. One measure is the **half-life** $T_{1/2}$ of a radionuclide, which is the time at which both N and R have been reduced to one-half their initial values. The other measure is the **mean life** τ, which is the time at which both N and R have been reduced to e^{-1} of their initial values.

To relate $T_{1/2}$ to the disintegration constant λ, we put $R = \frac{1}{2}R_0$ in Eq. 42-16 and substitute $T_{1/2}$ for t. We obtain

$$\tfrac{1}{2}R_0 = R_0 e^{-\lambda T_{1/2}}.$$

Taking the natural logarithm of both sides and solving for $T_{1/2}$, we find

$$T_{1/2} = \frac{\ln 2}{\lambda}.$$

Similarly, to relate τ to λ, we put $R = e^{-1}R_0$ in Eq. 42-16, substitute τ for t, and solve for τ, finding

$$\tau = \frac{1}{\lambda}.$$

We summarize these results with the following:

$$T_{1/2} = \frac{\ln 2}{\lambda} = \tau \ln 2. \tag{42-18}$$

CHECKPOINT 2 The nuclide ^{131}I is radioactive, with a half-life of 8.04 days. At noon on January 1, the activity of a certain sample is 600 Bq. Using the concept of half-life, without written calculation, determine whether the activity at noon on January 24 will be a little less than 200 Bq, a little more than 200 Bq, a little less than 75 Bq, or a little more than 75 Bq.

Sample Problem 42-4

The table that follows shows some measurements of the decay rate of a sample of ^{128}I, a radionuclide often used medically as a tracer to measure the rate at which iodine is absorbed by the thyroid gland.

Time (min)	R (counts/s)	Time (min)	R (counts/s)
4	392.2	132	10.9
36	161.4	164	4.56
68	65.5	196	1.86
100	26.8	218	1.00

Find the disintegration constant λ and the half-life $T_{1/2}$ for this radionuclide.

Solution: One **Key Idea** here is that the disintegration constant λ determines the exponential rate at which the decay rate R decreases with time t (as indicated by Eq. 42-16, $R = R_0 e^{-\lambda t}$). Therefore, we should be able to determine λ by plotting the measurements of R against the measurement times t.

However, obtaining λ from a plot of R versus t is difficult because R decreases exponentially with t, according to Eq. 42-16. Thus, a second **Key Idea** is to transform Eq. 42-16 into a linear function of t, so that we can easily find λ. To do so, we take the natural logarithms of both sides of Eq. 42-16. We obtain

$$\ln R = \ln(R_0 e^{-\lambda t}) = \ln R_0 + \ln(e^{-\lambda t}) \\ = \ln R_0 - \lambda t. \tag{42-19}$$

Because Eq. 42-19 is of the form $y = b + mx$, with b and m constants, it is a linear equation giving the quantity $\ln R$ as a function of t. Thus, if we plot $\ln R$ (instead of R) versus t, we should get a straight line. Further, the slope of the line should be equal to $-\lambda$.

Figure 42-8 shows a plot of $\ln R$ versus time t for the given measurements. The slope of the straight line that fits through the plotted points is

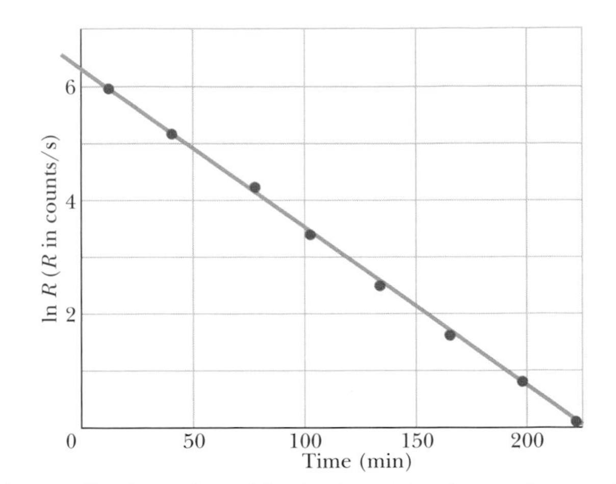

Fig. 42-8 A semilogarithmic plot of the decay of a sample of ^{128}I, based on the data in the table.

$$\text{slope} = \frac{0 - 6.2}{225 \text{ min} - 0} = -0.0275 \text{ min}^{-1}.$$

Thus,

$$-\lambda = -0.0275 \text{ min}^{-1}$$

or

$$\lambda = 0.0275 \text{ min}^{-1} \approx 1.7 \text{ h}^{-1}. \quad \text{(Answer)}$$

To find the half-life $T_{1/2}$ of the radionuclide, we use the **Key Idea** that the time for the decay rate R to decrease by 1/2 is related to the disintegration constant λ via Eq. 42-18 ($T_{1/2} = (\ln 2)/\lambda$). From that equation, we find

$$T_{1/2} = \frac{\ln 2}{\lambda} = \frac{\ln 2}{0.0275 \text{ min}^{-1}} \approx 25 \text{ min}. \quad \text{(Answer)}$$

Sample Problem 42-5

A 2.71 g sample of KCl from the chemistry stockroom is found to be radioactive, and it is decaying at a constant rate of 4490 Bq. The decays are traced to the element potassium and in particular to the isotope ^{40}K, which constitutes 1.17% of normal potassium. Calculate the half-life of this nuclide.

Solution: One **Key Idea** here is that because the activity R of the sample is apparently constant, we cannot find the half-life $T_{1/2}$ by plotting $\ln R$ versus time t as we did in Sample Problem 42-4. (We would just get a horizontal plot.) However, we can use the following two **Key Ideas**:

1. We can relate the half-life $T_{1/2}$ to the disintegration constant λ via Eq. 42-18 ($T_{1/2} = (\ln 2)/\lambda$).
2. We can then relate λ to the given activity R of 4490 Bq by means of Eq. 42-17 ($R = \lambda N$), where N is the number of ^{40}K nuclei (and thus atoms) in the sample.

Combining Eqs. 42-18 and 42-17 yields

$$T_{1/2} = \frac{N \ln 2}{R}. \tag{42-20}$$

We know that N in this equation is 1.17% of the total number N_K of potassium atoms in the sample. We also know that N_K must equal the number N_{KCl} of molecules in the sample. We can obtain N_{KCl} from the molar mass M_{KCl} of KCl (the mass of one mole of KCl) and the given mass M_{sam} of the sample by combining Eqs. 19-2 ($n = N/N_A$) and 19-3 ($n = M_{sam}/M$) to write

$$N_{KCl} = \left(\begin{matrix}\text{number of moles}\\ \text{in sample}\end{matrix}\right) N_A = \frac{M_{sam}}{M_{KCl}} N_A, \tag{42-21}$$

where N_A is Avogadro's number (6.02×10^{23} mol^{-1}). From Appendix F, we see that the molar mass of potassium is 39.102 g/mol and the molar mass of chlorine is 35.453 g/mol; thus, the molar mass of KCl is 74.555 g/mol. Equation 42-21 then gives us

$$N_{KCl} = \frac{(2.71 \text{ g})(6.02 \times 10^{23} \text{ mol}^{-1})}{74.555 \text{ g/mol}} = 2.188 \times 10^{22}$$

as the number of KCl molecules in the sample. Thus, the total number N_K of potassium atoms is also 2.188×10^{22}, and the number of ^{40}K in the sample must be

$$N = 0.0117 N_K = (0.0117)(2.188 \times 10^{22}) \\ = 2.560 \times 10^{20}.$$

Substituting this value for N and the given activity of 4490 Bq ($= 4490\ s^{-1}$) for R into Eq. 42-20 leads to

$$T_{1/2} = \frac{(2.560 \times 10^{20}) \ln 2}{4490\ s^{-1}}$$
$$= 3.95 \times 10^{16}\ s = 1.25 \times 10^9\ y. \quad \text{(Answer)}$$

This half-life of ^{40}K turns out to have the same order of magnitude as the age of the universe. Thus, the activity of ^{40}K in the stockroom sample decreases *very* slowly, too slowly for us to detect during a few days of observation or even an entire lifetime. A portion of the potassium in our bodies consists of this radioisotope, which means that we are all slightly radioactive.

42-5 Alpha Decay

When a nucleus undergoes **alpha decay,** it transforms to a different nuclide by emitting an alpha particle (a helium nucleus, ^{4}He). For example, when uranium ^{238}U undergoes alpha decay, it transforms to thorium ^{234}Th:

$$^{238}\text{U} \rightarrow {}^{234}\text{Th} + {}^{4}\text{He}. \tag{42-22}$$

This alpha decay of ^{238}U can occur spontaneously (without an external source of energy) because the total mass of the decay products ^{234}Th and ^{4}He is less than the mass of the original ^{238}U. Thus, the total mass energy of the decay products is less than the mass energy of the original nuclide. As defined by Eq. 37-50 ($Q = -\Delta M\, c^2$), in such a process the difference between the initial mass energy and the total final mass energy is called the Q of the process.

For a nuclear decay, we say that the difference in mass energy is the decay's *disintegration energy* Q. The Q for the decay in Eq. 42-22 is 4.25 MeV—that amount of energy is said to be released by the alpha decay of ^{238}U, with the energy transferred from mass energy to the kinetic energy of the two products.

The half-life of ^{238}U for this decay process is 4.5×10^9 y. Why so long? If ^{238}U can decay in this way, why doesn't every ^{238}U nuclide in a sample of ^{238}U atoms simply decay at once? To answer the questions, we must examine the process of alpha decay.

We choose a model in which the alpha particle is imagined to exist (already formed) inside the nucleus before it escapes from the nucleus. Figure 42-9 shows the approximate potential energy $U(r)$ of the system consisting of the alpha particle and the residual ^{234}Th nucleus, as a function of their separation r. This energy is a combination of (1) the potential energy associated with the (attractive) strong nuclear force that acts in the nuclear interior and (2) a Coulomb potential associated with the (repulsive) electric force that acts between the two particles before and after the decay has occurred.

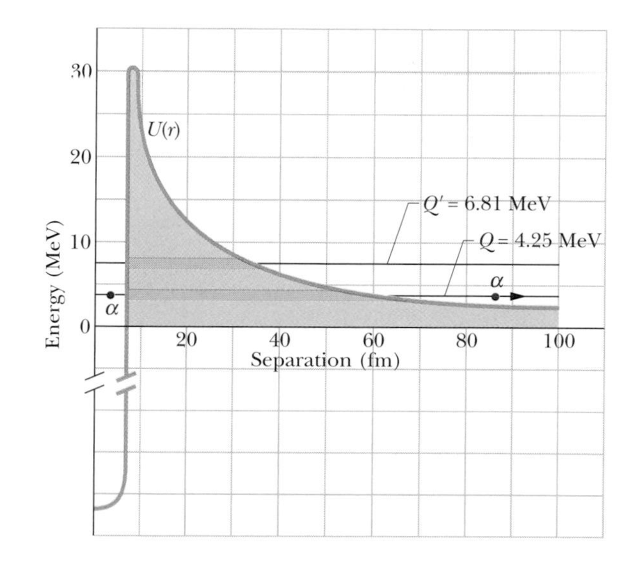

Fig. 42-9 A potential energy function for the emission of an alpha particle by ^{238}U. The horizontal black line marked $Q = 4.25$ MeV shows the disintegration energy for the process. The thick gray portion of this line represents separations r that are classically forbidden to the alpha particle. The alpha particle is represented by a dot, both inside this potential energy barrier (at the left) and outside it (at the right), after the particle has tunneled through. The horizontal black line marked $Q' = 6.81$ MeV shows the disintegration energy for the alpha decay of ^{228}U. (Both isotopes have the same potential energy function because they have the same nuclear charge.)

TABLE 42-2
Two Alpha Emitters Compared

Radionuclide	Q	Half-Life
^{238}U	4.25 MeV	4.5×10^9 y
^{228}U	6.81 MeV	9.1 min

The horizontal black line marked $Q = 4.25$ MeV shows the disintegration energy for the process. If we assume that this represents the total energy of the alpha particle during the decay process, then the part of the $U(r)$ curve above this line constitutes a potential energy barrier like that in Fig. 38-15. This barrier cannot be surmounted. If the alpha particle were able to be at some separation r within the barrier, its potential energy U would exceed its total energy E. This would mean, classically, that its kinetic energy K (which equals $E - U$) would be negative, an impossible situation.

We can see now why the alpha particle is not immediately emitted from the ^{238}U nucleus. That nucleus is surrounded by an impressive potential barrier, occupying—if you think of it in three dimensions—the volume lying between two spherical shells (of radii about 8 and 60 fm). This argument is so convincing that we now change our last question and ask: Since the particle seems permanently trapped inside the nucleus by the barrier, how can the ^{238}U nucleus *ever* emit an alpha particle? The answer is that, as you learned in Section 38-9, there is a finite probability that a particle can tunnel through an energy barrier that is classically insurmountable. In fact, alpha decay occurs as a result of barrier tunneling.

The very long half-life of ^{238}U tells us that the barrier is apparently not very "leaky." The alpha particle, presumed to be rattling back and forth within the nucleus, must arrive at the inner surface of the barrier about 10^{38} times before it succeeds in tunneling through the barrier. This is about 10^{21} times per second for about 4×10^9 years (the age of Earth)! We, of course, are waiting on the outside, able to count only the alpha particles that *do* manage to escape.

We can test this explanation of alpha decay by examining other alpha emitters. For an extreme contrast, consider the alpha decay of another uranium isotope, ^{228}U, which has a disintegration energy Q' of 6.81 MeV, about 60% higher than that of ^{238}U. (The value of Q' is also shown as a horizontal black line in Fig. 42-9.) Recall from Section 38-9 that the transmission coefficient of a barrier is very sensitive to small changes in the total energy of the particle seeking to penetrate it. Thus, we expect alpha decay to occur more readily for this nuclide than for ^{238}U. Indeed it does. As Table 42-2 shows, its half-life is only 9.1 min! An increase in Q by a factor of only 1.6 produces a decrease in half-life (that is, in the effectiveness of the barrier) by a factor of 3×10^{14}. This is sensitivity indeed.

Sample Problem 42-6

We are given the following atomic masses:

^{238}U	238.050 79 u	^4He	4.002 60 u
^{234}Th	234.043 63 u	^1H	1.007 83 u
^{237}Pa	237.051 21 u		

Here Pa is the symbol for the element protactinium ($Z = 91$).

(a) Calculate the energy released during the alpha decay of ^{238}U. The decay process is

$$^{238}U \rightarrow {}^{234}Th + {}^4He.$$

Note, incidentally, how nuclear charge is conserved in this equation: The atomic numbers of thorium (90) and helium (2) add up to the atomic number of uranium (92). The number of nucleons is also conserved: 238 = 234 + 4.

Solution: The **Key Idea** here is that the energy released in the decay is the disintegration energy Q, which we can calculate from the change in mass ΔM due to the ^{238}U decay. We use Eq. 37-50,

$$Q = M_i c^2 - M_f c^2, \tag{42-23}$$

where the initial mass M_i is that of ^{238}U and the final mass M_f is the sum of the ^{234}Th and ^4He masses. As in Sample

Problem 42-3, we must do this calculation for neutral atoms—that is, with atomic masses. Using the atomic masses given in the problem statement, Eq. 42-23 becomes

$$Q = (238.050\,79\text{ u})c^2 - (234.043\,63\text{ u} + 4.002\,60\text{ u})c^2$$
$$= (0.004\,56\text{ u})c^2 = (0.004\,56\text{ u})(931.494\,013\text{ MeV/u})$$
$$= 4.25\text{ MeV}. \qquad \text{(Answer)}$$

Note that using atomic masses instead of nuclear masses does not affect the result because the total mass of the electrons in the products subtracts out from the mass of the nucleons + electrons in the original ^{238}U.

(b) Show that ^{238}U cannot spontaneously emit a proton.

Solution: If this happened, the decay process would be

$$^{238}\text{U} \rightarrow {}^{237}\text{Pa} + {}^{1}\text{H}.$$

(You should verify that both nuclear charge and the number of nucleons are conserved in this process.) Using the same **Key Idea** as in part (a) and proceeding as we did there, we would find that the mass of the two decay products (= 237.051 21 u + 1.007 83 u) would *exceed* the mass of ^{238}U by $\Delta m = 0.008\,25$ u, with disintegration energy $Q = -7.68$ MeV. The minus sign indicates that we must *add* 7.68 MeV to a ^{238}U nucleus before it will emit a proton; it will certainly not do so spontaneously.

42-6 Beta Decay

A nucleus that decays spontaneously by emitting an electron or a positron (a positively charged particle with the mass of an electron) is said to undergo **beta decay.** Like alpha decay, this is a spontaneous process, with a definite disintegration energy and half-life. Again like alpha decay, beta decay is a statistical process, governed by Eqs. 42-15 and 42-16. In *beta-minus* (β^-) decay, an electron is emitted by a nucleus, as in the decay

$$^{32}\text{P} \rightarrow {}^{32}\text{S} + \text{e}^- + \nu \quad (T_{1/2} = 14.3\text{ d}). \qquad (42\text{-}24)$$

In *beta-plus* (β^+) decay, a positron is emitted by a nucleus, as in the decay

$$^{64}\text{Cu} \rightarrow {}^{64}\text{Ni} + \text{e}^+ + \nu \quad (T_{1/2} = 12.7\text{ h}). \qquad (42\text{-}25)$$

The symbol ν represents a **neutrino,** a neutral particle which has a very small mass, that is emitted from the nucleus along with the electron or positron during the decay process. Neutrinos interact only very weakly with matter and—for that reason—are so extremely difficult to detect that their presence long went unnoticed.*

Both charge and nucleon number are conserved in the above two processes. In the decay of Eq. 42-24, for example, we can write for charge conservation

$$(+15e) = (+16e) + (-e) + (0),$$

because ^{32}P has 15 protons, ^{32}S has 16 protons, and the neutrino ν has zero charge. Similarly, for nucleon conservation, we can write

$$(32) = (32) + (0) + (0),$$

because ^{32}P and ^{32}S each have 32 nucleons and neither the electron nor the neutrino is a nucleon.

It may seem surprising that nuclei can emit electrons, positrons, and neutrinos, since we have said that nuclei are made up of neutrons and protons only. However, we saw earlier that atoms emit photons, and we certainly do not say that atoms "contain" photons. We say that the photons are created during the emission process.

It is the same with the electrons, positrons, and neutrinos emitted from nuclei during beta decay. They are created during the emission process. For beta-minus decay, a neutron transforms into a proton within the nucleus according to

$$\text{n} \rightarrow \text{p} + \text{e}^- + \nu. \qquad (42\text{-}26)$$

*Beta decay also includes *electron capture,* in which a nucleus decays by absorbing one of its atomic electrons, emitting a neutrino in the process. We do not consider that process here. Also, the neutral particle emitted in the decay process of Eq. 42-24 is actually an *antineutrino,* a distinction we shall not make in this introductory treatment.

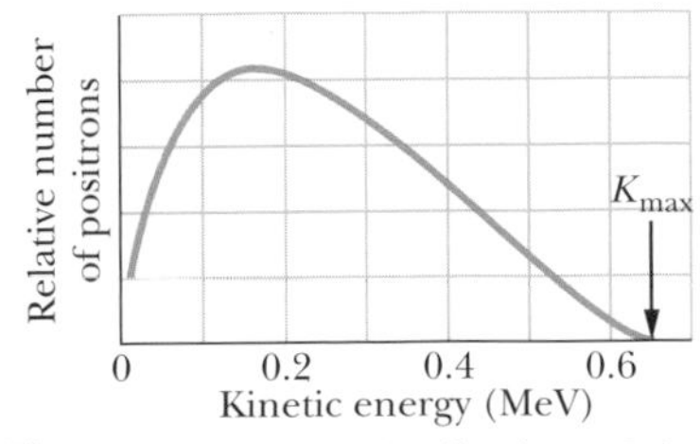

Fig. 42-10 The distribution of the kinetic energies of positrons emitted in the beta decay of ^{64}Cu. The maximum kinetic energy of the distribution (K_{max}) is 0.653 MeV. In all ^{64}Cu decay events, this energy is shared between the positron and the neutrino, in varying proportions. The *most probable* energy for an emitted positron is about 0.15 MeV.

For beta-plus decay, a proton transforms into a neutron via

$$p \rightarrow n + e^+ + \nu. \tag{42-27}$$

Both of these beta-decay processes provide evidence that—as was pointed out—neutrons and protons are not truly fundamental particles. These processes show why the mass number A of a nuclide undergoing beta decay does not change; one of its constituent nucleons simply changes its character according to Eq. 42-26 or 42-27.

In both alpha decay and beta decay, the same amount of energy is released in every individual decay of a particular radionuclide. In the alpha decay of a particular radionuclide, every emitted alpha particle has the same sharply defined kinetic energy. However, in the beta-minus decay of Eq. 42-26 with electron emission, the disintegration energy Q is shared—in varying proportions—between the emitted electron and neutrino. Sometimes the electron gets nearly all the energy, sometimes the neutrino does. In every case, however, the sum of the electron's energy and the neutrino's energy gives the same value Q. A similar sharing of energy, with a sum equal to Q, occurs in beta-plus decay (Eq. 42-27).

Thus, in beta decay the energy of the emitted electrons or positrons may range from zero up to a certain maximum K_{max}. Figure 42-10 shows the distribution of positron energies for the beta decay of ^{64}Cu (see Eq. 42-25). The maximum positron energy K_{max} must equal the disintegration energy Q because the neutrino has approximately zero energy when the positron has K_{max}:

$$Q = K_{max}. \tag{42-28}$$

The Neutrino

Wolfgang Pauli first suggested the existence of neutrinos in 1930. His neutrino hypothesis not only permitted an understanding of the energy distribution of electrons or positrons in beta decay but also solved another early beta-decay puzzle involving "missing" angular momentum.

The neutrino is a truly elusive particle; the mean free path of an energetic neutrino in water has been calculated as no less than several thousand light-years. At the same time, neutrinos left over from the big bang that presumably marked the creation of the universe are the most abundant particles of physics. Billions of them pass through our bodies every second, leaving no trace.

In spite of their elusive character, neutrinos have been detected in the laboratory. This was first done in 1953 by F. Reines and C. L. Cowan, using neutrinos generated in a high-power nuclear reactor. (In 1995, Reines received a Nobel Prize for this work.) In spite of the difficulties of detection, experimental neutrino physics is now a well-developed branch of experimental physics, with avid practitioners at laboratories throughout the world.

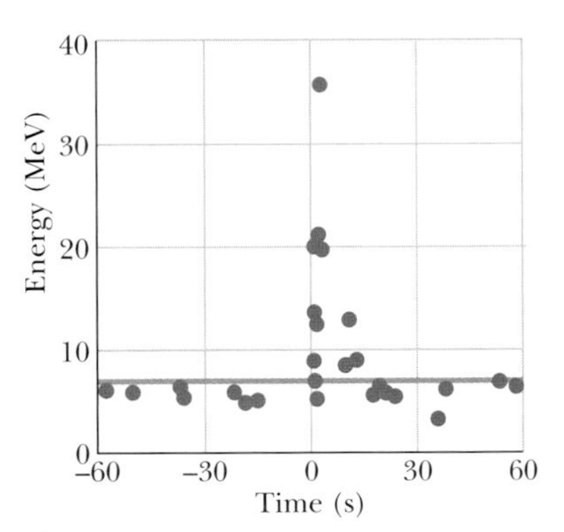

Fig. 42-11 A burst of neutrinos from the supernova SN 1987A, which occurred at (relative) time 0, stands out from the usual *background* of neutrinos. (For neutrinos, 10 is a "burst.") The particles were detected by an elaborate detector housed deep in a mine in Japan. The supernova was visible only in the Southern Hemisphere; so the neutrinos had to penetrate Earth (a trifling barrier for them) to reach the detector.

The Sun emits neutrinos copiously from the nuclear furnace at its core, and at night these messengers from the center of the Sun come up at us from below, Earth being almost totally transparent to them. In February 1987, light from an exploding star in the Large Magellanic Cloud (a nearby galaxy) reached Earth after having traveled for 170 000 years. Enormous numbers of neutrinos were generated in this explosion, and about 10 of them were picked up by a sensitive neutrino detector in Japan; Fig. 42-11 shows a record of their passage.

Radioactivity and the Nuclidic Chart

We can increase the amount of information obtainable from the nuclidic chart of Fig. 42-4 by including a third axis showing the mass excess Δ expressed in the unit MeV/c^2. The inclusion of such an axis gives Fig. 42-12, which reveals the degree of nuclear stability of the nuclides. For the low-mass nuclides, we find a "valley of the nuclides," with the stability band of Fig. 42-4 running along its

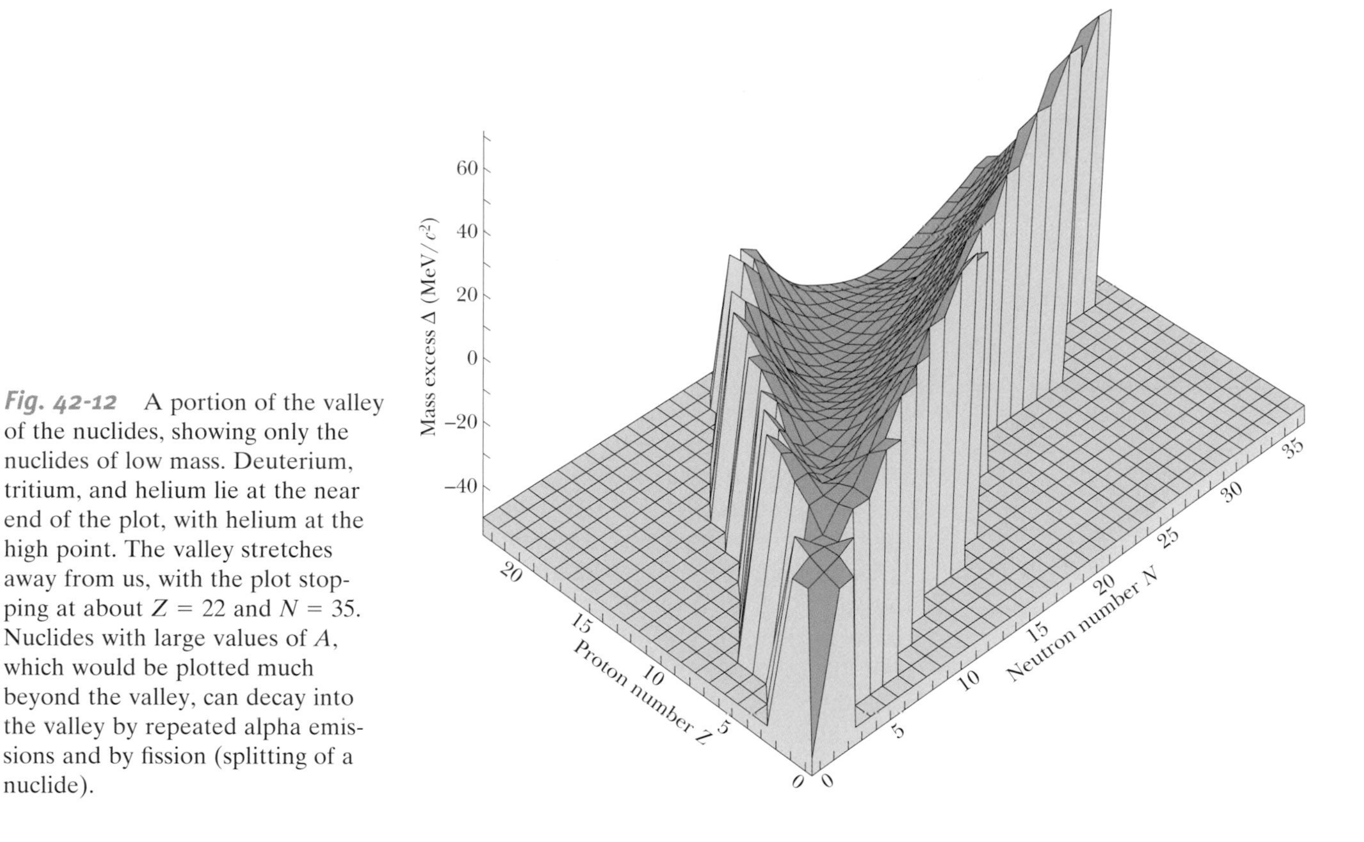

Fig. 42-12 A portion of the valley of the nuclides, showing only the nuclides of low mass. Deuterium, tritium, and helium lie at the near end of the plot, with helium at the high point. The valley stretches away from us, with the plot stopping at about $Z = 22$ and $N = 35$. Nuclides with large values of A, which would be plotted much beyond the valley, can decay into the valley by repeated alpha emissions and by fission (splitting of a nuclide).

bottom. Nuclides on the proton-rich side of the valley decay into it by emitting positrons, and those on the neutron-rich side do so by emitting electrons.

CHECKPOINT 3 ^{238}U decays to ^{234}Th by the emission of an alpha particle. There follows a chain of further radioactive decays, either by alpha decay or by beta decay. Eventually a stable nuclide is reached and, after that, no further radioactive decay is possible. Which of the following stable nuclides is the end product of the ^{238}U radioactive decay chain: ^{206}Pb, ^{207}Pb, ^{208}Pb, or ^{209}Pb? (*Hint:* You can decide by considering the changes in mass number A for the two types of decay.)

Sample Problem 42-7

Calculate the disintegration energy Q for the beta decay of ^{32}P, as described by Eq. 42-24. The needed atomic masses are 31.973 91 u for ^{32}P and 31.972 07 u for ^{32}S.

Solution: The **Key Idea** here is that the disintegration energy Q for the beta decay is the amount by which the mass energy is changed by the decay. Q is given by Eq. 37-50 ($Q = -\Delta M\,c^2$). However, we must be careful to distinguish between nuclear masses (which we do not know) and atomic masses (which we do know). Let the boldface symbols $\mathbf{m}_P$ and $\mathbf{m}_S$ represent the nuclear masses of ^{32}P and ^{32}S, and let the italic symbols m_P and m_S represent their atomic masses. Then we can write the change in mass for the decay of Eq. 42-24 as

$$\Delta m = (\mathbf{m}_S + m_e) - \mathbf{m}_P,$$

in which m_e is the mass of the electron. If we add and subtract $15m_e$ on the right side of this equation, we obtain

$$\Delta m = (\mathbf{m}_S + 16m_e) - (\mathbf{m}_P + 15m_e).$$

The quantities in parentheses are the atomic masses of ^{32}S and ^{32}P; so

$$\Delta m = m_S - m_P.$$

We thus see that if we subtract only the atomic masses, the mass of the emitted electron is automatically taken into account. (This procedure will not work for positron emission.)

The disintegration energy for the ^{32}P decay is then

$$\begin{aligned} Q &= -\Delta m\,c^2 \\ &= -(31.972\,07\text{ u} - 31.973\,91\text{ u})(931.494\,013\text{ MeV/u}) \\ &= 1.71\text{ MeV}. \end{aligned} \qquad \text{(Answer)}$$

Experimentally, this calculated quantity proves to be equal to K_{max}, the maximum energy the emitted electrons can have. Although 1.71 MeV is released every time a ^{32}P nucleus decays, in essentially every case the electron carries away less energy than this. The neutrino gets all the rest, carrying it stealthily out of the laboratory.

A fragment of the Dead Sea scrolls and the caves from which the scrolls were recovered.

42-7 Radioactive Dating

If you know the half-life of a given radionuclide, you can in principle use the decay of that radionuclide as a clock to measure time intervals. The decay of very long-lived nuclides, for example, can be used to measure the age of rocks—that is, the time that has elapsed since they were formed. Such measurements for rocks from Earth and the Moon, and for meteorites, yield a consistent maximum age of about 4.5×10^9 y for these bodies.

The radionuclide ^{40}K, for example, decays to ^{40}Ar, a stable isotope of the noble gas argon. The half-life for this decay is 1.25×10^9 y. A measurement of the ratio of ^{40}K to ^{40}Ar, as found in the rock in question, can be used to calculate the age of that rock. Other long-lived decays, such as that of ^{235}U to ^{207}Pb (involving a number of intermediate stages), can be used to verify this calculation.

For measuring shorter time intervals, in the range of historical interest, radiocarbon dating has proved invaluable. The radionuclide ^{14}C (with $T_{1/2} =$ 5730 y) is produced at a constant rate in the upper atmosphere as atmospheric nitrogen is bombarded by cosmic rays. This radiocarbon mixes with the carbon that is normally present in the atmosphere (as CO_2) so that there is about one atom of ^{14}C for every 10^{13} atoms of ordinary stable ^{12}C. Through biological activity such as photosynthesis and breathing, the atoms of atmospheric carbon trade places randomly, one atom at a time, with the atoms of carbon in every living thing, including broccoli, mushrooms, penguins, and humans. Eventually an exchange equilibrium is reached at which the carbon atoms of every living thing contain a fixed small fraction of the radioactive nuclide ^{14}C.

This equilibrium persists as long as the organism is alive. When the organism dies, the exchange with the atmosphere stops and the amount of radiocarbon trapped in the organism, since it is no longer being replenished, dwindles away with a half-life of 5730 y. By measuring the amount of radiocarbon per gram of organic matter, it is possible to measure the time that has elapsed since the organism died. Charcoal from ancient campfires, the Dead Sea scrolls, and many prehistoric artifacts have been dated in this way. The age of the scrolls was determined by radiocarbon dating a sample of the cloth used to plug the jars in which the scrolls were sealed.

Sample Problem 42-8

Mass spectrometric analysis of potassium and argon atoms in a Moon rock sample shows that the ratio of the number of (stable) ^{40}Ar atoms present to the number of (radioactive) ^{40}K atoms is 10.3. Assume that all the argon atoms were produced by the decay of potassium atoms, with a half-life of 1.25×10^9 y. (We neglect here several complications about the decay of potassium atoms.) How old is the rock?

Solution: The **Key Idea** here is that if N_0 potassium atoms were present at the time the rock was formed by solidification from a molten form, the number of potassium atoms remaining at the time of analysis is, from Eq. 42-15,

$$N_K = N_0 e^{-\lambda t}, \tag{42-29}$$

in which t is the age of the rock. For every potassium atom that decays, an argon atom is produced. Thus, the number of argon atoms present at the time of the analysis is

$$N_{Ar} = N_0 - N_K. \tag{42-30}$$

We cannot measure N_0; so let's eliminate it from Eqs. 42-29 and 42-30. We find, after some algebra, that

$$\lambda t = \ln\left(1 + \frac{N_{Ar}}{N_K}\right), \tag{42-31}$$

in which N_{Ar}/N_K *can* be measured. Solving for t and using Eq. 42-18 to replace λ with $(\ln 2)/T_{1/2}$ yield

$$t = \frac{T_{1/2} \ln(1 + N_{Ar}/N_K)}{\ln 2}$$

$$= \frac{(1.25 \times 10^9 \text{ y})[\ln(1 + 10.3)]}{\ln 2}$$

$$= 4.37 \times 10^9 \text{ y}. \qquad \text{(Answer)}$$

Lesser ages may be found for other lunar or terrestrial rock samples, but no substantially greater ones. Thus, the oldest rocks were formed soon after the solar system formed, and the solar system must be about 4 billion years old.

42-8 Measuring Radiation Dosage

The effect of radiation such as gamma rays, electrons, and alpha particles on living tissue (particularly our own) is a matter of public interest. Such radiation is found in nature in cosmic rays (from astronomical sources) and in the emissions by radioactive elements in Earth's crust. Radiation associated with some human activities, such as using x rays and radionuclides in medicine and in industry, also contributes.

Our task here is not to explore the various sources of radiation but simply to describe the units in which the properties and effects of such radiations are expressed. We have already discussed the *activity* of a radioactive source. There are two remaining quantities of interest.

1. *Absorbed Dose.* This is a measure of the radiation dose (as energy per unit mass) actually absorbed by a specific object, such as a patient's hand or chest. Its SI unit is the **gray** (Gy). An older unit, the **rad** (from **r**adiation **a**bsorbed **d**ose) is still in common use. The terms are defined and related as follows:

$$1 \text{ Gy} = 1 \text{ J/kg} = 100 \text{ rad}. \tag{42-32}$$

   A typical dose-related statement is: "A whole-body, short-term gamma-ray dose of 3 Gy (= 300 rad) will cause death in 50% of the population exposed to it." Thankfully, our present average absorbed dose per year, from sources of both natural and human origin, is only about 2 mGy (= 0.2 rad).

2. *Dose Equivalent.* Although different types of radiation (gamma rays and neutrons, say) may deliver the same amount of energy to the body, they do not have the same biological effect. The dose equivalent allows us to express the biological effect by multiplying the absorbed dose (in grays or rads) by a numerical **RBE** factor (from **r**elative **b**iological **e**ffectiveness). For x rays and electrons, for example, RBE = 1; for slow neutrons, RBE = 5; for alpha particles, RBE = 10; and so on. Personnel-monitoring devices such as film badges register the dose equivalent.

   The SI unit of dose equivalent is the **sievert** (Sv). An earlier unit, the **rem,** is still in common use. Their relationship is

$$1 \text{ Sv} = 100 \text{ rem}. \tag{42-33}$$

   An example of the correct use of these terms is: "The recommendation of the National Council on Radiation Protection is that no individual who is (non-occupationally) exposed to radiation should receive a dose equivalent greater than 5 mSv (= 0.5 rem) in any one year." This includes radiation of all kinds; of course the appropriate RBE factor must be used for each kind.

Cosmic rays consist primarily of protons ejected from the Sun. We are largely protected from this stream of high-speed protons by Earth's atmosphere and magnetic field. However, anyone flying at high altitude has less overhead atmosphere to stop the protons and thus intercepts more of them. Also, anyone flying at high latitude is located where Earth's magnetic field redirects the incoming solar protons (see Section 28-6). Thus, anyone flying at both high altitude and high latitude is at increased exposure to the solar protons.

Passengers who infrequently fly are probably not at risk, not even if they are in flight when a large influx of solar protons hits Earth. However, passengers who frequently fly a polar route (high-latitude flight path) between, say, Los Angeles and London are at greater risk because of the exposure over many flights. An air crew member would be at even greater risk were the allowed number of flight hours per year not restricted by the airline companies. For example, a crew member can be exposed to about 0.006 mSv per flight hour. If the crew member accumulates 900 flight hours per year (a reasonable workload), the net exposure would be 5.4 mSv per year, which exceeds the safety limit of 5 mSv we just discussed.

You might think that the worst radiation risk was on Concorde flights because, when that type of airplane was still flying, its supersonic speed required it to fly much higher than all other (slower) airplanes. In fact, the risk was less on Concorde flights because the flight times were so much shorter.

Sample Problem 42-9

We have seen that a gamma-ray dose of 3 Gy is lethal to half the people exposed to it. If the equivalent energy were absorbed as heat, what rise in body temperature would result?

Solution: One **Key Idea** here is that we can relate an absorbed energy Q and the resulting temperature increase ΔT with Eq. 18-14 ($Q = cm\,\Delta T$). In that equation, m is the mass of the material absorbing the energy and c is the specific heat of that material (*not* the speed of light). Another **Key Idea** is that an absorbed dose of 3 Gy corresponds to an absorbed energy per unit mass of 3 J/kg. Let us assume that c, the specific heat of the human body, is the same as that of water, 4180 J/kg · K. Then we find that

$$\Delta T = \frac{Q/m}{c} = \frac{3\text{ J/kg}}{4180\text{ J/kg}\cdot\text{K}} = 7.2 \times 10^{-4}\text{ K} \approx 0.7\text{ mK}. \qquad \text{(Answer)}$$

Obviously the damage done by ionizing radiation has nothing to do with thermal heating. The harmful effects arise because the radiation damages DNA and thus interferes with the normal functioning of tissues.

42-9 Nuclear Models

Nuclei are more complicated than atoms. For atoms, the basic force law (Coulomb's law) is simple in form and there is a natural force center, the nucleus. For nuclei, the force law is complicated and cannot, in fact, be written down explicitly in full detail. Furthermore, the nucleus—a jumble of protons and neutrons—has no natural force center to simplify the calculations.

In the absence of a comprehensive nuclear *theory,* we turn to the construction of nuclear *models.* A nuclear model is simply a way of looking at the nucleus that gives a physical insight into as wide a range of its properties as possible. The usefulness of a model is tested by its ability to provide predictions that can be verified experimentally in the laboratory.

Two models of the nucleus have proved useful. Although based on assumptions that seem flatly to exclude each other, each accounts very well for a selected group of nuclear properties. After describing them separately, we shall see how these two models may be combined to form a single coherent picture of the atomic nucleus.

The Collective Model

In the *collective model,* formulated by Niels Bohr, the nucleons, moving around within the nucleus at random, are imagined to interact strongly with each other, like the molecules in a drop of liquid. A given nucleon collides frequently with other nucleons in the nuclear interior, its mean free path as it moves about being substantially less than the nuclear radius.

The collective model permits us to correlate many facts about nuclear masses and binding energies; it is useful (as you will see later) in explaining nuclear fission. It is also useful for understanding a large class of nuclear reactions.

Consider, for example, a generalized nuclear reaction of the form

$$X + a \rightarrow C \rightarrow Y + b. \qquad (42\text{-}34)$$

We imagine that projectile a enters target nucleus X, forming a **compound nucleus** C and conveying to it a certain amount of excitation energy. The projectile, perhaps a neutron, is at once caught up by the random motions that characterize the nuclear interior. It quickly loses its identity—so to speak—and the excitation energy it carried into the nucleus is quickly shared with all the other nucleons in C.

The quasi-stable state represented by C in Eq. 42-34 may have a mean life of 10^{-16} s before it decays to Y and b. By nuclear standards, this is a very long time, being about one million times longer than the time required for a nucleon with a few million electron-volts of energy to travel across a nucleus.

The central feature of this compound-nucleus concept is that the formation of the compound nucleus and its eventual decay are totally independent events. At the time of its decay, the compound nucleus has "forgotten" how it was formed. Hence, its mode of decay is not influenced by its mode of formation. As an example, Fig. 42-13 shows three possible ways in which the compound nucleus ^{20}Ne might be formed and three in which it might decay. Any of the three formation modes can lead to any of the three decay modes.

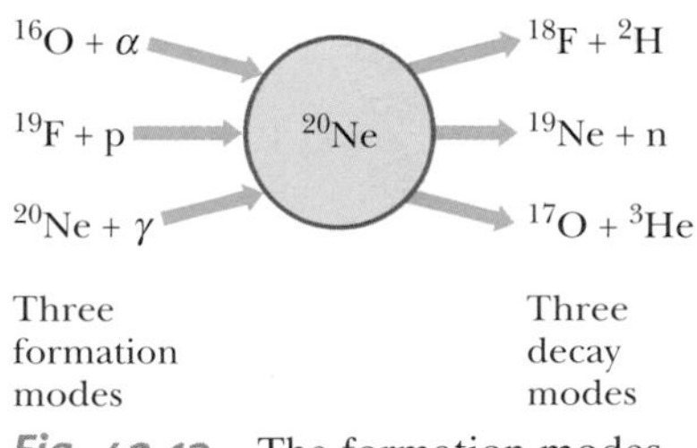

Fig. 42-13 The formation modes and the decay modes of the compound nucleus ^{20}Ne.

The Independent Particle Model

In the collective model, we assume that the nucleons move around at random and bump into one another frequently. The *independent particle model,* however, is based on just the opposite assumption—namely, that each nucleon remains in a well-defined quantum state within the nucleus and makes hardly any collisions at all! The nucleus, unlike the atom, has no fixed center of charge; we assume in this model that each nucleon moves in a potential well that is determined by the smeared-out (time-averaged) motions of all the other nucleons.

A nucleon in a nucleus, like an electron in an atom, has a set of quantum numbers that defines its state of motion. Also, nucleons obey the Pauli exclusion principle, just as electrons do; that is, no two nucleons in a nucleus may occupy the same quantum state at the same time. In this regard, the neutrons and the protons are treated separately, each particle type with its own set of quantum states.

The fact that nucleons obey the Pauli exclusion principle helps us to understand the relative stability of nucleon states. If two nucleons within the nucleus are to collide, the energy of each of them after the collision must correspond to the energy of an *unoccupied* state. If no such state is available, the collision simply cannot occur. Thus, any given nucleon experiencing repeated "frustrated collision opportunities" will maintain its state of motion long enough to give meaning to the statement that it exists in a quantum state with a well-defined energy.

In the atomic realm, the repetitions of physical and chemical properties that we find in the periodic table are associated with a property of atomic electrons—namely, they arrange themselves in shells that have a special stability when fully occupied. We can take the atomic numbers of the noble gases,

$$2, 10, 18, 36, 54, 86, \ldots,$$

as *magic electron numbers* that mark the completion (or closure) of such shells.

Nuclei also show such closed-shell effects, associated with certain **magic nucleon numbers:**

$$2, 8, 20, 28, 50, 82, 126, \ldots.$$

Any nuclide whose proton number Z or neutron number N has one of these values turns out to have a special stability that may be made apparent in a variety of ways.

Examples of "magic" nuclides are ^{18}O ($Z = 8$), ^{40}Ca ($Z = 20, N = 20$), ^{92}Mo ($N = 50$), and ^{208}Pb ($Z = 82$, $N = 126$). Both ^{40}Ca and ^{208}Pb are said to be "doubly magic" because they contain both filled shells of protons *and* filled shells of neutrons.

The magic number 2 shows up in the exceptional stability of the alpha particle (^{4}He), which, with $Z = N = 2$, is doubly magic. For example, on the binding energy curve of Fig. 42-6, the binding energy per nucleon for this nuclide stands well above those of its periodic-table neighbors hydrogen, lithium, and berylium. The neutrons and protons making up the alpha particle are so tightly bound to one another, in fact, that it is impossible to add another proton or neutron to it; there is no stable nuclide with $A = 5$.

The central idea of a closed shell is that a single particle outside a closed shell can be relatively easily removed, but considerably more energy must be expended to remove a particle from the shell itself. The sodium atom, for example, has one (valence) electron outside a closed electron shell. Only about 5 eV is required to strip the valence electron away from a sodium atom; however, to remove a *second* electron (which must be plucked out of a closed shell) requires 22 eV. As a nuclear case, consider ^{121}Sb ($Z = 51$), which contains a single proton outside a closed shell of 50 protons. To remove this lone proton requires 5.8 MeV; to remove a *second* proton, however, requires an energy of 11 MeV. There is much additional experimental evidence that the nucleons in a nucleus form closed shells and that these shells exhibit stable properties.

We have seen that quantum theory can account beautifully for the magic electron numbers—that is, for the populations of the subshells into which atomic electrons are grouped. It turns out that, under certain assumptions, quantum theory can account equally well for the magic nucleon numbers! The 1963 Nobel Prize in physics was, in fact, awarded to Maria Mayer and Hans Jensen "for their discoveries concerning nuclear shell structure."

A Combined Model

Consider a nucleus in which a small number of neutrons (or protons) exist outside a core of closed shells that contains magic numbers of neutrons or protons. The outside nucleons occupy quantized states in a potential well established by the central core, thus preserving the central feature of the independent-particle model. These outside nucleons also interact with the core, deforming it and setting up "tidal wave" motions of rotation or vibration within it. These collective motions of the core preserve the central feature of the collective model. Such a model of nuclear structure thus succeeds in combining the seemingly irreconcilable points of view of the collective and independent-particle models. It has been remarkably successful in explaining observed nuclear properties.

Sample Problem 42-10

Consider the neutron capture reaction

$$^{109}\text{Ag} + \text{n} \rightarrow {}^{110}\text{Ag} \rightarrow {}^{110}\text{Ag} + \gamma, \qquad (42\text{-}35)$$

in which a compound nucleus (^{110}Ag) is formed. Figure 42-14 shows the relative rate at which such events take place, plotted against the energy of the incoming neutron. Find the mean lifetime of this compound nucleus by using the uncertainty principle in the form

$$\Delta E \cdot \Delta t \approx \hbar. \qquad (42\text{-}36)$$

Here ΔE is a measure of the uncertainty with which the energy of a state can be defined. The quantity Δt is a measure of the time available to measure this energy. In fact, here Δt is just t_{avg}, the average life of the compound nucleus before it decays to its ground state.

Solution: We see that the relative reaction rate peaks sharply at a neutron energy of about 5.2 eV. This suggests that we are dealing with a single excited energy level of the compound nucleus ^{110}Ag. When the available energy (of the incoming neutron) just matches the energy of this level above

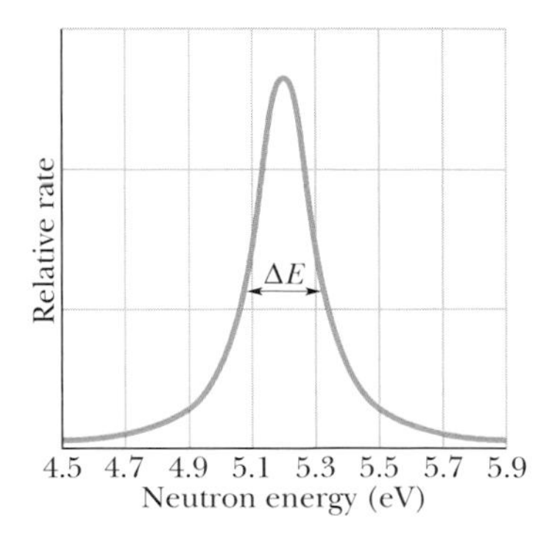

Fig. 42-14 A plot of the relative number of reaction events of the type described by Eq. 42-35 as a function of the energy of the incident neutron. The half-width ΔE of the resonance peak is about 0.20 eV.

the ^{110}Ag ground state, we have "resonance" and the reaction of Eq. 42-35 really "goes."

However, the resonance peak is not infinitely sharp but has an approximate half-width (ΔE in the figure) of about 0.20 eV. The **Key Idea** here is that we can account for this resonance-peak width by saying that the excited level is not sharply defined in energy but has an energy uncertainty ΔE of about 0.20 eV. Thus, Eq. 42-36 gives us

$$\Delta t = t_{avg} \approx \frac{\hbar}{\Delta E} = \frac{(4.14 \times 10^{-15}\ \text{eV}\cdot\text{s})/2\pi}{0.20\ \text{eV}}$$
$$\approx 3 \times 10^{-15}\ \text{s}. \quad \text{(Answer)}$$

This is several hundred times greater than the time a 0.20 eV neutron takes to cross the diameter of a ^{109}Ag nucleus. Therefore, the neutron is spending this time of 3×10^{-15} s *as part of* the nucleus.

Review & Summary

The Nuclides Approximately 2000 **nuclides** are known to exist. Each is characterized by an **atomic number** Z (the number of protons), a **neutron number** N, and a **mass number** A (the total number of **nucleons**—protons and neutrons). Thus, $A = Z + N$. Nuclides with the same atomic number but different neutron numbers are **isotopes** of one another. Nuclei have a mean radius r given by

$$r = r_0 A^{1/3}, \quad (42\text{-}3)$$

where $r_0 \approx 1.2$ fm.

Mass and Binding Energy Atomic masses are often reported in terms of *mass excess*

$$\Delta = M - A \quad \text{(mass excess)}, \quad (42\text{-}6)$$

where M is the actual mass of an atom in atomic mass units and A is the mass number for that atom's nucleus. The **binding energy** of a nucleus is the difference

$$\Delta E_{be} = \Sigma(mc^2) - Mc^2 \quad \text{(binding energy)}, \quad (42\text{-}7)$$

where $\Sigma(mc^2)$ is the total mass energy of the *individual* protons and neutrons. The **binding energy per nucleon** is

$$\Delta E_{ben} = \frac{\Delta E_{be}}{A} \quad \text{(binding energy per nucleon)}. \quad (42\text{-}8)$$

Mass–Energy Exchanges The energy equivalent of one mass unit (u) is 931.494 013 MeV. The binding energy curve shows that middle-mass nuclides are the most stable and that energy can be released both by fission of high-mass nuclei and by fusion of low-mass nuclei.

The Nuclear Force Nuclei are held together by an attractive force acting among the nucleons. It is thought to be a secondary effect of the **strong force** acting between the quarks that make up the nucleons. Nuclei can exist in a number of discrete energy states, each with a characteristic intrinsic angular momentum and magnetic moment.

Radioactive Decay Most known nuclides are radioactive; they spontaneously decay at a rate $R\ (= -dN/dt)$ that is proportional to the number N of radioactive atoms present, the proportionality constant being the **disintegration constant** λ. This leads to the law of exponential decay:

$$N = N_0 e^{-\lambda t}, \quad R = \lambda N = R_0 e^{-\lambda t}$$
$$\text{(radioactive decay)}. \quad (42\text{-}15, 42\text{-}17, 42\text{-}16)$$

The **half-life** $T_{1/2} = (\ln 2)/\lambda$ of a radioactive nuclide is the time required for the decay rate R (or the number N) in a sample to drop to half its initial value.

Alpha Decay Some nuclides decay by emitting an alpha particle (a helium nucleus, ^{4}He). Such decay is inhibited by a potential energy barrier that cannot be penetrated according to classical physics but is subject to tunneling according to quantum physics. The barrier penetrability, and thus the half-life for alpha decay, is very sensitive to the energy of the emitted alpha particle.

Beta Decay In **beta decay** either an electron or a positron is emitted by a nucleus, along with a neutrino. The emitted particles share the available disintegration energy. The electrons and positrons emitted in beta decay have a continuous spectrum of energies from near zero up to a limit K_{max} $(= Q = -\Delta m\, c^2)$.

Radioactive Dating Naturally occurring radioactive nuclides provide a means for estimating the dates of historic and prehistoric events. For example, the ages of organic materials can often be found by measuring their ^{14}C content; rock samples can be dated using the radioactive isotope ^{40}K.

Radiation Dosage Three units are used to describe exposure to ionizing radiation. The **becquerel** (1 Bq = 1 decay per second) measures the **activity** of a source. The amount of energy actually absorbed is measured in **grays,** with 1 Gy corresponding to 1 J/kg. The estimated biological effect of the absorbed energy is measured in **sieverts;** a dose equivalent of 1 Sv causes the same biological effect regardless of the radiation type by which it was acquired.

Nuclear Models The **collective** model of nuclear structure assumes that nucleons collide constantly with one another and that relatively long-lived **compound nuclei** are formed when a projectile is captured. The formation and eventual decay of a compound nucleus are totally independent events.

The **independent particle** model of nuclear structure assumes that each nucleon moves, essentially without collisions, in a quantized state within the nucleus. The model predicts nucleon levels and **magic nucleon numbers** (2, 8, 20, 28, 50, 82, and 126) associated with closed shells of nucleons; nuclides with any of these numbers of neutrons or protons are particularly stable.

The **combined** model, in which extra nucleons occupy quantized states outside a central core of closed shells, is highly successful in predicting many nuclear properties.

Questions

1 Suppose the alpha particle of Sample Problem 42-1 is replaced with a proton of the same initial kinetic energy and also headed directly toward the nucleus of the gold atom. Will the distance from the center of the nucleus at which the proton stops be greater than, less than, or the same as that of the alpha particle?

2 In your body are there more protons than neutrons, more neutrons than protons, or about the same number of each?

3 A certain nuclide is said to be particularly stable. Does its binding energy per nucleon lie slightly above or slightly below the binding energy curve of Fig. 42-6?

4 The nuclide ^{244}Pu ($Z = 94$) is an alpha-emitter. Into which of the following nuclides does it decay: ^{240}Np ($Z = 93$), ^{240}U ($Z = 92$), ^{248}Cm ($Z = 96$), or ^{244}Am ($Z = 95$)?

5 The radionuclide ^{196}Ir decays by emitting an electron. (a) Into which square in Fig. 42-5 is it transformed? (b) Do further decays then occur?

6 Is the mass excess of an alpha particle (use a straightedge on Fig. 42-12) greater than or less than the particle's total binding energy (use the binding energy per nucleon from Fig. 42-6)?

7 A lead nuclide contains 82 protons. (a) If it also contained 82 neutrons, where would it be located on the plot of Fig. 42-4? (b) If such a nucleus could be formed, would it emit positrons, emit electrons, or be stable? (c) From Fig. 42-4, about how many neutrons do you expect to find in a stable lead nuclide?

8 The nuclide ^{238}U ($Z = 92$) can fission into two parts that have identical atomic numbers and mass numbers. (a) Is the nuclide ^{238}U above or below the $N = Z$ line of Fig. 42-4? (b) Are the two fragments above or below this line? (c) Are these fragments stable or radioactive?

9 At $t = 0$, a sample of radionuclide A has the same decay rate as a sample of radionuclide B has at $t = 30$ min. The disintegration constants are λ_A and λ_B, with $\lambda_A < \lambda_B$. Will the two samples ever have (simultaneously) the same decay rate? (*Hint:* Sketch a graph of their activities.)

10 At $t = 0$, a sample of radionuclide A has twice the decay rate as a sample of radionuclide B. The disintegration constants are λ_A and λ_B, with $\lambda_A > \lambda_B$. Will the two samples ever have (simultaneously) the same decay rate?

11 Radionuclides decay exponentially, as in Eq. 42-16. Batteries, stars, and even students also decay, where "decay" stands for "burn out." Do these items decay exponentially?

12 Figure 42-15 gives the activities of three radioactive samples versus time. Rank the samples according to their (a) half-life and (b) disintegration constant, greatest first. (*Hint:* For (a), use a straightedge on the graph.)

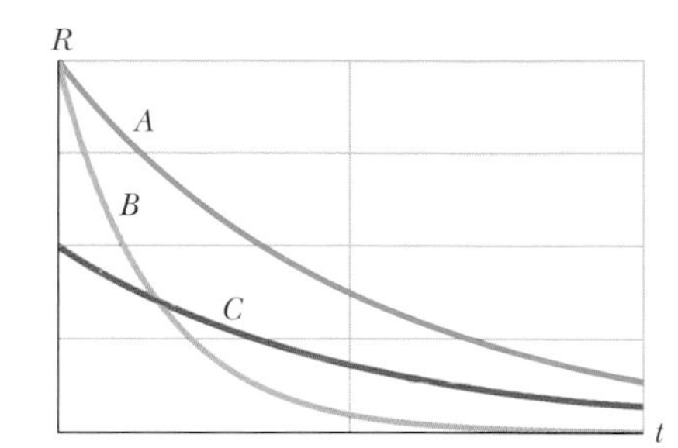

Fig. 42-15 Question 12.

13 At $t = 0$ we begin to observe two identical radioactive nuclei that have a half-life of 5 min. At $t = 1$ min, one of the nuclei decays. Does that event increase or decrease the chance that the second nucleus will decay in the next 4 min, or is there no effect on the second nucleus?

14 If the mass of a radioactive sample is doubled, do (a) the activity of the sample and (b) the disintegration constant of the sample increase, decrease, or remain the same?

15 The radionuclides ^{209}At and ^{209}Po emit alpha particles with energies of 5.65 and 4.88 MeV, respectively. Which nuclide has the longer half-life?

16 The radionuclide ^{49}Sc has a half-life of 57.0 min. At $t = 0$, the counting rate of a sample of it is 6000 counts/min above the general background activity, which is 30 counts/min. Without computation, determine whether the counting rate of the sample will be about equal to the background rate in 3 h, 7 h, 10 h, or a time much longer than 10 h.

17 The magic nucleon numbers for nuclei are given in Section 42-9 as 2, 8, 20, 28, 50, 82, and 126. Are nuclides magic (that is, especially stable) when (a) only the mass number A, (b) only the atomic number Z, (c) only the neutron number N, or (d) either Z or N (or both) is equal to one of these numbers? Pick all correct phrases.

18 (a) Which of the following nuclides are magic: ^{122}Sn, ^{132}Sn, ^{98}Cd, ^{198}Au, ^{208}Pb? (b) Which, if any, are doubly magic?

Problems

SSM Solution is in the Student Solutions Manual.
WWW Solution is at http://www.wiley.com/college/halliday
• – ••• Number of dots indicates level of problem difficulty.

sec. 42-2 Discovering the Nucleus

•1 Calculate the distance of closest approach for a head-on collision between a 5.30 MeV alpha particle and the nucleus of a copper atom.

•2 A ^{7}Li nucleus with a kinetic energy of 3.00 MeV is sent toward a ^{232}Th nucleus. What is the least center-to-center separation between the two nuclei, assuming that the (more massive) ^{232}Th nucleus does not move?

••3 When an alpha particle collides elastically with a nucleus, the nucleus recoils. Suppose a 5.00 MeV alpha particle has a head-on elastic collision with a gold nucleus that is initially at rest. What is the kinetic energy of (a) the recoiling nucleus and (b) the rebounding alpha particle?

sec. 42-3 Some Nuclear Properties

•4 The nuclide ^{14}C contains (a) how many protons and (b) how many neutrons?

•5 What is the mass excess Δ_1 of ^1H (whose actual mass is 1.007 825 u) in (a) atomic mass units and (b) MeV/c^2? What is the mass excess Δ_n of a neutron (actual mass is 1.008 664 u) in (c) atomic mass units and (d) MeV/c^2? What is the mass excess Δ_{120} of ^{120}Sn (actual mass is 119.902 197 u) in (e) atomic mass units and (f) MeV/c^2?

•6 The strong neutron excess (defined as $N - Z$) of high-mass nuclei is illustrated by noting that most high-mass nuclides could never fission into two stable nuclei without neutrons being left over. For example, consider the spontaneous fission of a ^{235}U nucleus into two stable *daughter nuclei* with atomic numbers 39 and 53. From Appendix F, determine the name of the (a) first and (b) second daughter nucleus. From Fig. 42-4, approximately how many neutrons are in the (c) first and (d) second? (e) Approximately how many neutrons are left over?

•7 A neutron star is a stellar object whose density is about that of nuclear matter, as calculated in Sample Problem 42-2. Suppose that the Sun were to collapse and become such a star without losing any of its present mass. What would be its radius?

•8 The electric potential energy of a uniform sphere of charge q and radius r is given by

$$U = \frac{3q^2}{20\pi\varepsilon_0 r}.$$

(a) Does the energy represent a tendency for the sphere to bind together or blow apart? The nuclide ^{239}Pu is spherical with radius 6.64 fm. For this nuclide, what are (b) the electric potential energy U according to the equation, (c) the electric potential energy per proton, and (d) the electric potential energy per nucleon? The binding energy per nucleon is 7.56 MeV. (e) Why is the nuclide bound so well when the answers to (c) and (d) are large and positive?

•9 What is the nuclear mass density ρ_m of (a) the fairly low-mass nuclide ^{55}Mn and (b) the fairly high-mass nuclide ^{209}Bi? (c) Compare the two answers, with an explanation. What is the nuclear charge density ρ_q of (d) ^{55}Mn and (e) ^{209}Bi? (f) Compare the two answers, with an explanation.

•10 (a) Show that the mass M of an atom is given approximately by $M_{app} = Am_p$, where A is the mass number and m_p is the proton mass. For (b) ^1H, (c) ^{31}P, (d) ^{120}Sn, (e) ^{197}Au, and (f) ^{239}Ps, use the mass values given in Table 42-1 to find the percentage deviation between M_{app} and M:

$$\text{percentage deviation} = \frac{M_{app} - M}{M} 100.$$

(g) Is a value of M_{app} accurate enough to be used in a calculation of a nuclear binding energy?

•11 Nuclear radii may be measured by scattering high-energy electrons from nuclei. (a) What is the de Broglie wavelength for 200 MeV electrons? (b) Are these electrons suitable probes for this purpose? SSM

••12 What is the binding energy per nucleon of ^{262}Bh? The mass of the atom is 262.1231 u.

••13 A periodic table might list the average atomic mass of magnesium as being 24.312 u. That average value is the result of *weighting* the atomic masses of the magnesium isotopes according to their natural abundances on Earth. The three isotopes and their masses are ^{24}Mg (23.985 04 u), ^{25}Mg (24.985 84 u), and ^{26}Mg (25.982 59 u). The natural abundance of ^{24}Mg is 78.99% by mass (that is, 78.99% of the mass of a naturally occurring sample of magnesium is due to the presence of ^{24}Mg). What is the abundance of (a) ^{25}Mg and (b) ^{26}Mg?

••14 An α particle (^4He nucleus) is to be taken apart in the following steps. Give the energy (work) required for each step: (a) remove a proton, (b) remove a neutron, and (c) separate the remaining proton and neutron. For an α particle, what are (d) the total binding energy and (e) the binding energy per nucleon? (f) Does either match an answer to (a), (b), or (c)? Some needed atomic masses are

^4He	4.002 60 u	^2H	2.014 10 u
^3H	3.016 05 u	^1H	1.007 83 u
n	1.008 67 u		

••15 Verify the binding energy per nucleon given in Table 42-1 for ^{239}Pu. The mass of the atom is 239.052 16 u. SSM

••16 A penny has a mass of 3.0 g. Calculate the energy that would be required to separate all the neutrons and protons in this coin from one another. For simplicity, assume that the penny is made entirely of ^{63}Cu atoms (of mass 62.929 60 u). The masses of the proton and the neutron are 1.007 83 u and 1.008 66 u, respectively.

••17 (a) Show that the energy associated with the strong force between nucleons in a nucleus is proportional to A, the mass number of the nucleus in question. (b) Show that the energy associated with the Coulomb force between protons in a nucleus is proportional to $Z(Z - 1)$. (c) Show that, as we move to larger and larger nuclei (see Fig. 42-4), the importance of the Coulomb force increases more rapidly than does that of the strong force.

••18 Because the neutron has no charge, its mass must be found in some way other than by using a mass spectrometer. When a neutron and a proton meet (assume both to be almost stationary), they combine and form a deuteron, emitting a gamma ray whose energy is 2.2233 MeV. The masses of the proton and the deuteron are 1.007 276 467 u and 2.013 553 212 u, respectively. Find the mass of the neutron from these data.

••19 (a) Show that the total binding energy E_{be} of a given nuclide is

$$E_{be} = Z\Delta_H + N\Delta_n - \Delta,$$

where Δ_H is the mass excess of ^1H, Δ_n is the mass excess of a neutron, and Δ is the mass excess of the given nuclide. (b) Using this method, calculate the binding energy per nucleon for ^{197}Au. Compare your result with the value listed in Table 42-1. The needed mass excesses, rounded to three significant figures, are $\Delta_H = +7.29$ MeV, $\Delta_n = +8.07$ MeV, and $\Delta_{197} = -31.2$ MeV. Note the economy of calculation that results when mass excesses are used in place of the actual masses. SSM WWW

sec. 42-4 Radioactive Decay

•20 When aboveground nuclear tests were conducted, the explosions shot radioactive dust into the upper atmosphere. Global air circulations then spread the dust worldwide before it settled out on ground and water. One such test was conducted in October 1976. What fraction of the ^{90}Sr produced by that explosion will still exist in October 2006? The half-life of ^{90}Sr is 29 y.

•21 A radioactive nuclide has a half-life of 30.0 y. What fraction of an initially pure sample of this nuclide will remain undecayed at the end of (a) 60.0 y and (b) 90.0 y?

•22 The half-life of a radioactive isotope is 140 d. How many days would it take for the decay rate of a sample of this isotope to fall to one-fourth of its initial value?

•23 Consider an initially pure 3.4 g sample of ^{67}Ga, an isotope that has a half-life of 78 h. (a) What is its initial decay rate? (b) What is its decay rate 48 h later?

•24 The half-life of a particular radioactive isotope is 6.5 h. If there are initially 48×10^{19} atoms of this isotope, how many remain at the end of 26 h?

•25 A radioactive isotope of mercury, ^{197}Hg, decays to gold, ^{197}Au, with a disintegration constant of $0.0108\ h^{-1}$. (a) Calculate the half-life of the ^{197}Hg. What fraction of a sample will remain at the end of (b) three half-lives and (c) 10.0 days? SSM WWW

•26 The plutonium isotope ^{239}Pu is produced as a by-product in nuclear reactors and hence is accumulating in our environment. It is radioactive, decaying with a half-life of 2.41×10^4 y. (a) How many nuclei of Pu constitute a chemically lethal dose of 2.00 mg? (b) What is the decay rate of this amount?

•27 Cancer cells are more vulnerable to x and gamma radiation than are healthy cells. In the past, the standard source for radiation therapy was radioactive ^{60}Co, which decays, with a half-life of 5.27 y, into an excited nuclear state of ^{60}Ni. That nickel isotope then immediately emits two gamma-ray photons, each with an approximate energy of 1.2 MeV. How many radioactive ^{60}Co nuclei are present in a 6000 Ci source of the type used in hospitals? (Energetic particles from linear accelerators are now used in radiation therapy.)

••28 The air in some caves includes a significant amount of radon gas, which can lead to lung cancer if breathed over a prolonged time. In British caves, the air in the cave with the greatest amount of the gas has an activity per volume of $1.55 \times 10^5\ Bq/m^3$. Suppose that you spend two full days exploring (and sleeping in) that cave. Approximately how many ^{222}Rn atoms would you take in and out of your lungs during your two-day stay? The radionuclide ^{222}Rn in radon gas has a half-life of 3.82 days. You need to estimate your lung capacity and average breathing rate.

••29 A radioactive sample intended for irradiation of a hospital patient is prepared at a nearby laboratory. The sample has a half-life of 83.61 h. What should its initial activity be if its activity is to be 7.4×10^8 Bq when it is used to irradiate the patient 24 h later?

••30 In October 1992, Swiss police arrested two men who were attempting to smuggle osmium out of Eastern Europe for a clandestine sale. However, by error, the smugglers had picked up ^{137}Cs. Reportedly, each smuggler was carrying a 1.0 g sample of ^{137}Cs in a pocket. In (a) bequerels and (b) curies, what was the activity of each sample? The isotope ^{137}Cs has a half-life of 30.2 y. (The activities of radioisotopes commonly used in hospitals range up to a few millicuries.)

••31 After long effort, in 1902 Marie and Pierre Curie succeeded in separating from uranium ore the first substantial quantity of radium, one decigram of pure $RaCl_2$. The radium was the radioactive isotope ^{226}Ra, which has a half-life of 1600 y. (a) How many radium nuclei had the Curies isolated? (b) What was the decay rate of their sample, in disintegrations per second? SSM

••32 The radioactive nuclide ^{99}Tc can be injected into a patient's bloodstream in order to monitor the blood flow, measure the blood volume, or find a tumor, among other goals. The nuclide is produced in a hospital by a "cow" containing ^{99}Mo, a radioactive nuclide that decays to ^{99}Tc with a half-life of 67 h. Once a day, the cow is "milked" for its ^{99}Tc, which is produced in an excited state by the ^{99}Mo; the ^{99}Tc de-excites to its lowest energy state by emitting a gamma-ray photon, which is recorded by detectors placed around the patient. The de-excitation has a half-life of 6.0 h. (a) By what process does ^{99}Mo decay to ^{99}Tc? (b) If a patient is injected with an 8.2×10^7 Bq sample of ^{99}Tc, how many gamma-ray photons are initially produced within the patient each second? (c) If the emission rate of gamma-ray photons from a small tumor that has collected ^{99}Tc is 38 per second at a certain time, how many excited-state ^{99}Tc are located in the tumor at that time?

••33 The radionuclide ^{64}Cu has a half-life of 12.7 h. If a sample contains 5.50 g of initially pure ^{64}Cu at $t = 0$, how much of it will decay between $t = 14.0$ h and $t = 16.0$ h?

••34 A source contains two phosphorus radionuclides, ^{32}P ($T_{1/2} = 14.3$ d) and ^{33}P ($T_{1/2} = 25.3$ d). Initially, 10.0% of the decays come from ^{33}P. How long must one wait until 90.0% do so?

••35 A 1.00 g sample of samarium emits alpha particles at a rate of 120 particles/s. The responsible isotope is ^{147}Sm, whose natural abundance in bulk samarium is 15.0%. Calculate the half-life for the decay process.

••36 Plutonium isotope ^{239}Pu decays by alpha decay with a half-life of 24 100 y. How many milligrams of helium are produced by an initially pure 12.0 g sample of ^{239}Pu at the end of 20 000 y? (Consider only the helium produced directly by the plutonium and not by any by-products of the decay process.)

••37 A certain radionuclide is being manufactured in a cyclotron at a constant rate R. It is also decaying with disintegration constant λ. Assume that the production process has been going on for a time that is much longer than the half-life of the radionuclide. (a) Show that the number of radioactive nuclei present after such time remains constant and is given by $N = R/\lambda$. (b) Now show that this result holds no matter how many radioactive nuclei were present initially. The nuclide is said to be in *secular equilibrium* with its source; in this state its decay rate is just equal to its production rate. SSM

••38 Calculate the mass of a sample of (initially pure) ^{40}K that has an initial decay rate of 1.70×10^5 disintegrations/s. The isotope has a half-life of 1.28×10^9 y.

••39 The radionuclide ^{56}Mn has a half-life of 2.58 h and is produced in a cyclotron by bombarding a manganese target with deuterons. The target contains only the stable manganese isotope ^{55}Mn, and the manganese–deuteron reaction that produces ^{56}Mn is

$$^{55}\text{Mn} + \text{d} \rightarrow {}^{56}\text{Mn} + \text{p}.$$

If the bombardment lasts much longer than the half-life of ^{56}Mn, the activity of the ^{56}Mn produced in the target reaches a final value of 8.88×10^{10} Bq. (a) At what rate is ^{56}Mn being produced? (b) How many ^{56}Mn nuclei are then in the target? (c) What is their total mass?

sec. 42-5 Alpha Decay

•40 How much energy is released when a ^{238}U nucleus decays by emitting (a) an alpha particle and (b) a sequence of neutron, proton, neutron, proton? (c) Convince yourself both by reasoned argument and by direct calculation that the difference between these two numbers is just the total binding energy of the alpha particle. (d) Find that binding energy. Some needed atomic and particle masses are

^{238}U	238.050 79 u	^{234}Th	234.043 63 u
^{237}U	237.048 73 u	^{4}He	4.002 60 u
^{236}Pa	236.048 91 u	^{1}H	1.007 83 u
^{235}Pa	235.045 44 u	n	1.008 66 u

•41 Generally, more massive nuclides tend to be more unstable to alpha decay. For example, the most stable isotope of uranium, ^{238}U, has an alpha decay half-life of 4.5×10^9 y. The most stable isotope of plutonium is ^{244}Pu with an 8.0×10^7 y half-life, and for curium we have ^{248}Cm and 3.4×10^5 y. When half of an original sample of ^{238}U has decayed, what fraction of the original sample of (a) plutonium and (b) curium is left? SSM

••42 Large radionuclides emit an alpha particle rather than other combinations of nucleons because the alpha particle has such a stable, tightly bound structure. To confirm this statement, calculate the disintegration energies for these hypothetical decay processes and discuss the meaning of your findings:

(a) $^{235}U \rightarrow {}^{232}Th + {}^{3}He$,
(b) $^{235}U \rightarrow {}^{231}Th + {}^{4}He$,
(c) $^{235}U \rightarrow {}^{230}Th + {}^{5}He$.

The needed atomic masses are

^{232}Th	232.0381 u	^{3}He	3.0160 u
^{231}Th	231.0363 u	^{4}He	4.0026 u
^{230}Th	230.0331 u	^{5}He	5.0122 u
^{235}U	235.0429 u		

••43 A ^{238}U nucleus emits a 4.196 MeV alpha particle. Calculate the disintegration energy Q for this process, taking the recoil energy of the residual ^{234}Th nucleus into account.

••44 Under certain rare circumstances, a nucleus can decay by emitting a particle more massive than an alpha particle. Consider the decays

$$^{223}\text{Ra} \rightarrow {}^{209}\text{Pb} + {}^{14}\text{C}$$

and

$$^{223}\text{Ra} \rightarrow {}^{219}\text{Rn} + {}^{4}\text{He}.$$

Calculate the Q value for the (a) first and (b) second decay and determine that both are energetically possible. (c) The Coulomb barrier height for alpha-particle emission is 30.0 MeV. What is the barrier height for ^{14}C emission? The needed atomic masses are

^{223}Ra	223.018 50 u	^{14}C	14.003 24 u
^{209}Pb	208.981 07 u	^{4}He	4.002 60 u
^{219}Rn	219.009 48 u		

sec. 42-6 Beta Decay

•45 The cesium isotope ^{137}Cs is present in the fallout from aboveground detonations of nuclear bombs. Because it decays with a slow (30.2 y) half-life into ^{137}Ba, releasing considerable energy in the process, it is of environmental concern. The atomic masses of the Cs and Ba are 136.9071 and 136.9058 u, respectively; calculate the total energy released in such a decay. SSM

•46 An electron is emitted from a middle-mass nuclide ($A = 150$, say) with a kinetic energy of 1.0 MeV. (a) What is its de Broglie wavelength? (b) Calculate the radius of the emitting nucleus. (c) Can such an electron be confined as a standing wave in a "box" of such dimensions? (d) Can you use these numbers to disprove the (abandoned) argument that electrons actually exist in nuclei?

•47 A free neutron decays according to Eq. 42-26. If the neutron–hydrogen atom mass difference is 840 μu, what is the maximum kinetic energy K_{max} possible for the electron produced in a neutron decay?

•48 Some radionuclides decay by capturing one of their own atomic electrons, a K-shell electron, say. An example is

$$^{49}\text{V} + \text{e}^- \rightarrow {}^{49}\text{Ti} + \nu, \qquad T_{1/2} = 331 \text{ d}.$$

Show that the disintegration energy Q for this process is given by

$$Q = (m_V - m_{Ti})c^2 - E_K,$$

where m_V and m_{Ti} are the atomic masses of ^{49}V and ^{49}Ti, respectively, and E_K is the binding energy of the vanadium K-shell electron. (*Hint:* Put $\mathbf{m}_V$ and $\mathbf{m}_{Ti}$ as the corresponding nuclear masses and proceed as in Sample Problem 42-7.)

••49 The radionuclide ^{11}C decays according to

$$^{11}\text{C} \rightarrow {}^{11}\text{B} + \text{e}^+ + \nu, \qquad T_{1/2} = 20.3 \text{ min}.$$

The maximum energy of the emitted positrons is 0.960 MeV. (a) Show that the disintegration energy Q for this process is given by

$$Q = (m_C - m_B - 2m_e)c^2,$$

where m_C and m_B are the atomic masses of ^{11}C and ^{11}B, respectively, and m_e is the mass of a positron. (b) Given the mass values $m_C = 11.011\,424$ u, $m_B = 11.009\,305$ u, and $m_e = 0.000\,548\,6$ u, calculate Q and compare it with the maximum energy of the emitted positron given above. (*Hint:* Let $\mathbf{m}_C$ and $\mathbf{m}_B$ be the nuclear masses and proceed as in Sample Problem 42-7 for beta decay. Note that beta-plus decay is an exception to the general rule that if atomic masses are used in nuclear decay calculations, the mass of the emitted electron is automatically taken care of.)

••50 Two radioactive materials that alpha decay, ^{238}U and ^{232}Th, and one that beta decays, ^{40}K, are sufficiently abundant in granite to contribute significantly to the heating of Earth through the decay energy produced. The alpha-decay isotopes give rise to decay chains that stop when stable lead isotopes are formed. The isotope ^{40}K has a single beta decay. (Assume this is the only possible decay of that isotope.) Here is the information:

Parent	Decay Mode	Half-Life (y)	Stable End Point	Q (MeV)	f (ppm)
^{238}U	α	4.47×10^9	^{206}Pb	51.7	4
^{232}Th	α	1.41×10^{10}	^{208}Pb	42.7	13
^{40}K	β	1.28×10^9	^{40}Ca	1.31	4

In the table Q is the *total* energy released in the decay of one parent nucleus to the *final* stable end point and f is the abundance of the isotope in kilograms per kilogram of granite; ppm means parts per million. (a) Show that these materials produce energy as heat at the rate of 1.0×10^{-9} W for each kilogram of granite. (b) Assuming that there is 2.7×10^{22} kg of granite in a 20-km-thick spherical shell at the surface of Earth, estimate the power of this decay process over all of Earth. Compare this power with the total solar power intercepted by Earth, 1.7×10^{17} W.

•••51 The radionuclide ^{32}P decays to ^{32}S as described by Eq. 42-24. In a particular decay event, a 1.71 MeV electron is emitted, the maximum possible value. What is the kinetic energy of the recoiling ^{32}S atom in this event? (*Hint:* For the electron it is necessary to use the relativistic expressions for kinetic energy and linear momentum. The ^{32}S atom is nonrelativistic.) SSM WWW

sec. 42-7 Radioactive Dating

•52 A 5.00 g charcoal sample from an ancient fire pit has a ^{14}C activity of 63.0 disintegrations/min. A living tree has a ^{14}C activity of 15.3 disintegrations/min per 1.00 g. The half-life of ^{14}C is 5730 y. How old is the charcoal sample?

•53 The isotope ^{238}U decays to ^{206}Pb with a half-life of 4.47×10^9 y. Although the decay occurs in many individual steps, the first step has by far the longest half-life; therefore, one can often consider the decay to go directly to lead. That is,

$$^{238}\text{U} \rightarrow {}^{206}\text{Pb} + \text{various decay products.}$$

A rock is found to contain 4.20 mg of ^{238}U and 2.135 mg of ^{206}Pb. Assume that the rock contained no lead at formation, so all the lead now present arose from the decay of uranium. How many atoms of (a) ^{238}U and (b) ^{206}Pb does the rock now contain? (c) How many atoms of ^{238}U did the rock contain at formation? (d) What is the age of the rock?

••54 A particular rock is thought to be 260 million years old. If it contains 3.70 mg of ^{238}U, how much ^{206}Pb should it contain? See Problem 53.

••55 A rock recovered from far underground is found to contain 0.86 mg of ^{238}U, 0.15 mg of ^{206}Pb, and 1.6 mg of ^{40}Ar. How much ^{40}K will it likely contain? Assume that ^{40}K decays to only ^{40}Ar with a half-life of 1.25×10^9 y. Also assume that ^{238}U has a half-life of 4.47×10^9 y.

•••56 The isotope ^{40}K can decay to either ^{40}Ca or ^{40}Ar; assume both decays have a half-life of 1.26×10^9 y. The ratio of the Ca produced to the Ar produced is $8.54/1 = 8.54$. A sample originally had only ^{40}K. It now has equal amounts of ^{40}K and ^{40}Ar; that is, the ratio of K to Ar is $1/1 = 1$. How old is the sample? (*Hint:* Work this like other radioactive-dating problems, except that this decay has two products.)

sec. 42-8 Measuring Radiation Dosage

•57 The nuclide ^{198}Au, with a half-life of 2.70 d, is used in cancer therapy. What mass of this nuclide is required to produce an activity of 250 Ci? SSM

•58 A radiation detector records 8700 counts in 1.00 min. Assuming that the detector records all decays, what is the activity of the radiation source in (a) becquerels and (b) curies?

••59 A typical chest x-ray radiation dose is 250 μSv, delivered by x rays with an RBE factor of 0.85. Assuming that the mass of the exposed tissue is one-half the patient's mass of 88 kg, calculate the energy absorbed in joules.

••60 A 75 kg person receives a whole-body radiation dose of 2.4×10^{-4} Gy, delivered by alpha particles for which the RBE factor is 12. Calculate (a) the absorbed energy in joules and the dose equivalent in (b) sieverts and (c) rem.

••61 An 85 kg worker at a breeder reactor plant accidentally ingests 2.5 mg of ^{239}Pu dust. This isotope has a half-life of 24 100 y, decaying by alpha decay. The energy of the emitted alpha particles is 5.2 MeV, with an RBE factor of 13. Assume that the plutonium resides in the worker's body for 12 h and that 95% of the emitted alpha particles are stopped within the body. Calculate (a) the number of plutonium atoms ingested, (b) the number that decay during the 12 h, (c) the energy absorbed by the body, (d) the resulting physical dose in grays, and (e) the dose equivalent in sieverts.

sec. 42-9 Nuclear Models

•62 A typical kinetic energy for a nucleon in a middle-mass nucleus may be taken as 5.00 MeV. To what effective nuclear temperature does this correspond, based on the assumptions of the collective model of nuclear structure?

•63 An intermediate nucleus in a particular nuclear reaction decays within 10^{-22} s of its formation. (a) What is the uncertainty ΔE in our knowledge of this intermediate state? (b) Can this state be called a compound nucleus? (See Sample Problem 42-10.)

•64 In the following list of nuclides, identify (a) those with filled nucleon shells, (b) those with one nucleon outside a filled shell, and (c) those with one vacancy in an otherwise filled shell: ^{13}C, ^{18}O, ^{40}K, ^{49}Ti, ^{60}Ni, ^{91}Zr, ^{92}Mo, ^{121}Sb, ^{143}Nd, ^{144}Sm, ^{205}Tl, and ^{207}Pb.

••65 Consider the three formation processes shown for the compound nucleus ^{20}Ne in Fig. 42-13. Here are some of the masses:

^{20}Ne	19.992 44 u	α	4.002 60 u
^{19}F	18.998 40 u	p	1.007 83 u
^{16}O	15.994 91 u		

What energy must (a) the alpha particle, (b) the proton, and (c) the γ-ray photon have to provide 25.0 MeV of excitation energy to the compound nucleus? SSM

Additional Problems

66 At the end of World War II, Dutch authorities arrested Dutch artist Hans van Meegeren for treason because, during the war, he had sold a masterpiece painting to the Nazi Hermann Goering. The painting, *Christ and His Disciples at Emmaus* by Dutch master Johannes Vermeer (1632–1675), had been discovered in 1937 by van Meegeren, after it had been lost for almost 300 years. Soon after the discovery, art experts

proclaimed that *Emmaus* was possibly the best Vermeer ever seen. Selling such a Dutch national treasure to the enemy was unthinkable treason.

However, shortly after being imprisoned, van Meegeren suddenly announced that he, not Vermeer, had painted *Emmaus.* He explained that he had carefully mimicked Vermeer's style, using a 300-year-old canvas and Vermeer's choice of pigments; he had then signed Vermeer's name to the work and baked the painting to give it an authentically old look.

Was van Meegeren lying to avoid a conviction of treason, hoping to be convicted of only the lesser crime of fraud? To art experts, *Emmaus* certainly looked like a Vermeer but, at the time of van Meegeren's trial in 1947, there was no scientific way to answer the question. However, in 1968 Bernard Keisch of Carnegie-Mellon University was able to answer the question with newly developed techniques of radioactive analysis.

Specifically, he analyzed a small sample of white lead-bearing pigment removed from *Emmaus.* This pigment is refined from lead ore, in which the lead is produced by a long radioactive decay series that starts with unstable ^{238}U and ends with stable ^{206}Pb. To follow the spirit of Keisch's analysis, focus on the following abbreviated portion of that decay series, in which intermediate, relatively short-lived radionuclides have been omitted:

$$^{230}\mathrm{Th} \xrightarrow[75.4\ \mathrm{ky}]{} {}^{226}\mathrm{Ra} \xrightarrow[1.60\ \mathrm{ky}]{} {}^{210}\mathrm{Pb} \xrightarrow[22.6\ \mathrm{y}]{} {}^{206}\mathrm{Pb}.$$

The longer and more important half-lives in this portion of the decay series are indicated.

(a) Show that in a sample of lead ore, the rate at which the number of ^{210}Pb nuclei changes is given by

$$\frac{dN_{210}}{dt} = \lambda_{226}N_{226} - \lambda_{210}N_{210},$$

where N_{210} and N_{226} are the numbers of ^{210}Pb nuclei and ^{226}Ra nuclei in the sample and λ_{210} and λ_{226} are the corresponding disintegration constants.

Because the decay series has been active for billions of years and because the half-life of ^{210}Pb is much less than that of ^{226}Ra, the nuclides ^{226}Ra and ^{210}Pb are in *equilibrium;* that is, the numbers of these nuclides (and thus their concentrations) in the sample do not change. (b) What is the ratio R_{226}/R_{210} of the activities of these nuclides in the sample of lead ore? (c) What is the ratio N_{226}/N_{210} of their numbers?

When lead pigment is refined from the ore, most of the ^{226}Ra is eliminated. Assume that only 1.00% remains. Just after the pigment is produced, what are the ratios (d) R_{226}/R_{210} and (e) N_{226}/N_{210}?

Keisch realized that with time the ratio R_{226}/R_{210} of the pigment would gradually change from the value in freshly refined pigment back to the value in the ore, as equilibrium between the ^{210}Pb and the remaining ^{226}Ra is established in the pigment. If *Emmaus* were painted by Vermeer and the sample of pigment taken from it were 300 years old when examined in 1968, the ratio would be close to the answer of (b). If *Emmaus* were painted by van Meegeren in the 1930s and the sample were only about 30 years old, the ratio would be close to the answer of (d). Keisch found a ratio of 0.09. (f) Is *Emmaus* a Vermeer?

67 The radionuclide ^{32}P ($T_{1/2}$ = 14.28 d) is often used as a tracer to follow the course of biochemical reactions involving phosphorus. (a) If the counting rate in a particular experimental setup is initially 3050 counts/s, how much time will the rate take to fall to 170 counts/s? (b) A solution containing ^{32}P is fed to the root system of an experimental tomato plant, and the ^{32}P activity in a leaf is measured 3.48 days later. By what factor must this reading be multiplied to correct for the decay that has occurred since the experiment began?

68 Because of the 1986 explosion and fire in a reactor at the Chernobyl nuclear power plant in northern Ukraine, part of Ukraine is contaminated with ^{137}Cs, which undergoes beta-minus decay with a half-life of 30.2 y. In 1996, the total activity of this contamination over an area of 2.6×10^5 km^2 was estimated to be 1×10^{16} Bq. Assume that the ^{137}Cs is uniformly spread over that area and that the beta-decay electrons travel either directly upward or directly downward. How many beta-decay electrons would you intercept were you to lie on the ground in that area for 1 h (a) in 1996 and (b) today? (You need to estimate your cross-sectional area that intercepts those electrons.)

69 One of the dangers of radioactive fallout from a nuclear bomb is its ^{90}Sr, which decays with a 29 year half-life. Because it has chemical properties much like those of calcium, the strontium, if ingested by a cow, becomes concentrated in the cow's milk. Some of the ^{90}Sr ends up in the bones of whoever drinks the milk. The energetic electrons emitted in the beta decay of ^{90}Sr damage the bone marrow and thus impair the production of red blood cells. A 1 megaton bomb produces approximately 400 g of ^{90}Sr. If the fallout spreads uniformly over a 2000 km^2 area, what ground area would hold an amount of radioactivity equal to the "allowed" limit for one person, which is 74 000 counts/s?

70 A radium source contains 1.00 mg of ^{226}Ra, which decays with a half-life of 1600 y to produce ^{222}Rn, a noble gas. This radon isotope in turn decays by alpha emission with a half-life of 3.82 d. If this process continues for a time much longer than the half-life of ^{222}Rn, the ^{222}Rn decay rate reaches a limiting value that matches the rate at which ^{222}Rn is being produced, which is approximately constant because of the relatively long half-life of ^{226}Ra. For the source under this limiting condition, what are (a) the activity of ^{226}Ra, (b) the activity of ^{222}Rn, and (c) the total mass of ^{222}Rn?

71 Find the disintegration energy Q for the decay of ^{49}V by K-electron capture (see Problem 48). The needed data are m_V = 48.948 52 u, m_{Ti} = 48.947 87 u, and E_K = 5.47 keV.

72 Assume that a gold nucleus has a radius of 6.23 fm and an alpha particle has a radius of 1.80 fm. What energy must an incident alpha particle have in order to "touch" the gold nucleus according to the type of calculation in Sample Problem 42-1?

73 Because a nucleon is confined to a nucleus, we can take the uncertainty in its position to be approximately the nuclear radius r. Use the uncertainty principle to determine the uncertainty Δp in the linear momentum of the nucleon. Using the approximation $p \approx \Delta p$ and the fact that the nucleon is nonrelativistic, calculate the kinetic energy of the nucleon in a nucleus with A = 100.

74 What is the activity of a 20 ng sample of ^{92}Kr, which has a half-life of 1.84 s?

75 How many years are needed to reduce the activity of ^{14}C to 0.020 of its original activity? The half-life of ^{14}C is 5730 y.

76 High-mass radionuclides, which may be either alpha or beta emitters, belong to one of four decay chains, depending on whether their mass number A is of the form $4n$, $4n + 1$, $4n + 2$, or $4n + 3$, where n is a positive integer. (a) Justify this statement and show that if a nuclide belongs to one of these families, all its decay products belong to the same family. Classify the following nuclides as to family: (b) ^{235}U, (c) ^{236}U, (d) ^{238}U, (e) ^{239}Pu, (f) ^{240}Pu, (g) ^{245}Cm, (h) ^{246}Cm, (i) ^{249}Cf, and (j) ^{253}Fm.

77 A certain stable nuclide, after absorbing a neutron, emits an electron, and the new nuclide splits spontaneously into two alpha particles. Identify the nuclide.

78 After a brief neutron irradiation of silver, two isotopes are present: ^{108}Ag ($T_{1/2}$ = 2.42 min) with an initial decay rate of 3.1×10^5/s, and ^{110}Ag ($T_{1/2}$ = 24.6 s) with an initial decay rate of 4.1×10^6/s. Make a semilog plot similar to Fig. 42-8 showing the total combined decay rate of the two isotopes as a function of time from $t = 0$ until $t = 10$ min. We used Fig. 42-8 to illustrate the extraction of the half-life for simple (one isotope) decays. Given only your plot of total decay rate for the two-isotope system here, suggest a way to analyze it in order to find the half-lives of both isotopes.

79 Figure 42-16 shows part of the decay scheme of ^{237}Np on a plot of mass number A versus proton number Z; five lines that represent either alpha decay or beta-minus decay connect dots that represent isotopes. What is the isotope at the end of the five decays (as marked with a question mark in Fig. 42-16)?

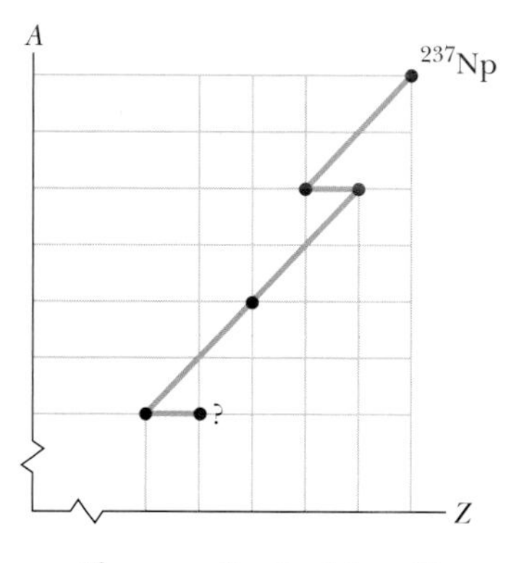

Fig. 42-16 Problem 79.

80 From data presented in the first few paragraphs of Section 42-4, find (a) the disintegration constant λ and (b) the half-life of ^{238}U.

81 Consider a ^{238}U nucleus to be made up of an alpha particle (^{4}He) and a residual nucleus (^{234}Th). Plot the electrostatic potential energy $U(r)$, where r is the distance between these particles. Cover the approximate range 10 fm $< r <$ 100 fm and compare your plot with that of Fig. 42-9.

82 Locate the nuclides displayed in Table 42-1 on the nuclidic chart of Fig. 42-4. Verify that they lie in the stability zone.

83 Make a nuclidic chart similar to Fig. 42-5 for the 25 nuclides $^{118-122}$Te, $^{117-121}$Sb, $^{116-120}$Sn, $^{115-119}$In, and $^{114-118}$Cd. Draw in and label (a) all isobaric (constant A) lines and (b) all lines of constant neutron excess, defined as $N - Z$.

84 Using a nuclidic chart, write the symbols for (a) all stable isotopes with $Z = 60$, (b) all radioactive nuclides with $N = 60$, and (c) all nuclides with $A = 60$.

85 If the unit for atomic mass were defined so that the mass of ^{1}H were exactly 1.000 000 u, what would be the mass of (a) ^{12}C (actual mass 12.000 000 u) and (b) ^{238}U (actual mass 238.050 785 u)?

86 Characteristic nuclear time is a useful but loosely defined quantity, taken to be the time required for a nucleon with a few million electron-volts of kinetic energy to travel a distance equal to the diameter of a middle-mass nuclide. What is the order of magnitude of this quantity? Consider 5 MeV neutrons traversing a nuclear diameter of ^{197}Au; use Eq. 42-3.

87 The radius of a spherical nucleus is measured, by electron-scattering methods, to be 3.6 fm. What is the likely mass number of the nucleus?

Energy from the Nucleus

This image has transfixed the world since World War II. When Robert Oppenheimer, the head of the scientific team that developed the atomic bomb, witnessed the first atomic explosion, he quoted from a sacred Hindu text: "Now I am become Death, the destroyer of worlds."

What is the physics behind this image that has so horrified the world?

The answer is in this chapter.

43-1 What Is Physics?

Now that we have discussed some of the properties of nuclei, let's turn to a central concern of physics and certain types of engineering: Can we get useful energy from nuclear sources, as people have done for thousands of years from atomic sources by burning materials like wood and coal?

As you already know, the answer is yes, but there are major differences between the two energy sources. When we get energy from wood and coal by burning them, we are tinkering with atoms of carbon and oxygen, rearranging their outer *electrons* into more stable combinations. When we get energy from uranium in a nuclear reactor, we are again burning a fuel, but now we are tinkering with the uranium nucleus, rearranging its *nucleons* into more stable combinations.

Electrons are held in atoms by the electromagnetic Coulomb force, and it takes only a few electron-volts to pull one of them out. On the other hand, nucleons are held in nuclei by the strong force, and it takes a few *million* electron-volts to pull one of *them* out. This factor of a few million is reflected in the fact that we can extract a few million times more energy from a kilogram of uranium than we can from a kilogram of coal.

In both atomic and nuclear burning, the release of energy is accompanied by a decrease in mass, according to the equation $Q = -\Delta m\, c^2$. The central difference between burning uranium and burning coal is that, in the former case, a much larger fraction of the available mass (again, by a factor of a few million) is consumed.

The different processes that can be used for atomic or nuclear burning provide different levels of power, or rates at which the energy is delivered. In the nuclear case, we can burn a kilogram of uranium explosively in a bomb or slowly in a power reactor. In the atomic case, we might consider exploding a stick of dynamite or digesting a jelly doughnut.

Table 43-1 shows how much energy can be extracted from 1 kg of matter by doing various things to it. Instead of reporting the energy directly, the table shows how long the extracted energy could operate a 100 W lightbulb. Only processes in the first three rows of the table have actually been carried out; the remaining three represent theoretical limits that may not be attainable in practice. The bottom row, the total mutual annihilation of matter and antimatter, is an ultimate energy production goal. In that process, *all* the mass energy is transferred to other forms of energy.

The comparisons of Table 43-1 are computed on a per-unit-mass basis. Kilogram for kilogram, you get several million times more energy from uranium than you do from coal or from falling water. On the other hand, there is a lot of coal in Earth's crust, and water is easily backed up behind a dam.

TABLE 43-1
Energy Released by 1 kg of Matter

Form of Matter	Process	Time[a]
Water	A 50 m waterfall	5 s
Coal	Burning	8 h
Enriched UO_2	Fission in a reactor	690 y
^{235}U	Complete fission	3×10^4 y
Hot deuterium gas	Complete fusion	3×10^4 y
Matter and antimatter	Complete annihilation	3×10^7 y

[a]This column shows the time interval for which the generated energy could power a 100 W lightbulb.

43-2 Nuclear Fission: The Basic Process

In 1932 English physicist James Chadwick discovered the neutron. A few years later Enrico Fermi in Rome found that when various elements are bombarded by neutrons, new radioactive elements are produced. Fermi had predicted that the neutron, being uncharged, would be a useful nuclear projectile; unlike the proton or the alpha particle, it experiences no repulsive Coulomb force when it nears a nuclear surface. Even *thermal neutrons,* which are slowly moving neutrons in thermal equilibrium with the surrounding matter at room temperature, with a kinetic energy of only about 0.04 eV, are useful projectiles in nuclear studies.

In the late 1930s physicist Lise Meitner and chemists Otto Hahn and Fritz Strassmann, working in Berlin and following up on the work of Fermi and his co-workers, bombarded solutions of uranium salts with such thermal neutrons. They found that after the bombardment a number of new radionuclides were present. In 1939 one of the radionuclides produced in this way was positively identified, by repeated tests, as barium. But how, Hahn and Strassmann wondered, could this middle-mass element ($Z = 56$) be produced by bombarding uranium ($Z = 92$) with neutrons?

The puzzle was solved within a few weeks by Meitner and her nephew Otto Frisch. They suggested the mechanism by which a uranium nucleus, having absorbed a thermal neutron, could split, with the release of energy, into two roughly equal parts, one of which might well be barium. Frisch named the process **fission.**

Meitner's central role in the discovery of fission was not fully recognized until recent historical research brought it to light. She did not share in the Nobel Prize in chemistry that was awarded to Otto Hahn in 1944. However, both Hahn and Meitner have been honored by having elements named after them: hahnium (symbol Ha, $Z = 105$) and meitnerium (symbol Mt, $Z = 109$).

A Closer Look at Fission

Figure 43-1 shows the distribution by mass number of the fragments produced when ^{235}U is bombarded with thermal neutrons. The most probable mass numbers, occurring in about 7% of the events, are centered around $A \approx 95$ and $A \approx 140$. Curiously, the "double-peaked" character of Fig. 43-1 is still not understood.

In a typical ^{235}U fission event, a ^{235}U nucleus absorbs a thermal neutron, producing a compound nucleus ^{236}U in a highly excited state. It is *this* nucleus that actually undergoes fission, splitting into two fragments. These fragments—between them—rapidly emit two neutrons, leaving (in a typical case) ^{140}Xe ($Z = 54$) and ^{94}Sr ($Z = 38$) as fission fragments. Thus, the stepwise fission equa-

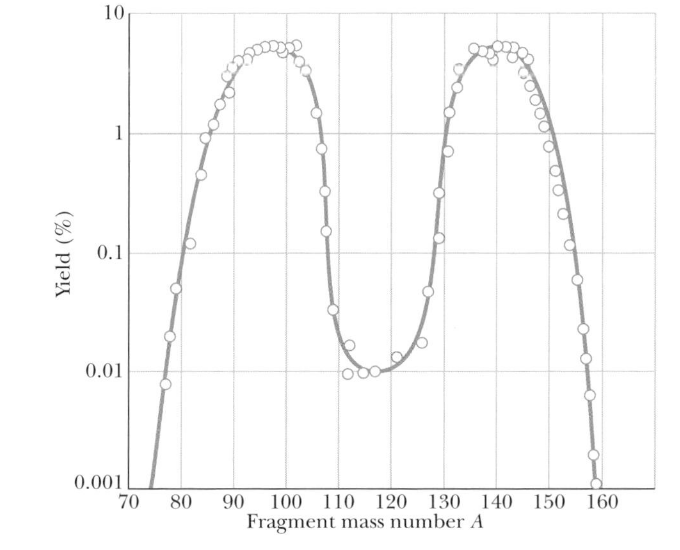

Fig. 43-1 The distribution by mass number of the fragments that are found when many fission events of ^{235}U are examined. Note that the vertical scale is logarithmic.

tion for this event is

$$^{235}\text{U} + \text{n} \rightarrow {}^{236}\text{U} \rightarrow {}^{140}\text{Xe} + {}^{94}\text{Sr} + 2\text{n}. \tag{43-1}$$

Note that during the formation and fission of the compound nucleus, there is conservation of the number of protons and of the number of neutrons involved in the process (and thus conservation of their total number and the net charge).

In Eq. 43-1, the fragments ^{140}Xe and ^{94}Sr are both highly unstable, undergoing beta decay (with the conversion of a neutron to a proton and the emission of an electron and a neutrino) until each reaches a stable end product. For xenon, the decay chain is

$$^{140}\text{Xe} \rightarrow {}^{140}\text{Cs} \rightarrow {}^{140}\text{Ba} \rightarrow {}^{140}\text{La} \rightarrow {}^{140}\text{Ce} \tag{43-2}$$

$T_{1/2}$	14 s	64 s	13 d	40 h	Stable
Z	54	55	56	57	58

For strontium, it is

$$^{94}\text{Sr} \rightarrow {}^{94}\text{Y} \rightarrow {}^{94}\text{Zr} \tag{43-3}$$

$T_{1/2}$	75 s	19 min	Stable
Z	38	39	40

As we should expect from Section 42-6, the mass numbers (140 and 94) of the fragments remain unchanged during these beta-decay processes and the atomic numbers (initially 54 and 38) increase by unity at each step.

Inspection of the stability band on the nuclidic chart of Fig. 42-4 shows why the fission fragments are unstable. The nuclide ^{236}U, which is the fissioning nucleus in the reaction of Eq. 43-1, has 92 protons and 236 − 92, or 143, neutrons, for a neutron/proton ratio of about 1.6. The primary fragments formed immediately after the fission reaction have about this same neutron/proton ratio. However, stable nuclides in the middle-mass region have smaller neutron/proton ratios, in the range of 1.3 to 1.4. The primary fragments are thus *neutron rich* (they have too many neutrons) and will eject a few neutrons, two in the case of the reaction of Eq. 43-1. The fragments that remain are still too neutron rich to be stable. Beta decay offers a mechanism for getting rid of the excess neutrons—namely, by changing them into protons within the nucleus.

We can estimate the energy released by the fission of a high-mass nuclide by examining the total binding energy per nucleon ΔE_{ben} before and after the fission. The idea is that fission can occur because the total mass energy will decrease; that is, ΔE_{ben} will *increase* so that the products of the fission are *more* tightly bound. Thus, the energy Q released by the fission is

$$Q = \begin{pmatrix}\text{total final}\\ \text{binding energy}\end{pmatrix} - \begin{pmatrix}\text{initial}\\ \text{binding energy}\end{pmatrix}. \tag{43-4}$$

For our estimate, let us assume that fission transforms an initial high-mass nucleus to two middle-mass nuclei with the same number of nucleons. Then we have

$$Q = \begin{pmatrix}\text{final}\\ \Delta E_{\text{ben}}\end{pmatrix}\begin{pmatrix}\text{final number}\\ \text{of nucleons}\end{pmatrix} - \begin{pmatrix}\text{initial}\\ \Delta E_{\text{ben}}\end{pmatrix}\begin{pmatrix}\text{initial number}\\ \text{of nucleons}\end{pmatrix}. \tag{43-5}$$

From Fig. 42-6, we see that for a high-mass nuclide ($A \approx 240$), the binding energy per nucleon is about 7.6 MeV/nucleon. For middle-mass nuclides ($A \approx 120$), it is about 8.5 MeV/nucleon. Thus, the energy released by fission of a high-mass nuclide to two middle-mass nuclides is

$$\begin{aligned} Q = {} & \left(8.5\,\frac{\text{MeV}}{\text{nucleon}}\right)(2\text{ nuclei})\left(120\,\frac{\text{nucleons}}{\text{nucleus}}\right) \\ & - \left(7.6\,\frac{\text{MeV}}{\text{nucleon}}\right)(240\text{ nucleons}) \approx 200\text{ MeV}. \end{aligned} \tag{43-6}$$

CHECKPOINT 1 A generic fission event is

$$^{235}\text{U} + \text{n} \rightarrow X + Y + 2\text{n}.$$

Which of the following pairs *cannot* represent X and Y: (a) ^{141}Xe and ^{93}Sr; (b) ^{139}Cs and ^{95}Rb; (c) ^{156}Nd and ^{79}Ge; (d) ^{121}In and ^{113}Ru?

Sample Problem 43-1

Find the disintegration energy Q for the fission event of Eq. 43-1, taking into account the decay of the fission fragments as displayed in Eqs. 43-2 and 43-3. Some needed atomic and particle masses are

^{235}U	235.0439 u	^{140}Ce	139.9054 u
n	1.008 66 u	^{94}Zr	93.9063 u

Solution: The **Key Ideas** here are (1) that the disintegration energy Q is the energy transferred from mass energy to kinetic energy of the decay products and (2) that $Q = -\Delta m\, c^2$, where Δm is the change in mass. Because we are to include the decay of the fission fragments, we combine Eqs. 43-1, 43-2, and 43-3 to write the overall transformation as

$$^{235}\text{U} \rightarrow {}^{140}\text{Ce} + {}^{94}\text{Zr} + \text{n}. \tag{43-7}$$

Only the single neutron appears here because the initiating neutron on the left side of Eq. 43-1 cancels one of the two neutrons on the right of that equation. The mass difference for the reaction of Eq. 43-7 is

$$\begin{aligned}\Delta m &= (139.9054\text{ u} + 93.9063\text{ u} + 1.008\,66\text{ u}) - (235.0439\text{ u})\\ &= -0.223\,54\text{ u},\end{aligned}$$

and the corresponding disintegration energy is

$$\begin{aligned}Q &= -\Delta m\, c^2 = (-0.223\,54\text{ u})(931.494\,013\text{ MeV/u})\\ &= 208\text{ MeV},\end{aligned} \qquad \text{(Answer)}$$

which is in good agreement with our estimate of Eq. 43-6.

If the fission event takes place in a bulk solid, most of this disintegration energy, which first goes into kinetic energy of the decay products, appears eventually as an increase in the internal energy of that body, revealing itself as a rise in temperature. Five or six percent or so of the disintegration energy, however, is associated with neutrinos that are emitted during the beta decay of the primary fission fragments. This energy is carried out of the system and is lost.

43-3 A Model for Nuclear Fission

Soon after the discovery of fission, Niels Bohr and John Wheeler used the collective model of the nucleus (Section 42-9), based on the analogy between a nucleus and a charged liquid drop, to explain the main nuclear features. Figure 43-2 suggests how the fission process proceeds from this point of view. When a high-mass nucleus—let us say ^{235}U—absorbs a slow (thermal) neutron, as in Fig. 43-2*a*, that neutron falls into the potential well associated with the strong forces that act in the nuclear interior. The neutron's potential energy is then transformed into internal excitation energy of the nucleus, as Fig. 43-2*b* suggests.

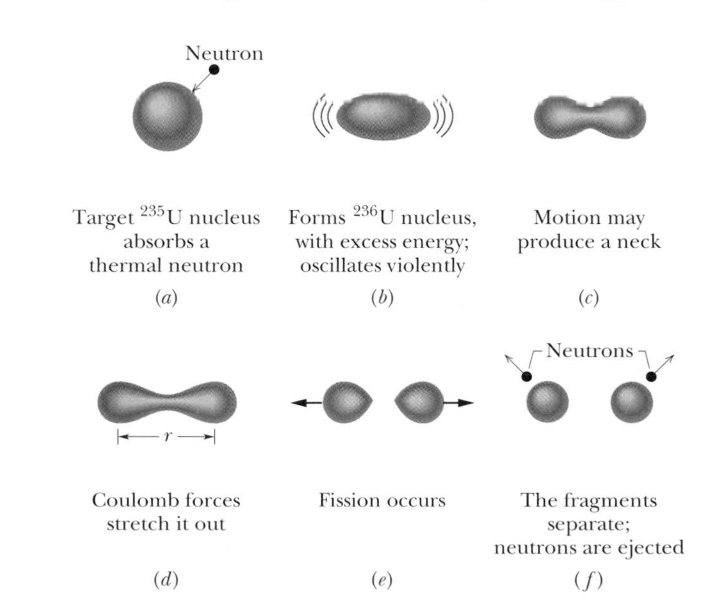
Neutron
Target 235U nucleus absorbs a thermal neutron
(a)
Forms 236U nucleus, with excess energy; oscillates violently
(b)
Motion may produce a neck
(c)
r
Coulomb forces stretch it out
(d)
Fission occurs
(e)
Neutrons
The fragments separate; neutrons are ejected
(f)

Fig. 43-2 The stages of a typical fission process, according to the collective model of Bohr and Wheeler.

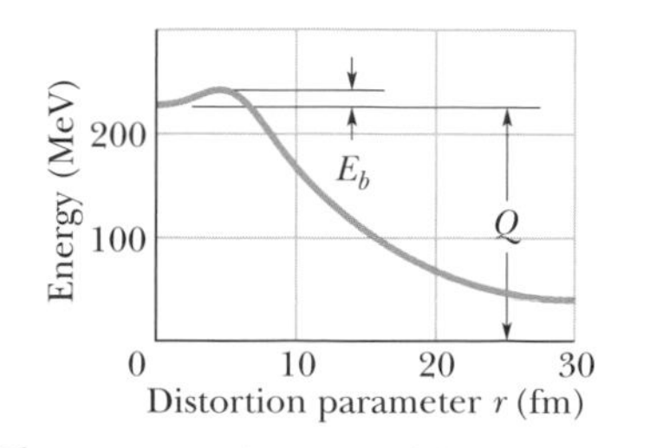

Fig. 43-3 The potential energy at various stages in the fission process, as predicted from the collective model of Bohr and Wheeler. The Q of the reaction (about 200 MeV) and the fission barrier height E_b are both indicated.

The amount of excitation energy that a slow neutron carries into a nucleus is equal to the binding energy E_n of the neutron in that nucleus, which is the change in mass energy of the neutron–nucleus system due to the neutron's capture.

Figures 43-2*c* and *d* show that the nucleus, behaving like an energetically oscillating charged liquid drop, will sooner or later develop a short "neck" and will begin to separate into two charged "globs." If the electric repulsion between these two globs forces them far enough apart to break the neck, the two fragments, each still carrying some residual excitation energy, will fly apart (Figs. 43-2*e* and *f*). Fission has occurred.

This model gave a good qualitative picture of the fission process. What remained to be seen, however, was whether it could answer a hard question: Why are some high-mass nuclides (^{235}U and ^{239}Pu, say) readily fissionable by thermal neutrons when other, equally massive nuclides (^{238}U and ^{243}Am, say) are not?

Bohr and Wheeler were able to answer this question. Figure 43-3 shows a graph of the potential energy of the fissioning nucleus at various stages, derived from their model for the fission process. This energy is plotted against the *distortion parameter r*, which is a rough measure of the extent to which the oscillating nucleus departs from a spherical shape. Figure 43-2*d* suggests how this parameter is defined just before fission occurs. When the fragments are far apart, this parameter is simply the distance between their centers.

The energy difference between the initial state ($r = 0$) and the final state ($r = \infty$) of the fissioning nucleus—that is, the disintegration energy Q—is labeled in Fig. 43-3. The central feature of that figure, however, is that the potential energy curve passes through a maximum at a certain value of r. Thus, there is a *potential barrier* of height E_b that must be surmounted (or tunneled through) before fission can occur. This reminds us of alpha decay (Fig. 42-9), which is also a process that is inhibited by a potential barrier.

We see then that fission will occur only if the absorbed neutron provides an excitation energy E_n great enough to overcome the barrier. This energy E_n need not be *quite* as great as the barrier height E_b because of the possibility of quantum-physics tunneling.

Table 43-2 shows, for four high-mass nuclides, this test of whether capture of a thermal neutron can cause fissioning. For each nuclide, the table shows both the barrier height E_b of the nucleus that is formed by the neutron capture and the excitation energy E_n due to the capture. The values of E_b are calculated from the theory of Bohr and Wheeler. The values of E_n are calculated from the change in mass energy due to the neutron capture.

For an example of the calculation of E_n, we can go to the first line in the table, which represents the neutron capture process

$$^{235}\text{U} + \text{n} \rightarrow {}^{236}\text{U}.$$

The masses involved are 235.043 922 u for ^{235}U, 1.008 664 u for the neutron, and 236.045 562 u for ^{236}U. It is easy to show that, because of the neutron capture, the mass decreases by 7.024×10^{-3} u. Thus, energy is transferred from mass energy to excitation energy E_n. Multiplying the change in mass by

TABLE 43-2
Test of the Fissionability of Four Nuclides

Target Nuclide	Nuclide Being Fissioned	E_n (MeV)	E_b (MeV)	Fission by Thermal Neutrons?
^{235}U	^{236}U	6.5	5.2	Yes
^{238}U	^{239}U	4.8	5.7	No
^{239}Pu	^{240}Pu	6.4	4.8	Yes
^{243}Am	^{244}Am	5.5	5.8	No

c^2 (= 931.494 013 MeV/u) gives us E_n = 6.5 MeV, which is listed on the first line of the table.

The first and third results in Table 43-2 are historically profound because they are the reasons the two atomic bombs used in World War II contained ^{235}U (first bomb) and ^{239}Pu (second bomb). That is, for ^{235}U and ^{239}Pu, $E_n > E_b$. This means that fission by absorption of a thermal neutron is predicted to occur for these nuclides. For the other two nuclides in Table 43-2 (^{238}U and ^{243}Am), we have $E_n < E_b$; thus, there is not enough energy from a thermal neutron for the excited nucleus to surmount the barrier or to tunnel through it effectively. Instead of fissioning, the nucleus gets rid of its excitation energy by emitting a gamma-ray photon.

The nuclides ^{238}U and ^{243}Am *can* be made to fission, however, if they absorb a substantially energetic (rather than a thermal) neutron. A ^{238}U nucleus, for example, might fission if it happens to absorb a neutron of at least 1.3 MeV in a so-called *fast fission* process ("fast" because the neutron is fast).

The two atomic bombs used in World War II depended on the ability of thermal neutrons to cause many high-mass nuclides in the cores of the bombs to fission nearly all at once, so that the fissioning would result in an explosive and devastating output of energy. The first bomb used ^{235}U because enough of it had been refined from uranium ore to make that bomb and a test bomb. (The ore consists mainly of ^{238}U, which, as we have seen, is not caused to fission by thermal neutrons.) The second bomb used ^{239}Pu, based only on theoretical calculations as summarized in Table 43-2, because not enough additional ^{235}U was available when the second bomb was ordered.

43-4 The Nuclear Reactor

For large-scale energy release due to fission, one fission event must trigger others, so that the process spreads throughout the nuclear fuel like flame through a log. The fact that more neutrons are produced in fission than are consumed raises the possibility of just such a **chain reaction,** with each neutron that is produced potentially triggering another fission. The reaction can be either rapid (as in a nuclear bomb) or controlled (as in a nuclear reactor).

Suppose that we wish to design a reactor based on the fission of ^{235}U by thermal neutrons. Natural uranium contains 0.7% of this isotope, the remaining 99.3% being ^{238}U, which is not fissionable by thermal neutrons. Let us give ourselves an edge by artificially *enriching* the uranium fuel so that it contains perhaps 3% ^{235}U. Three difficulties still stand in the way of a working reactor.

The scene 20 m from the Chernobyl reactor unit 4 (near Kiev) after it exploded in April 1986. Nearly all the volatile radionuclides inside the reactor were released into the air.

1. *The Neutron Leakage Problem.* Some of the neutrons produced by fission will leak out of the reactor and so not be part of the chain reaction. Leakage is a surface effect; its magnitude is proportional to the square of a typical reactor dimension (the surface area of a cube of edge length a is $6a^2$). Neutron production, however, occurs throughout the volume of the fuel and is thus proportional to the cube of a typical dimension (the volume of the same cube is a^3). We can make the fraction of neutrons lost by leakage as small as we wish by making the reactor core large enough, thereby reducing the surface-to-volume ratio (= $6/a$ for a cube).
2. *The Neutron Energy Problem.* The neutrons produced by fission are fast, with kinetic energies of about 2 MeV. However, fission is induced most effectively by thermal neutrons. The fast neutrons can be slowed down by mixing the uranium fuel with a substance—called a **moderator**—that has two properties: It is effective in slowing down neutrons via elastic collisions, and it does not remove neutrons from the core by absorbing them so that they do not result in fission. Most power reactors in North America use water as a moderator; the hydrogen nuclei (protons) in the water are the effective component.

A reactor vessel from a nuclear power plant is being buried (note the dirt falling from the red scoop) at the Hanford Nuclear Reservation near Richland, Washington.

We saw in Chapter 9 that if a moving particle has a head-on elastic collision with a stationary particle, the moving particle loses *all* its kinetic energy if the two particles have the same mass. Thus, protons form an effective moderator because they have approximately the same mass as the fast neutrons whose speed we wish to reduce.

3. *The Neutron Capture Problem.* As the fast (2 MeV) neutrons generated by fission are slowed down in the moderator to thermal energies (about 0.04 eV), they must pass through a critical energy interval (from 1 to 100 eV) in which they are particularly susceptible to nonfission capture by ^{238}U nuclei. Such *resonance capture,* which results in the emission of a gamma ray, removes the neutron from the fission chain. To minimize such nonfission capture, the uranium fuel and the moderator are not intimately mixed but rather are placed in different regions of the reactor volume.

   In a typical reactor, the uranium fuel is in the form of uranium oxide pellets, which are inserted end to end into long, hollow metal tubes. The liquid moderator surrounds bundles of these **fuel rods,** forming the reactor **core.** This geometric arrangement increases the probability that a fast neutron, produced in a fuel rod, will find itself in the moderator when it passes through the critical energy interval. Once the neutron has reached thermal energies, it may *still* be captured in ways that do not result in fission (called *thermal capture*). However, it is much more likely that the thermal neutron will wander back into a fuel rod and produce a fission event.

Figure 43-4 shows the neutron balance in a typical power reactor operating at constant power. Let us trace a sample of 1000 thermal neutrons through one complete cycle, or *generation,* in the reactor core. They produce 1330 neutrons by fission in the ^{235}U fuel and 40 neutrons by fast fission in ^{238}U, which gives 370 neutrons more than the original 1000, all of them fast. When the reactor is operating at a steady power level, exactly the same number of neutrons (370) is then lost by leakage from the core and by nonfission capture, leaving 1000 thermal neutrons to start the next generation. In this cycle, of course, each of the 370 neutrons produced by fission events represents a deposit of energy in the reactor core, heating up the core.

The *multiplication factor* k—an important reactor parameter—is the ratio of the number of neutrons present at the beginning of a particular generation to the number present at the beginning of the next generation. In Fig. 43-4, the multiplication factor is 1000/1000, or exactly unity. For $k = 1$, the operation of

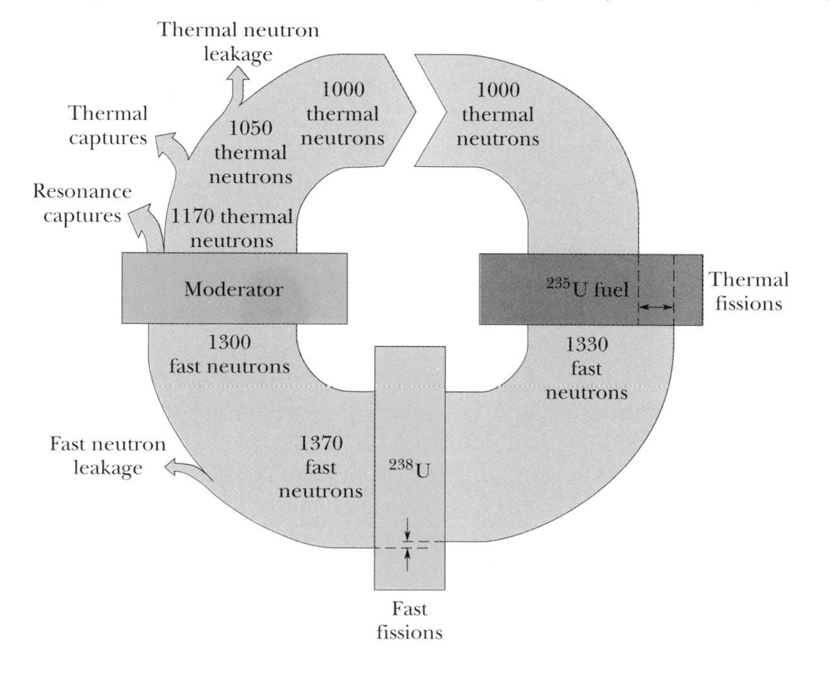

Fig. 43-4 Neutron bookkeeping in a reactor. A generation of 1000 thermal neutrons interacts with the ^{235}U fuel, the ^{238}U matrix, and the moderator. They produce 1370 neutrons by fission, but 370 of these are lost by nonfission capture or by leakage, meaning that 1000 thermal neutrons are left to form the next generation. The figure is drawn for a reactor running at a steady power level.

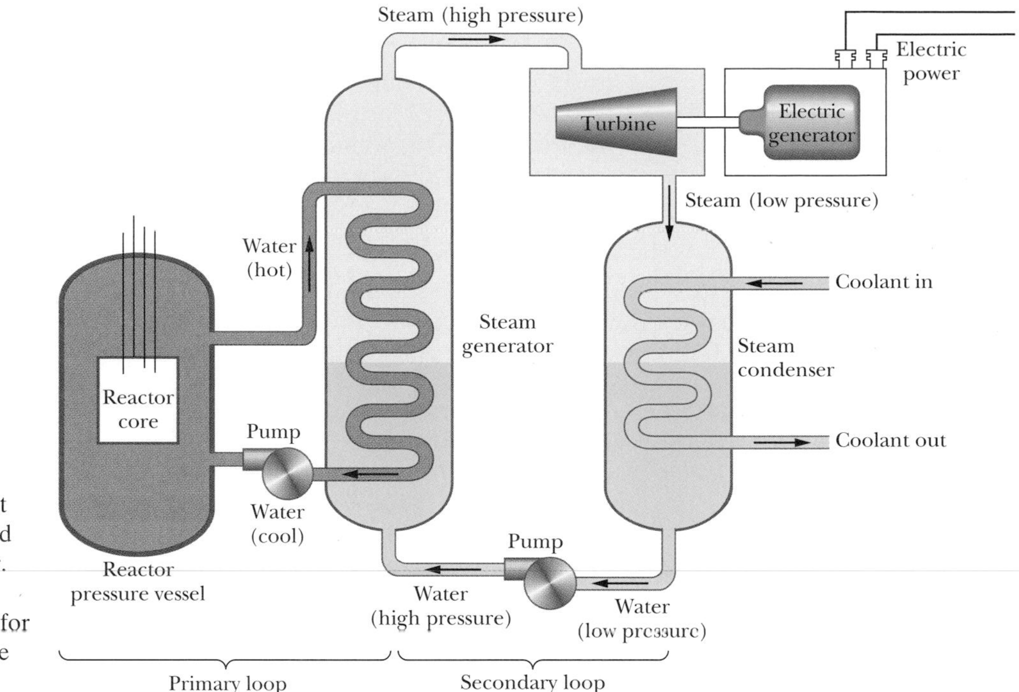

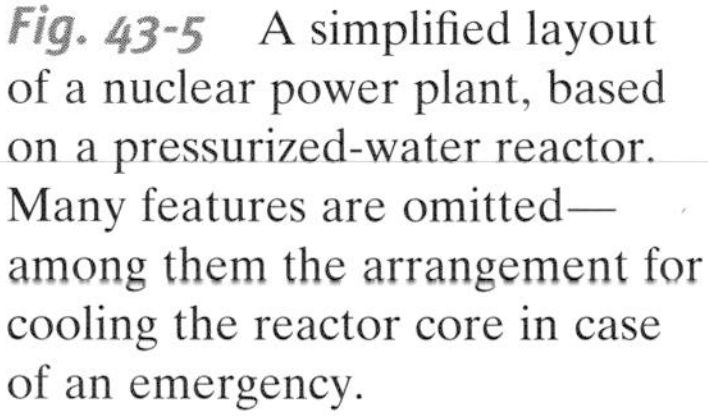

Fig. 43-5 A simplified layout of a nuclear power plant, based on a pressurized-water reactor. Many features are omitted—among them the arrangement for cooling the reactor core in case of an emergency.

the reactor is said to be exactly *critical,* which is what we wish it to be for steady-power operation. Reactors are actually designed so that they are inherently *supercritical* ($k > 1$); the multiplication factor is then adjusted to critical operation ($k = 1$) by inserting **control rods** into the reactor core. These rods, containing a material such as cadmium that absorbs neutrons readily, can be inserted farther to reduce the operating power level and withdrawn to increase the power level or to compensate for the tendency of reactors to go *subcritical* as (neutron-absorbing) fission products build up in the core during continued operation.

If you pulled out one of the control rods rapidly, how fast would the reactor power level increase? This *response time* is controlled by the fascinating circumstance that a small fraction of the neutrons generated by fission do not escape promptly from the newly formed fission fragments but are emitted from these fragments later, as the fragments decay by beta emission. Of the 370 "new" neutrons produced in Fig. 43-4, for example, perhaps 16 are delayed, being emitted from fragments following beta decays whose half-lives range from 0.2 to 55 s. These delayed neutrons are few in number, but they serve the essential purpose of slowing the reactor response time to match practical mechanical reaction times.

Figure 43-5 shows the broad outlines of an electrical power plant based on a *pressurized-water reactor* (PWR), a type in common use in North America. In such a reactor, water is used both as the moderator and as the heat transfer medium. In the *primary loop,* water is circulated through the reactor vessel and transfers energy at high temperature and pressure (possibly 600 K and 150 atm) from the hot reactor core to the steam generator, which is part of the *secondary loop.* In the steam generator, evaporation provides high-pressure steam to operate the turbine that drives the electric generator. To complete the secondary loop, low-pressure steam from the turbine is cooled and condensed to water and forced back into the steam generator by a pump. To give some idea of scale, a typical reactor vessel for a 1000 MW (electric) plant may be 12 m high and weigh 4 MN. Water flows through the primary loop at a rate of about 1 ML/min.

An unavoidable feature of reactor operation is the accumulation of radioactive wastes, including both fission products and heavy *transuranic* nuclides such as plutonium and americium. One measure of their radioactivity is the rate at which they release energy in thermal form. Figure 43-6 shows the thermal power

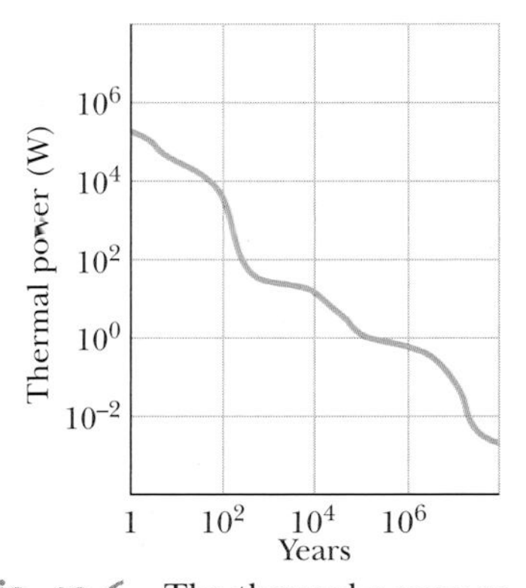

Fig. 43-6 The thermal power released by the radioactive wastes from one year's operation of a typical large nuclear power plant, shown as a function of time. The curve is the superposition of the effects of many radionuclides, with a wide variety of half-lives. Note that both scales are logarithmic.

generated by such wastes from one year's operation of a typical large nuclear plant. Note that both scales are logarithmic. Most "spent" fuel rods from power reactor operation are stored on site, immersed in water; permanent secure storage facilities for reactor waste have yet to be completed. Much weapons-derived radioactive waste accumulated during World War II and in subsequent years is also still in on-site storage. Radioactive waste and contaminated components do not just disappear—they will remain a concern long into the future.

Sample Problem 43-2

A large electric generating station is powered by a pressurized-water nuclear reactor. The thermal power produced in the reactor core is 3400 MW, and 1100 MW of electricity is generated by the station. The *fuel charge* is 8.60×10^4 kg of uranium, in the form of uranium oxide, distributed among 5.70×10^4 fuel rods. The uranium is enriched to 3.0% ^{235}U.

(a) What is the station's efficiency?

Solution: The **Key Idea** here is the definition of efficiency for this power plant or any other energy device: Efficiency is the ratio of the output power (rate at which useful energy is provided) to the input power (rate at which energy must be supplied). Here the efficiency (eff) is

$$\text{eff} = \frac{\text{useful output}}{\text{energy input}} = \frac{1100 \text{ MW (electric)}}{3400 \text{ MW (thermal)}}$$
$$= 0.32, \text{ or } 32\%. \qquad \text{(Answer)}$$

The efficiency—as for all power plants—is controlled by the second law of thermodynamics. To run this plant, energy at the rate of 3400 MW − 1100 MW, or 2300 MW, must be discharged as thermal energy to the environment.

(b) At what rate R do fission events occur in the reactor core?

Solution: The **Key Ideas** here are (1) that the fission events provide the input power P of 3400 MW ($= 3.4 \times 10^9$ J/s) and (2) from Eq. 43-6, the energy Q released by each event is about 200 MeV. Thus, for steady-state operation (P is constant), we find

$$R = \frac{P}{Q} = \left(\frac{3.4 \times 10^9 \text{ J/s}}{200 \text{ MeV/fission}}\right)\left(\frac{1 \text{ MeV}}{1.60 \times 10^{-13} \text{ J}}\right)$$
$$= 1.06 \times 10^{20} \text{ fissions/s}$$
$$\approx 1.1 \times 10^{20} \text{ fissions/s}. \qquad \text{(Answer)}$$

(c) At what rate (in kilograms per day) is the ^{235}U fuel disappearing? Assume conditions at start-up.

Solution: Here the **Key Idea** is that ^{235}U disappears due to two processes: (1) the fission process with the rate calculated in part (b) and (2) the nonfission capture of neutrons at about one-fourth that rate. Thus, the total rate at which ^{235}U disappears is

$$(1 + 0.25)(1.06 \times 10^{20} \text{ atoms/s}) = 1.33 \times 10^{20} \text{ atoms/s}.$$

We next need the mass of each ^{235}U atom. We cannot use the molar mass for uranium listed in Appendix F because that molar mass is for ^{238}U, the most common uranium isotope. Instead, we shall assume that the mass of each ^{235}U atom in atomic mass units is equal to the mass number A. Thus, the mass of each ^{235}U atom is 235 u ($= 3.90 \times 10^{-25}$ kg). Then the rate at which the ^{235}U fuel disappears is

$$\frac{dM}{dt} = (1.33 \times 10^{20} \text{ atoms/s})(3.90 \times 10^{-25} \text{ kg/atom})$$
$$= 5.19 \times 10^{-5} \text{ kg/s} \approx 4.5 \text{ kg/d}. \qquad \text{(Answer)}$$

(d) At this rate of fuel consumption, how long would the fuel supply of ^{235}U last?

Solution: At start-up, we know that the total mass of ^{235}U is 3.0% of the 8.60×10^4 kg of uranium oxide. Then the **Key Idea** here is that the time T required to consume this total mass of ^{235}U at the steady rate of 4.5 kg/d is

$$T = \frac{(0.030)(8.60 \times 10^4 \text{ kg})}{4.5 \text{ kg/d}} \approx 570 \text{ d}. \qquad \text{(Answer)}$$

In practice, the fuel rods must be replaced (usually in batches) before their ^{235}U content is entirely consumed.

(e) At what rate is mass being converted to other forms of energy by the fission of ^{235}U in the reactor core?

Solution: The **Key Idea** here is that the conversion of mass energy to other forms of energy is linked only to the fissioning that produces the input power (3400 MW) and not to the nonfission capture of neutrons (although both these processes affect the rate at which ^{235}U is consumed). Thus, from Einstein's relation $E = mc^2$, we can write

$$\frac{dm}{dt} = \frac{dE/dt}{c^2} = \frac{3.4 \times 10^9 \text{ W}}{(3.00 \times 10^8 \text{ m/s})^2}$$
$$= 3.8 \times 10^{-8} \text{ kg/s} = 3.3 \text{ g/d}. \qquad \text{(Answer)}$$

We see that the mass conversion rate is about the mass of one common coin per day, considerably less than the fuel consumption rate calculated in (c).

CHECKPOINT 2 In this sample problem, we saw that the generated power of the nuclear power plant ($P_{gen} = 1100$ MW) was less than the power discharged to the environment ($P_{dis} = 2300$ MW). Does the second law of thermodynamics (a) require that P_{gen} always be less than P_{dis}, (b) permit P_{gen} to be greater than P_{dis}, or (c) permit P_{dis} to be zero, assuming optimum reactor design?

43-5 A Natural Nuclear Reactor

On December 2, 1942, when their reactor first became operational (Fig. 43-7), Enrico Fermi and his associates had every right to assume that they had put into operation the first fission reactor that had ever existed on this planet. About 30 years later it was discovered that, if they did in fact think that, they were wrong.

Some two billion years ago, in a uranium deposit now being mined in Gabon, West Africa, a natural fission reactor apparently went into operation and ran for perhaps several hundred thousand years before shutting down. We can test whether this could actually have happened by considering two questions:

1. *Was There Enough Fuel?* The fuel for a uranium-based fission reactor must be the easily fissionable isotope ^{235}U, which, as noted earlier, constitutes only 0.72% of natural uranium. This isotopic ratio has been measured for terrestrial samples, in Moon rocks, and in meteorites; in all cases the abundance values are the same. The clue to the discovery in West Africa was that the uranium in that deposit was deficient in ^{235}U, some samples having abundances as low as 0.44%. Investigation led to the speculation that this deficit in ^{235}U could be accounted for if, at some earlier time, the ^{235}U was partially consumed by the operation of a natural fission reactor.

The serious problem remains that, with an isotopic abundance of only 0.72%, a reactor can be assembled (as Fermi and his team learned) only after thoughtful design and with scrupulous attention to detail. There seems no chance that a nuclear reactor could go critical "naturally."

However, things were different in the distant past. Both ^{235}U and ^{238}U are radioactive, with half-lives of 7.04×10^8 y and 44.7×10^8 y, respectively. Thus, the half-life of the readily fissionable ^{235}U is about 6.4 times shorter than that of ^{238}U. Because ^{235}U decays faster, there was more of it, relative to ^{238}U, in the past. Two billion years ago, in fact, this abundance was not 0.72%, as it is now, but 3.8%. This abundance happens to be just about the abundance to which natural uranium is artificially enriched to serve as fuel in modern power reactors.

With this readily fissionable fuel available, the presence of a natural reactor (provided certain other conditions are met) is less surprising. The fuel was there. Two billion years ago, incidentally, the highest order of life-form to have evolved was the blue-green alga.

2. *What Is the Evidence?* The mere depletion of ^{235}U in an ore deposit does not prove the existence of a natural fission reactor. One looks for more convincing evidence.

If there was a reactor, there must now be fission products. Of the 30 or so elements whose stable isotopes are produced in a reactor, some must still remain. Study of their isotopic abundances could provide the evidence we need.

Fig. 43-7 A painting of the first nuclear reactor, assembled during World War II on a squash court at the University of Chicago by a team headed by Enrico Fermi. This reactor, which went critical on December 2, 1942, was built of lumps of uranium embedded in blocks of graphite. It served as a prototype for later reactors whose purpose was to manufacture plutonium for the construction of nuclear weapons.

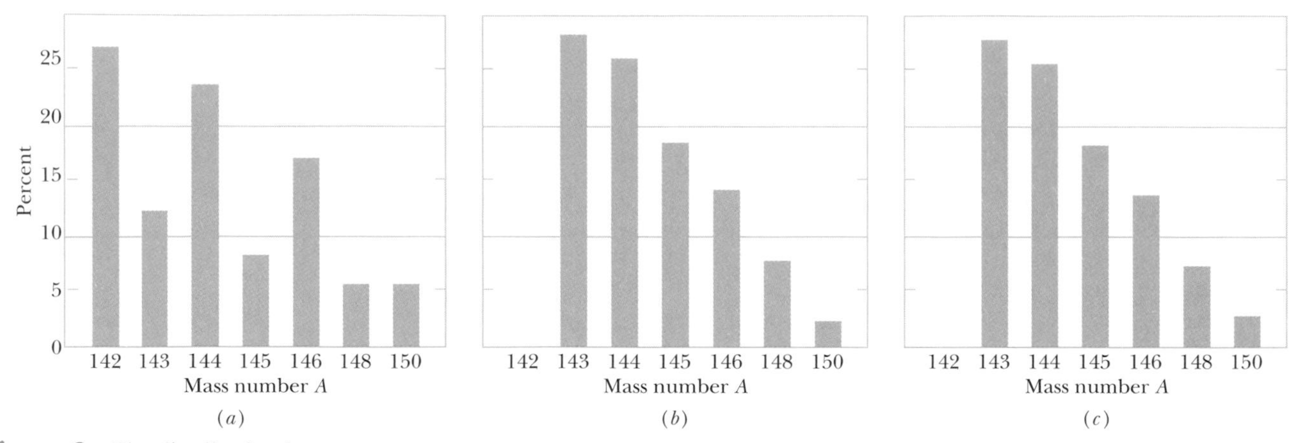

Fig. 43-8 The distribution by mass number of the isotopes of neodymium as they occur in (*a*) natural terrestrial deposits of the ores of this element and (*b*) the spent fuel of a power reactor. (*c*) The distribution (after several corrections) found for neodymium from the uranium mine in Gabon, West Africa. Note that (*b*) and (*c*) are virtually identical and are quite different from (*a*).

Of the several elements investigated, the case of neodymium is spectacularly convincing. Figure 43-8*a* shows the isotopic abundances of the seven stable neodymium isotopes as they are normally found in nature. Figure 43-8*b* shows these abundances as they appear among the ultimate stable fission products of the fission of ^{235}U. The clear differences are not surprising, considering the totally different origins of the two sets of isotopes. Note particularly that ^{142}Nd, the dominant isotope in the natural element, is absent from the fission products.

The big question is: What do the neodymium isotopes found in the uranium ore body in West Africa look like? If a natural reactor operated there, we would expect to find isotopes from *both* sources (that is, natural isotopes as well as fission-produced isotopes). Figure 43-8*c* shows the abundances after dual-source and other corrections have been made to the data. Comparison of Figs. 43-8*b* and *c* indicates that there was indeed a natural fission reactor at work.

The failure of the fission products of the West African natural reactor to migrate far from their region of production over about 2 billion years may support the idea of long-term storage of radioactive waste in suitably chosen geological environments.

Sample Problem 43-3

The ratio of ^{235}U to ^{238}U in natural uranium deposits today is 0.0072. What was this ratio 2.0×10^9 y ago? The half-lives of the two isotopes are 7.04×10^8 y and 44.7×10^8 y, respectively.

Solution: The **Key Idea** here is that the ratio N_5/N_8 of ^{235}U to ^{238}U at time $t = 0$ was not equal to 0.0072 (the ratio today, at time $t = 2.0 \times 10^9$ y), because those two isotopes have been decaying at different rates. Let $N_5(0)$ and $N_8(0)$ be the numbers of the isotopes in a sample of uranium at $t = 0$, and let $N_5(t)$ and $N_8(t)$ be the numbers of the isotopes at some later time t. Then, for each isotope, we can use Eq. 42-15, $N = N_0 e^{-\lambda t}$, to write the number at time t in terms of the number at time $t = 0$:

$$N_5(t) = N_5(0)e^{-\lambda_5 t} \quad \text{and} \quad N_8(t) = N_8(0)e^{-\lambda_8 t},$$

in which λ_5 and λ_8 are the corresponding disintegration constants. Dividing gives

$$\frac{N_5(t)}{N_8(t)} = \frac{N_5(0)}{N_8(0)} e^{-(\lambda_5 - \lambda_8)t}.$$

Because we want the ratio $N_5(0)/N_8(0)$, we rearrange this to

$$\frac{N_5(0)}{N_8(0)} = \frac{N_5(t)}{N_8(t)} e^{(\lambda_5 - \lambda_8)t}. \tag{43-8}$$

The disintegration constants are related to the half-lives by Eq. 42-18, $T_{1/2} = (\ln 2)/\lambda$, which yields

$$\lambda_5 = \frac{\ln 2}{7.04 \times 10^8 \text{ y}} = 9.85 \times 10^{-10} \text{ y}^{-1}$$

and

$$\lambda_8 = \frac{\ln 2}{44.7 \times 10^8 \text{ y}} = 1.55 \times 10^{-10} \text{ y}^{-1}.$$

The exponent in Eq. 43-8 is then

$$(\lambda_5 - \lambda_8)t = [(9.85 - 1.55) \times 10^{-10} \text{ y}^{-1}](2 \times 10^9 \text{ y}) = 1.66.$$

Equation 43-8 then yields

$$\frac{N_5(0)}{N_8(0)} = \frac{N_5(t)}{N_8(t)} e^{(\lambda_5 - \lambda_8)t} = (0.0072)(e^{1.66}) = 0.0379 \approx 3.8\%. \quad \text{(Answer)}$$

43-6 Thermonuclear Fusion: The Basic Process

The binding energy curve of Fig. 42-6 shows that energy can be released if two light nuclei combine to form a single larger nucleus, a process called nuclear **fusion.** That process is hindered by the Coulomb repulsion that acts to prevent the two positively charged particles from getting close enough to be within range of their attractive nuclear forces and thus "fusing." The height of this *Coulomb barrier* depends on the charges and the radii of the two interacting nuclei. We show in Sample Problem 43-4 that, for two protons ($Z = 1$), the barrier height is 400 keV. For more highly charged particles, of course, the barrier is correspondingly higher.

To generate useful amounts of energy, nuclear fusion must occur in bulk matter. The best hope for bringing this about is to raise the temperature of the material until the particles have enough energy—due to their thermal motions alone—to penetrate the Coulomb barrier. We call this process **thermonuclear fusion.**

In thermonuclear studies, temperatures are reported in terms of the kinetic energy K of interacting particles via the relation

$$K = kT, \tag{43-9}$$

in which K is the kinetic energy corresponding to the *most probable speed* of the interacting particles, k is the Boltzmann constant, and the temperature T is in kelvins. Thus, rather than saying, "The temperature at the center of the Sun is 1.5×10^7 K," it is more common to say, "The temperature at the center of the Sun is 1.3 keV."

Room temperature corresponds to $K \approx 0.03$ eV; a particle with only this amount of energy could not hope to overcome a barrier as high as, say, 400 keV. Even at the center of the Sun, where $kT = 1.3$ keV, the outlook for thermonuclear fusion does not seem promising at first glance. Yet we know that thermonuclear fusion not only occurs in the core of the Sun but is the dominant feature of that body and of all other stars.

The puzzle is solved when we realize two facts: (1) The energy calculated with Eq. 43-9 is that of the particles with the *most probable* speed, as defined in Section 19-7; there is a long tail of particles with much higher speeds and, correspondingly, much higher energies. (2) The barrier heights that we have calculated represent the *peaks* of the barriers. Barrier tunneling can occur at energies considerably below those peaks, as we saw with alpha decay in Section 42-5.

Figure 43-9 sums things up. The curve marked $n(K)$ in this figure is a Maxwell distribution curve for the protons in the Sun's core, drawn to correspond to the Sun's central temperature. This curve differs from the Maxwell distribution curve given in Fig. 19-7 in that here the curve is drawn in terms of energy and not of speed. Specifically, for any kinetic energy K, the expression $n(K)\,dK$ gives the probability that a proton will have a kinetic energy lying between the values K and $K + dK$. The value of kT in the core of the Sun is indicated by the vertical line in the figure; note that many of the Sun's core protons have energies greater than this value.

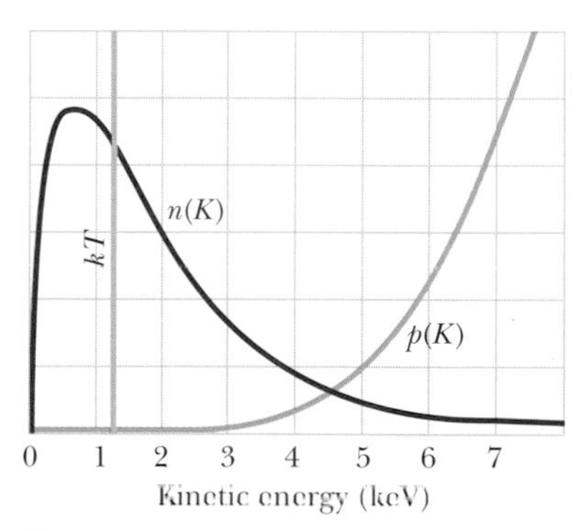

Fig. 43-9 The curve marked $n(K)$ gives the number density per unit energy for protons at the center of the Sun. The curve marked $p(K)$ gives the probability of barrier penetration (and hence fusion) for proton-proton collisions at the Sun's core temperature. The vertical line marks the value of kT at this temperature. Note that the two curves are drawn to (separate) arbitrary vertical scales.

The curve marked $p(K)$ in Fig. 43-9 is the probability of barrier penetration by two colliding protons. The two curves in Fig. 43-9 suggest that there is a particular proton energy at which proton-proton fusion events occur at a maximum rate. At energies much above this value, the barrier is transparent enough but too few protons have these energies, and so the fusion reaction cannot be sustained. At energies much below this value, plenty of protons have these energies but the Coulomb barrier is too formidable.

CHECKPOINT 3 Which of these potential fusion reactions will *not* result in the net release of energy: (a) ^{6}Li + ^{6}Li, (b) ^{4}He + ^{4}He, (c) ^{12}C + ^{12}C, (d) ^{20}Ne + ^{20}Ne, (e) ^{35}Cl + ^{35}Cl, and (f) ^{14}N + ^{35}Cl? (*Hint:* Consult the curve of Fig. 42-6.)

Sample Problem 43-4

Assume a proton is a sphere of radius $R \approx 1$ fm. Two protons are fired at each other with the same kinetic energy K.

(a) What must K be if the particles are brought to rest by their mutual Coulomb repulsion when they are just "touching" each other? We can take this value of K as a representative measure of the height of the Coulomb barrier.

Solution: The **Key Idea** here is that the mechanical energy E of the two-proton system is conserved as the protons move toward each other and momentarily stop. In particular, the initial mechanical energy E_i is equal to the mechanical energy E_f when they stop. The initial energy E_i consists only of the total kinetic energy $2K$ of the two protons. When the protons stop, energy E_f consists only of the electric potential energy U of the system, as given by Eq. 24-43 ($U = q_1q_2/4\pi\varepsilon_0 r$). Here the distance r between the protons when they stop is their center-to-center distance $2R$, and their charges q_1 and q_2 are both e. Then we can write the conservation of energy $E_i = E_f$ as

$$2K = \frac{1}{4\pi\varepsilon_0}\frac{e^2}{2R}.$$

This yields, with known values,

$$K = \frac{e^2}{16\pi\varepsilon_0 R}$$
$$= \frac{(1.60 \times 10^{-19}\ \text{C})^2}{(16\pi)(8.85 \times 10^{-12}\ \text{F/m})(1 \times 10^{-15}\ \text{m})}$$
$$= 5.75 \times 10^{-14}\ \text{J} = 360\ \text{keV} \approx 400\ \text{keV}. \qquad \text{(Answer)}$$

(b) At what temperature would a proton in a gas of protons have the average kinetic energy calculated in (a) and thus have energy equal to the height of the Coulomb barrier?

Solution: The **Key Idea** here is that if we treat the proton gas as an ideal gas, then from Eq. 19-24, the average energy of the protons is $K_{avg} = \frac{3}{2}kT$, where k is the Boltzmann constant. Solving this equation for T and using the result of (a) yield

$$T = \frac{2K_{avg}}{3k} = \frac{(2)(5.75 \times 10^{-14}\ \text{J})}{(3)(1.38 \times 10^{-23}\ \text{J/K})}$$
$$\approx 3 \times 10^9\ \text{K}. \qquad \text{(Answer)}$$

The temperature of the core of the Sun is only about 1.5×10^7 K; thus fusion in the Sun's core must involve protons whose energies are *far* above the average energy.

43-7 Thermonuclear Fusion in the Sun and Other Stars

The Sun radiates energy at the rate of 3.9×10^{26} W and has been doing so for several billion years. Where does all this energy come from? Chemical burning is ruled out; if the Sun had been made of coal and oxygen—in the right proportions for combustion—it would have lasted for only about 1000 y. Another possibility is that the Sun is slowly shrinking, under the action of its own gravitational forces. By transferring gravitational potential energy to thermal energy, the Sun might maintain its temperature and continue to radiate. Calculation shows, however, that this mechanism also fails; it produces a solar lifetime that is too short by a factor of at least 500. That leaves only thermonuclear fusion. The Sun, as you will see, burns not coal but hydrogen, and in a nuclear furnace, not an atomic or chemical one.

The fusion reaction in the Sun is a multistep process in which hydrogen is burned to form helium, hydrogen being the "fuel" and helium the "ashes." Figure 43-10 shows the **proton-proton** (p-p) **cycle** by which this occurs.

The p-p cycle starts with the collision of two protons (^{1}H + ^{1}H) to form a deuteron (^{2}H), with the simultaneous creation of a positron (e^+) and a neutrino

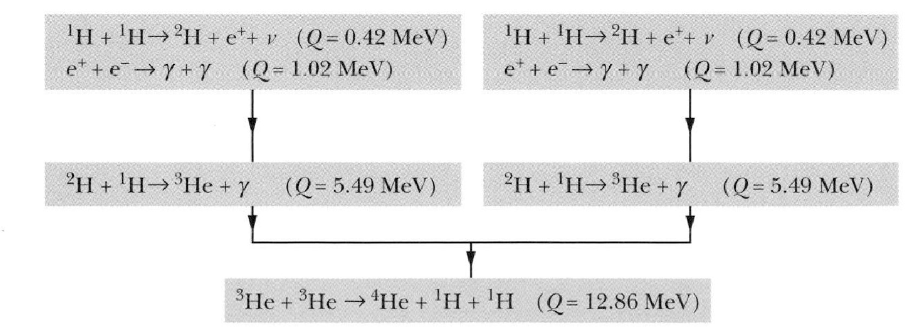

Fig. 43-10 The proton-proton mechanism that accounts for energy production in the Sun. In this process, protons fuse to form an alpha particle (^{4}He), with a net energy release of 26.7 MeV for each event.

(ν). The positron very quickly encounters a free electron (e^-) in the Sun, and both particles annihilate (see Section 21-6), their mass energy appearing as two gamma-ray photons (γ).

A pair of such events is shown in the top row of Fig. 43-10. These events are actually extremely rare. In fact, only once in about 10^{26} proton-proton collisions is a deuteron formed; in the vast majority of cases, the two protons simply rebound elastically from each other. It is the slowness of this "bottleneck" process that regulates the rate of energy production and keeps the Sun from exploding. In spite of this slowness, there are so very many protons in the huge and dense volume of the Sun's core that deuterium is produced in just this way at the rate of 10^{12} kg/s.

Once a deuteron has been produced, it quickly collides with another proton and forms a ^{3}He nucleus, as the middle row of Fig. 43-10 shows. Two such ^{3}He nuclei may eventually (within 10^5 y; there is plenty of time) find each other, forming an alpha particle (^{4}He) and two protons, as the bottom row in the figure shows.

Overall, we see from Fig. 43-10 that the p-p cycle amounts to the combination of four protons and two electrons to form an alpha particle, two neutrinos, and six gamma-ray photons. That is,

$$4\,{}^1\text{H} + 2e^- \rightarrow {}^4\text{He} + 2\nu + 6\gamma. \tag{43-10}$$

Let us now add two electrons to each side of Eq. 43-10, obtaining

$$(4\,{}^1\text{H} + 4e^-) \rightarrow ({}^4\text{He} + 2e^-) + 2\nu + 6\gamma. \tag{43-11}$$

The quantities in the two sets of parentheses then represent *atoms* (not bare nuclei) of hydrogen and of helium. That allows us to compute the energy release in the overall reaction of Eq. 43-10 (and Eq. 43-11) as

$$\begin{aligned} Q &= -\Delta m\, c^2 \\ &= -[4.002\,603\text{ u} - (4)(1.007\,825\text{ u})][931.5\text{ MeV/u}] \\ &= 26.7\text{ MeV}, \end{aligned}$$

in which 4.002 603 u is the mass of a helium atom and 1.007 825 u is the mass of a hydrogen atom. Neutrinos have a negligibly small mass, and gamma-ray photons have no mass; thus, they do not enter into the calculation of the disintegration energy.

This same value of Q follows (as it must) from adding up the Q values for the separate steps of the proton-proton cycle in Fig. 43-10. Thus,

$$\begin{aligned} Q &= (2)(0.42\text{ MeV}) + (2)(1.02\text{ MeV}) + (2)(5.49\text{ MeV}) + 12.86\text{ MeV} \\ &= 26.7\text{ MeV}. \end{aligned}$$

About 0.5 MeV of this energy is carried out of the Sun by the two neutrinos indicated in Eqs. 43-10 and 43-11; the rest (= 26.2 MeV) is deposited in the core of the Sun as thermal energy. That thermal energy is then gradually transported to the Sun's surface, where it is radiated away from the Sun as electromagnetic waves, including visible light.

The burning of hydrogen in the Sun's core is alchemy on a grand scale in the sense that one element is turned into another. The medieval alchemists, however, were more interested in changing lead into gold than in changing hydrogen into helium. In a sense, they were on the right track, except that their furnaces were not hot enough. Instead of being at a temperature of, say, 600 K, the ovens should have been at least as hot as 10^8 K.

Hydrogen burning has been going on in the Sun for about 5×10^9 y, and calculations show that there is enough hydrogen left to keep the Sun going for about the same length of time into the future. In 5 billion years, however, the Sun's core, which by that time will be largely helium, will begin to cool and the

Fig. 43-11 (*a*) The star known as Sanduleak, as it appeared until 1987. (*b*) We then began to intercept light from the star's supernova, designated SN1987a; the explosion was 100 million times brighter than our Sun and could be seen with the unaided eye. The explosion took place 155 000 light-years away and thus actually occurred 155 000 years ago.

Sun will start to collapse under its own gravity. This will raise the core temperature and cause the outer envelope to expand, turning the Sun into what is called a *red giant.*

If the core temperature increases to about 10^8 K again, energy can be produced through fusion once more—this time by burning helium to make carbon. As a star evolves further and becomes still hotter, other elements can be formed by other fusion reactions. However, elements more massive than those near the peak of the binding energy curve of Fig. 42-6 cannot be produced by further fusion processes.

Elements with mass numbers beyond the peak of that curve are thought to be formed by neutron capture during cataclysmic stellar explosions that we call *supernovas* (Fig. 43-11). In such an event the outer shell of the star is blown outward into space, where it mixes with—and becomes part of—the tenuous medium that fills the space between the stars. It is from this medium, continually enriched by debris from stellar explosions, that new stars form, by condensation under the influence of the gravitational force.

The abundance on Earth of elements heavier than hydrogen and helium suggests that our solar system has condensed out of interstellar material that contained the remnants of such explosions. Thus, all the elements around us—including those in our own bodies—were manufactured in the interiors of stars that no longer exist. As one scientist put it: "In truth, we are the children of the stars."

Sample Problem 43-5

At what rate dm/dt is hydrogen being consumed in the core of the Sun by the p-p cycle of Fig. 43-10?

Solution: The **Key Idea** is that the rate dE/dt at which energy is produced by hydrogen (proton) consumption within the Sun is equal to the rate P at which energy is radiated by the Sun:

$$P = \frac{dE}{dt}.$$

To bring the mass consumption rate dm/dt into this equation, we can rewrite it as

$$P = \frac{dE}{dt} = \frac{dE}{dm}\frac{dm}{dt} \approx \frac{\Delta E}{\Delta m}\frac{dm}{dt}, \tag{43-12}$$

where ΔE is the energy produced when protons of mass Δm are consumed. From our discussion in this section, we know that 26.2 MeV ($= 4.20 \times 10^{-12}$ J) of thermal energy is produced when four protons are consumed. That is, $\Delta E = 4.20 \times 10^{-12}$ J for a mass consumption of $\Delta m = 4(1.67 \times 10^{-27}$ kg). Substituting these data into Eq. 43-12 and using the power P of the Sun given in Appendix C, we find that

$$\frac{dm}{dt} = \frac{\Delta m}{\Delta E}P = \frac{4(1.67 \times 10^{-27}\text{ kg})}{4.20 \times 10^{-12}\text{ J}}(3.90 \times 10^{26}\text{ W})$$
$$= 6.2 \times 10^{11}\text{ kg/s}. \qquad \text{(Answer)}$$

Thus, a huge amount of hydrogen is consumed by the Sun every second. However, you need not worry too much about the Sun running out of hydrogen, because its mass of 2×10^{30} kg will keep it burning for a long, long time.

43-8 Controlled Thermonuclear Fusion

The first thermonuclear reaction on Earth occurred at Eniwetok Atoll on November 1, 1952, when the United States exploded a fusion device, generating an energy release equivalent to 10 million tons of TNT. The high temperatures and densities needed to initiate the reaction were provided by using a fission bomb as a trigger.

A sustained and controllable source of fusion power—a fusion reactor as part of, say, an electric generating plant—is considerably more difficult to achieve. That goal is nonetheless being pursued vigorously in many countries around the world, because many people look to the fusion reactor as the power source of the future, at least for the generation of electricity.

The p-p scheme displayed in Fig. 43-10 is not suitable for an Earth-bound fusion reactor because it is hopelessly slow. The process succeeds in the Sun only because of the enormous density of protons in the center of the Sun. The most attractive reactions for terrestrial use appear to be two deuteron-deuteron (d-d) reactions,

$$^2H + {^2H} \rightarrow {^3He} + n \qquad (Q = +3.27\ \text{MeV}), \tag{43-13}$$

$$^2H + {^2H} \rightarrow {^3H} + {^1H} \qquad (Q = +4.03\ \text{MeV}), \tag{43-14}$$

and the deuteron-triton (d-t) reaction*

$$^2H + {^3H} \rightarrow {^4He} + n \qquad (Q = +17.59\ \text{MeV}). \tag{43-15}$$

Deuterium, the source of deuterons for these reactions, has an isotopic abundance of only 1 part in 6700 but is available in unlimited quantities as a component of seawater. Proponents of power from the nucleus have described our ultimate power choice—after we have burned up all our fossil fuels—as either "burning rocks" (fission of uranium extracted from ores) or "burning water" (fusion of deuterium extracted from water).

There are three requirements for a successful thermonuclear reactor:

1. *A High Particle Density n.* The number density of interacting particles (the number of, say, deuterons per unit volume) must be great enough to ensure that the d-d collision rate is high enough. At the high temperatures required, the deuterium would be completely ionized, forming an electrically neutral **plasma** (ionized gas) of deuterons and electrons.
2. *A High Plasma Temperature T.* The plasma must be hot. Otherwise the colliding deuterons will not be energetic enough to penetrate the Coulomb barrier that tends to keep them apart. A plasma ion temperature of 35 keV, corresponding to 4×10^8 K, has been achieved in the laboratory. This is about 30 times higher than the Sun's central temperature.
3. *A Long Confinement Time τ.* A major problem is containing the hot plasma long enough to maintain it at a density and a temperature sufficiently high to ensure the fusion of enough of the fuel. Because it is clear that no solid container can withstand the high temperatures that are necessary, clever confining techniques are called for; we shall shortly discuss two of them.

It can be shown that, for the successful operation of a thermonuclear reactor using the d-t reaction, it is necessary to have

$$n\tau > 10^{20}\ \text{s/m}^3. \tag{43-16}$$

This condition, known as **Lawson's criterion,** tells us that we have a choice between confining a lot of particles for a short time or fewer particles for a longer

*The nucleus of the hydrogen isotope ^3H (tritium) is called the *triton.* It is a radionuclide with a half-life of 12.3 y.

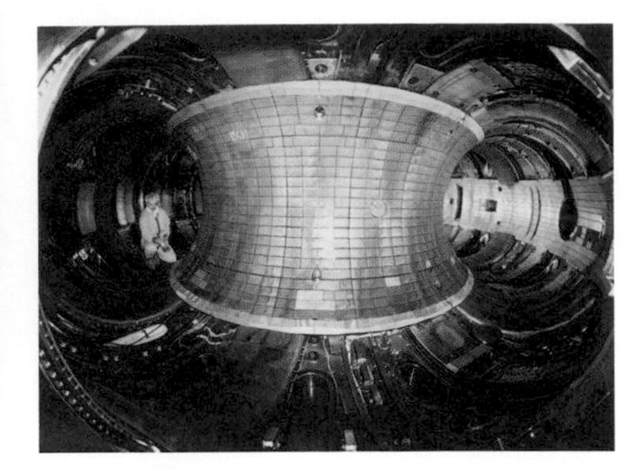

Fig. 43-12 The Tokamak Fusion Test Reactor at Princeton University.

time. Beyond meeting this criterion, it is still necessary that the plasma temperature be high enough.

Two approaches to controlled nuclear power generation are currently under study. Although neither approach has yet been successful, both are being pursued because of their promise and because of the potential importance of controlled fusion to the solving of the world's energy problems.

Magnetic Confinement

One avenue to controlled fusion is to contain the fusing material in a very strong magnetic field—hence the name **magnetic confinement.** In one version of this approach, a suitably shaped magnetic field is used to confine the hot plasma in an evacuated doughnut-shaped chamber called a **tokamak** (the name is an abbreviation consisting of parts of three Russian words). The magnetic forces acting on the charged particles that make up the hot plasma keep the plasma from touching the walls of the chamber. Figure 43-12 shows such a device at the Plasma Physics Laboratory of Princeton University.

The plasma is heated by inducing a current in it and by bombarding it with an externally accelerated beam of particles. The first goal of this approach is to achieve **breakeven,** which occurs when the Lawson criterion is met or exceeded. The ultimate goal is **ignition,** which corresponds to a self-sustaining thermonuclear reaction and a net generation of energy. As of 2004, ignition has not been achieved, either in tokamaks or in other magnetic confinement devices.

Inertial Confinement

A second approach, called **inertial confinement,** involves "zapping" a solid fuel pellet from all sides with intense laser beams, evaporating some material from the surface of the pellet. This boiled-off material causes an inward-moving shock wave that compresses the core of the pellet, increasing both its particle density and its temperature. The process is called inertial confinement because (a) the fuel is *confined* to the pellet and (b) the particles do not escape from the heated pellet during the very short zapping interval because of their *inertia* (their mass).

Fig. 43-13 The small spheres on the quarter are deuterium–tritium fuel pellets, designed to be used in a laser fusion chamber.

Laser fusion, using the inertial confinement approach, is being investigated in many laboratories in the United States and elsewhere. At the Lawrence Livermore Laboratory, for example, deuterium–tritium fuel pellets, each smaller than a grain of sand (Fig. 43-13), are to be zapped by 10 synchronized high-power laser pulses symmetrically arranged around the pellet. The laser pulses are designed to deliver, in total, some 200 kJ of energy to each fuel pellet in less than a nanosecond. This is a delivered power of about 2×10^{14} W during the pulse, which is roughly 100 times the total installed (sustained) electrical power generating capacity of the world!

In an operating thermonuclear reactor of the laser-fusion type, fuel pellets are to be exploded—like miniature hydrogen bombs—at a rate of perhaps 10 to 100 per second. The feasibility of laser fusion as the basis of a thermonuclear power reactor has not been demonstrated as of 2004, but research is continuing at a vigorous pace.

Sample Problem 43-6

Suppose a fuel pellet in a laser fusion device contains equal numbers of deuterium and tritium atoms (and no other material). The density $d = 200\ \text{kg/m}^3$ of the pellet is increased by a factor of 10^3 by the action of the laser pulses.

(a) How many particles per unit volume (both deuterons and tritons) does the pellet contain in its compressed state? The molar mass M_d of deuterium atoms is 2.0×10^{-3} kg/mol, and the molar mass M_t of tritium atoms is 3.0×10^{-3} kg/mol.

Solution: The **Key Idea** here is that, for a system consisting of only one type of particle, we can write the (mass) density of the system in terms of the particle masses and number density:

$$\begin{pmatrix}\text{density,}\\ \text{kg/m}^3\end{pmatrix} = \begin{pmatrix}\text{number density,}\\ \text{m}^{-3}\end{pmatrix}\begin{pmatrix}\text{particle mass,}\\ \text{kg}\end{pmatrix}. \quad (43\text{-}17)$$

Let n be the total number of particles per unit volume in the compressed pellet. Then the number of deuterium atoms per unit volume is $n/2$, and the number of tritium atoms per unit volume is also $n/2$.

Next, we can extend Eq. 43-17 to the system consisting of the two types of particles by writing the density d^* of the compressed pellet as the sum of the individual densities:

$$d^* = \frac{n}{2}m_d + \frac{n}{2}m_t, \quad (43\text{-}18)$$

where m_d and m_t are the masses of a deuterium atom and a tritium atom, respectively. We can replace those masses with the given molar masses by substituting

$$m_d = \frac{M_d}{N_A} \quad \text{and} \quad m_t = \frac{M_t}{N_A},$$

where N_A is Avogadro's number. After making those replacements and substituting $1000d$ for the compressed density d^*, we solve Eq. 43-18 for n to obtain

$$n = \frac{2000dN_A}{M_d + M_t},$$

which gives us

$$n = \frac{(2000)(200\ \text{kg/m}^3)(6.02 \times 10^{23}\ \text{mol}^{-1})}{2.0 \times 10^{-3}\ \text{kg/mol} + 3.0 \times 10^{-3}\ \text{kg/mol}}$$
$$= 4.8 \times 10^{31}\ \text{m}^{-3}. \quad \text{(Answer)}$$

(b) According to Lawson's criterion, how long must the pellet maintain this particle density if breakeven operation is to take place?

Solution: The **Key Idea** here is that, if breakeven operation is to occur, the compressed density must be maintained for a time period τ given by Eq. 43-16 ($n\tau > 10^{20}$ s/m^3). Thus, we have

$$\tau > \frac{10^{20}\ \text{s/m}^3}{4.8 \times 10^{31}\ \text{m}^{-3}} \approx 10^{-12}\ \text{s}. \quad \text{(Answer)}$$

(The plasma temperature must also be suitably high.)

Review & Summary

Energy from the Nucleus Nuclear processes are about a million times more effective, per unit mass, than chemical processes in transforming mass into other forms of energy.

Nuclear Fission Equation 43-1 shows a **fission** of ^{236}U induced by thermal neutrons bombarding ^{235}U. Equations 43-2 and 43-3 show the beta-decay chains of the primary fragments. The energy released in such a fission event is $Q \approx 200$ MeV.

Fission can be understood in terms of the collective model, in which a nucleus is likened to a charged liquid drop carrying a certain excitation energy. A potential barrier must be tunneled through if fission is to occur. Fissionability depends on the relationship between the barrier height E_b and the excitation energy E_n.

The neutrons released during fission make possible a fission **chain reaction.** Figure 43-4 shows the neutron balance for one cycle of a typical reactor. Figure 43-5 suggests the layout of a complete nuclear power plant.

Nuclear Fusion The release of energy by the **fusion** of two light nuclei is inhibited by their mutual Coulomb barrier. Fusion can occur in bulk matter only if the temperature is high enough (that is, if the particle energy is high enough) for appreciable barrier tunneling to occur.

The Sun's energy arises mainly from the thermonuclear burning of hydrogen to form helium by the **proton-proton cycle** outlined in Fig. 43-10. Elements up to $A \approx 56$ (the peak of the binding energy curve) can be built up by other fusion processes once the hydrogen fuel supply of a star has been exhausted.

Controlled Fusion Controlled **thermonuclear fusion** for energy generation has not yet been achieved. The d-d and d-t reactions are the most promising mechanisms. A successful fusion reactor must satisfy **Lawson's criterion,**

$$n\tau > 10^{20}\ \text{s/m}^3, \quad (43\text{-}16)$$

and must have a suitably high plasma temperature T.

In a **tokamak** the plasma is confined by a magnetic field. In **laser fusion** inertial confinement is used.

Questions

1 According to Fig. 43-1, does thermal-neutron fission of ^{235}U into two equally massive fragments occur in about one event in 10 000, 1000, 100, or 10?

2 In Table 43-1, does the relation $Q = -\Delta m\, c^2$ apply to all the processes, all the processes except the waterfall, the fission processes only, or the fission and fusion processes only?

3 For the fission reaction

$$^{235}\text{U} + \text{n} \rightarrow \text{X} + \text{Y} + 2\text{n},$$

rank the following possibilities for X (or Y), most likely first: ^{152}Nd, ^{140}I, ^{128}In, ^{115}Pd, ^{105}Mo. (*Hint:* See Fig. 43-1.)

4 Do the initial fragments formed by fission have more protons than neutrons, more neutrons than protons, or about the same number of each?

5 Suppose a ^{238}U nucleus "swallows" a neutron and then decays not by fission but by beta decay, emitting an electron and a neutrino. Which nuclide remains after this decay: ^{239}Pu, ^{238}Np, ^{239}Np, or ^{238}Pa?

6 Pick the most likely member of each pair to be one of the initial fragments formed by a fission event: (a) ^{93}Sr or ^{93}Ru, (b) ^{140}Gd or ^{140}I, (c) ^{155}Nd or ^{155}Lu. (*Hint:* See Fig. 42-4 and the periodic table.)

7 A nuclear reactor core should have the smallest possible surface-to-volume ratio. Order the following solids according to their surface-to-volume ratios, greatest first: (a) a cube of edge a, (b) a sphere of radius a, (c) a right circular cone of height a and base radius a, (d) a right cylinder of radius a and height a. (*Hint:* The area of the *curved* surface of the right circular cone is $\sqrt{2}\pi a^2$, and the volume of the cone is $\pi a^3/3$.)

8 A nuclear reactor is operating at a certain power level, with its multiplication factor k adjusted to unity. If the control rods are used to reduce the power output of the reactor to 25% of its former value, is the multiplication factor now a little less than unity, substantially less than unity, or still equal to unity?

9 Which of these elements is *not* "cooked up" by thermonuclear fusion processes in stellar interiors: carbon, silicon, chromium, bromine?

10 Figure 43-6 shows how the heat generated by the nuclear waste from one year's operation of a large nuclear power plant decays over time. By what approximate factor has this thermal energy output decreased at the end of 100 years: 20, 200, 2000, or more than 2000?

11 Lawson's criterion for the d-t reaction (Eq. 43-16) is $n\tau > 10^{20}$ s/m^3. For the d-d reaction, do you expect the number on the right-hand side to be the same, smaller, or larger?

12 About 2% of the energy generated in the Sun's core by the p-p reaction is carried out of the Sun by neutrinos. Is the energy associated with this neutrino flux equal to, greater than, or less than the energy radiated from the Sun's surface as electromagnetic radiation?

Problems

SSM Solution is in the Student Solutions Manual.
WWW Solution is at http://www.wiley.com/college/halliday
• – ••• Number of dots indicates level of problem difficulty.

sec. 43-2 Nuclear Fission: The Basic Process

•1 At what rate must ^{235}U nuclei undergo fission by neutron bombardment to generate energy at the rate of 1.0 W? Assume that Q = 200 MeV.

•2 (a)–(d) Complete the following table, which refers to the generalized fission reaction $^{235}\text{U} + \text{n} \rightarrow \text{X} + \text{Y} + b\text{n}$.

X	Y	b
^{140}Xe	(a)	1
^{139}I	(b)	2
(c)	^{100}Zr	2
^{141}Cs	^{92}Rb	(d)

•3 (a) How many atoms are contained in 1.0 kg of pure ^{235}U? (b) How much energy, in joules, is released by the complete fissioning of 1.0 kg of ^{235}U? Assume Q = 200 MeV. (c) For how long would this energy light a 100 W lamp? SSM

•4 The fission properties of the plutonium isotope ^{239}Pu are very similar to those of ^{235}U. The average energy released per fission is 180 MeV. How much energy, in MeV, is released if all the atoms in 1.00 kg of pure ^{239}Pu undergo fission?

•5 Calculate the disintegration energy Q for the fission of ^{52}Cr into two equal fragments. The masses you will need are 51.940 51 u for ^{52}Cr and 25.982 59 u for ^{26}Mg.

•6 (a) Calculate the disintegration energy Q for the fission of ^{98}Mo into two equal parts. The masses you will need are 97.905 41 u for ^{98}Mo and 48.950 02 u for ^{49}Sc. (b) If Q turns out to be positive, discuss why this process does not occur spontaneously.

•7 The isotope ^{235}U decays by alpha emission with a half-life of 7.0×10^8 y. It also decays (rarely) by spontaneous fission, and if the alpha decay did not occur, its half-life due to spontaneous fission alone would be 3.0×10^{17} y. (a) At what rate do spontaneous fission decays occur in 1.0 g of ^{235}U? (b) How many ^{235}U alpha-decay events are there for every spontaneous fission event?

•8 Calculate the energy released in the fission reaction

$$^{235}\text{U} + \text{n} \rightarrow {}^{141}\text{Cs} + {}^{93}\text{Rb} + 2\text{n}.$$

Needed atomic and particle masses are

^{235}U	235.043 92 u	^{93}Rb	92.921 57 u
^{141}Cs	140.919 63 u	n	1.008 66 u

••9 In a particular fission event in which ^{235}U is fissioned by slow neutrons, no neutron is emitted and one of the primary fission fragments is ^{83}Ge. (a) What is the other fragment? The disintegration energy is Q = 170 MeV. How much of this energy goes to (b) the ^{83}Ge fragment and (c) the other fragment? Just after the fission, what is the speed of (d) the ^{83}Ge fragment and (e) the other fragment? SSM WWW

••10 A ^{236}U nucleus undergoes fission and breaks into two middle-mass fragments, ^{140}Xe and ^{96}Sr. (a) By what percentage does the surface area of the fission products differ from that of the original ^{236}U nucleus? (b) By what percentage does the volume change? (c) By what percentage does the electric potential energy change? The electric potential energy of a uniformly charged sphere of radius r and charge Q is given by

$$U = \frac{3}{5}\left(\frac{Q^2}{4\pi\varepsilon_0 r}\right).$$

••11 Assume that immediately after the fission of ^{236}U according to Eq. 43-1, the resulting ^{140}Xe and ^{94}Sr nuclei are just touching at their surfaces. (a) Assuming the nuclei to be spherical, calculate the electric potential energy associated with the repulsion between the two fragments. (*Hint:* Use Eq. 42-3 to calculate the radii of the fragments.) (b) Compare this energy with the energy released in a typical fission event. SSM

••12 Consider the fission of ^{238}U by fast neutrons. In one fission event, no neutrons are emitted and the final stable end products, after the beta decay of the primary fission fragments, are ^{140}Ce and ^{99}Ru. (a) What is the total of the beta-decay events in the two beta-decay chains? (b) Calculate Q for this fission process. The relevant atomic and particle masses are

^{238}U	238.050 79 u	^{140}Ce	139.905 43 u
n	1.008 66 u	^{99}Ru	98.905 94 u

sec. 43-4 The Nuclear Reactor

•13 A 200 MW fission reactor consumes half its fuel in 3.00 y. How much ^{235}U did it contain initially? Assume that all the energy generated arises from the fission of ^{235}U and that this nuclide is consumed only by the fission process.

•14 The nuclide ^{238}Np requires 4.2 MeV for fission. To remove a neutron from this nuclide requires an energy expenditure of 5.0 MeV. Is ^{237}Np fissionable by thermal neutrons?

••15 The thermal energy generated when radiation from radionuclides is absorbed in matter can serve as the basis for a small power source for use in satellites, remote weather stations, and other isolated locations. Such radionuclides are manufactured in abundance in nuclear reactors and may be separated chemically from the spent fuel. One suitable radionuclide is ^{238}Pu ($T_{1/2} = 87.7$ y), which is an alpha emitter with $Q = 5.50$ MeV. At what rate is thermal energy generated in 1.00 kg of this material? SSM

••16 In an atomic bomb, energy release is due to the uncontrolled fission of plutonium ^{239}Pu (or ^{235}U). The bomb's rating is the magnitude of the released energy, specified in terms of the mass of TNT required to produce the same energy release. One megaton of TNT releases 2.6×10^{28} MeV of energy. (a) Calculate the rating, in tons of TNT, of an atomic bomb containing 95.0 kg of ^{239}Pu, of which 2.5 kg actually undergoes fission. (See Problem 4.) (b) Why is the other 92.5 kg of ^{239}Pu needed if it does not fission?

••17 (See Problem 15.) Among the many fission products that may be extracted chemically from the spent fuel of a nuclear reactor is ^{90}Sr ($T_{1/2} = 29$ y). This isotope is produced in typical large reactors at the rate of about 18 kg/y. By its radioactivity, the isotope generates thermal energy at the rate of 0.93 W/g. (a) Calculate the effective disintegration energy Q_{eff} associated with the decay of a ^{90}Sr nucleus. (This energy Q_{eff} includes contributions from the decay of the ^{90}Sr daughter products in its decay chain but not from neutrinos, which escape totally from the sample.) (b) It is desired to construct a power source generating 150 W (electric) to use in operating electronic equipment in an underwater acoustic beacon. If the power source is based on the thermal energy generated by ^{90}Sr and if the efficiency of the thermal–electric conversion process is 5.0%, how much ^{90}Sr is needed?

••18 The neutron generation time t_{gen} in a reactor is the average time needed for a fast neutron emitted in one fission event to be slowed to thermal energies by the moderator and then initiate another fission event. Suppose the power output of a reactor at time $t = 0$ is P_0. Show that the power output a time t later is $P(t)$, where $P(t) = P_0 k^{t/t_{gen}}$ and k is the multiplication factor. For constant power output, $k = 1$.

••19 A 66 kiloton atomic bomb is fueled with pure ^{235}U (Fig. 43-14), 4.0% of which actually undergoes fission. (a) What is the mass of the uranium in the bomb? (It is not 66 kilotons—that is the amount of released energy specified in terms of the mass of TNT required to produce the same amount of energy.) (b) How many primary fission fragments are produced? (c) How many fission neutrons generated are released to the environment? (On average, each fission produces 2.5 neutrons.)

Fig. 43-14 Problem 19. A "button" of ^{235}U ready to be recast and machined for a warhead.

••20 A reactor operates at 400 MW with a neutron generation time (see Problem 18) of 30.0 ms. If its power increases for 5.00 min with a multiplication factor of 1.0003, what is the power output at the end of the 5.00 min?

••21 (a) A neutron of mass m_n and kinetic energy K makes a head-on elastic collision with a stationary atom of mass m. Show that the fractional kinetic energy loss of the neutron is given by

$$\frac{\Delta K}{K} = \frac{4 m_n m}{(m + m_n)^2}.$$

Find $\Delta K/K$ for each of the following acting as the stationary atom: (b) hydrogen, (c) deuterium, (d) carbon, and (e) lead. (f) If $K = 1.00$ MeV initially, how many such head-on collisions would it take to reduce the neutron's kinetic energy to a thermal value (0.025 eV) if the stationary atoms it collides with are deuterium, a commonly used moderator? (In actual moderators, most collisions are not head-on.) SSM

••22 The neutron generation time t_{gen} (see Problem 18) in a particular reactor is 1.0 ms. If the reactor is operating at a power level of 500 MW, about how many free neutrons are present in the reactor at any moment?

••23 The neutron generation time (see Problem 18) of a particular reactor is 1.3 ms. The reactor is generating energy at the rate of 1200.0 MW. To perform certain maintenance checks, the power level must temporarily be reduced to 350.00 MW. It is desired that the transition to the reduced power level take 2.6000 s. To what (constant) value should the multiplication factor be set to effect the transition in the desired time? SSM WWW

sec. 43-5 A Natural Nuclear Reactor

•**24** The natural fission reactor discussed in Section 43-5 is estimated to have generated 15 gigawatt-years of energy during its lifetime. (a) If the reactor lasted for 200 000 y, at what average power level did it operate? (b) How many kilograms of ^{235}U did it consume during its lifetime?

•**25** How long ago was the ratio ^{235}U/^{238}U in natural uranium deposits equal to 0.15?

••**26** Some uranium samples from the natural reactor site described in Section 43-5 were found to be slightly *enriched* in ^{235}U, rather than depleted. Account for this in terms of neutron absorption by the abundant isotope ^{238}U and the subsequent beta and alpha decay of its products.

••**27** The uranium ore mined today contains only 0.72% of fissionable ^{235}U, too little to make reactor fuel for thermal-neutron fission. For this reason, the mined ore must be enriched with ^{235}U. Both ^{235}U ($T_{1/2} = 7.0 \times 10^8$ y) and ^{238}U ($T_{1/2} = 4.5 \times 10^9$ y) are radioactive. How far back in time would natural uranium ore have been a practical reactor fuel, with a ^{235}U/^{238}U ratio of 3.0%? SSM

sec. 43-6 Thermonuclear Fusion: The Basic Process

•**28** Verify that the fusion of 1.0 kg of deuterium by the reaction

$$^2\text{H} + ^2\text{H} \rightarrow ^3\text{He} + \text{n} \qquad (Q = +3.27 \text{ MeV})$$

could keep a 100 W lamp burning for 3×10^4 y.

•**29** Calculate the height of the Coulomb barrier for the head-on collision of two deuterons. Take the effective radius of a deuteron to be 2.1 fm. SSM

•**30** For overcoming the Coulomb barrier for fusion, methods other than heating the fusible material have been suggested. For example, if you were to use two particle accelerators to accelerate two beams of deuterons directly toward each other so as to collide head-on, (a) what voltage would each accelerator require in order for the colliding deuterons to overcome the Coulomb barrier? (b) Why do you suppose this method is not presently used?

••**31** Calculate the Coulomb barrier height for two ^{7}Li nuclei that are fired at each other with the same initial kinetic energy K. (*Hint:* Use Eq. 42-3 to calculate the radii of the nuclei.)

••**32** In Fig. 43-9, the equation for $n(K)$, the number density per unit energy for particles, is

$$n(K) = 1.13n \frac{K^{1/2}}{(kT)^{3/2}} e^{-K/kT},$$

where n is the total particle number density. At the center of the Sun, the temperature is 1.50×10^7 K and the average proton energy K_{avg} is 1.94 keV. Find the ratio of the proton number density at 5.00 keV to the number density at the average proton energy.

sec. 43-7 Thermonuclear Fusion in the Sun and Other Stars

•**33** Show that the energy released when three alpha particles fuse to form ^{12}C is 7.27 MeV. The atomic mass of ^{4}He is 4.0026 u, and that of ^{12}C is 12.0000 u.

•**34** We have seen that Q for the overall proton-proton fusion cycle is 26.7 MeV. How can you relate this number to the Q values for the reactions that make up this cycle, as displayed in Fig. 43-10?

•**35** The Sun has a mass of 2.0×10^{30} kg and radiates energy at the rate of 3.9×10^{26} W. (a) At what rate does the Sun transfer its mass to other forms of energy? (b) What fraction of its original mass has the Sun lost in this way since it began to burn hydrogen, about 4.5×10^9 y ago?

••**36** Verify the three Q values reported for the reactions given in Fig. 43-10. The needed atomic and particle masses are

^{1}H	1.007 825 u	^{4}He	4.002 603 u
^{2}H	2.014 102 u	$e^{\pm}$	0.000 548 6 u
^{3}He	3.016 029 u		

(*Hint:* Distinguish carefully between atomic and nuclear masses, and take the positrons properly into account.)

••**37** (a) Calculate the rate at which the Sun generates neutrinos. Assume that energy production is entirely by the proton-proton fusion cycle. (b) At what rate do solar neutrinos reach Earth?

••**38** Calculate and compare the energy released by (a) the fusion of 1.0 kg of hydrogen deep within the Sun and (b) the fission of 1.0 kg of ^{235}U in a fission reactor.

••**39** Coal burns according to the reaction $C + O_2 \rightarrow CO_2$. The heat of combustion is 3.3×10^7 J/kg of atomic carbon consumed. (a) Express this in terms of energy per carbon atom. (b) Express it in terms of energy per kilogram of the initial reactants, carbon and oxygen. (c) Suppose that the Sun (mass $= 2.0 \times 10^{30}$ kg) were made of carbon and oxygen in combustible proportions and that it continued to radiate energy at its present rate of 3.9×10^{26} W. How long would the Sun last? SSM WWW

••**40** In certain stars the *carbon cycle* is more effective than the proton-proton cycle in generating energy. This carbon cycle is

$$
\begin{aligned}
^{12}\text{C} + {^1\text{H}} &\rightarrow {^{13}\text{N}} + \gamma, & Q_1 &= 1.95 \text{ MeV},\\
^{13}\text{N} &\rightarrow {^{13}\text{C}} + e^+ + \nu, & Q_2 &= 1.19,\\
^{13}\text{C} + {^1\text{H}} &\rightarrow {^{14}\text{N}} + \gamma, & Q_3 &= 7.55,\\
^{14}\text{N} + {^1\text{H}} &\rightarrow {^{15}\text{O}} + \gamma, & Q_4 &= 7.30,\\
^{15}\text{O} &\rightarrow {^{15}\text{N}} + e^+ + \nu, & Q_5 &= 1.73,\\
^{15}\text{N} + {^1\text{H}} &\rightarrow {^{12}\text{C}} + {^4\text{He}}, & Q_6 &= 4.97.
\end{aligned}
$$

(a) Show that this cycle is exactly equivalent in its overall effects to the proton-proton cycle of Fig. 43-10. (b) Verify that the two cycles, as expected, have the same Q value.

••**41** A star converts all its hydrogen to helium, achieving a 100% helium composition. Next it converts the helium to carbon via the triple-alpha process,

$$^4\text{He} + {^4\text{He}} + {^4\text{He}} \rightarrow {^{12}\text{C}} + 7.27 \text{ MeV}.$$

The mass of the star is 4.6×10^{32} kg, and it generates energy at the rate of 5.3×10^{30} W. How long will it take to convert all the helium to carbon at this rate?

••**42** Assume that the core of the Sun has one-eighth of the Sun's mass and is compressed within a sphere whose radius is one-fourth of the solar radius. Assume further that the composition of the core is 35% hydrogen by mass and that essentially all the Sun's energy is generated there. If the Sun continues to burn hydrogen at the rate calculated in Sample Problem 43-5, how long will it be before the hydrogen is entirely consumed? The Sun's mass is 2.0×10^{30} kg.

••43 Figure 43-15 shows an early proposal for a hydrogen bomb. The fusion fuel is deuterium, ^2H. The high temperature and particle density needed for fusion are provided by an atomic bomb "trigger" that involves a ^{235}U or ^{239}Pu fission fuel arranged to impress an imploding, compressive shock wave on the deuterium. The fusion reaction is

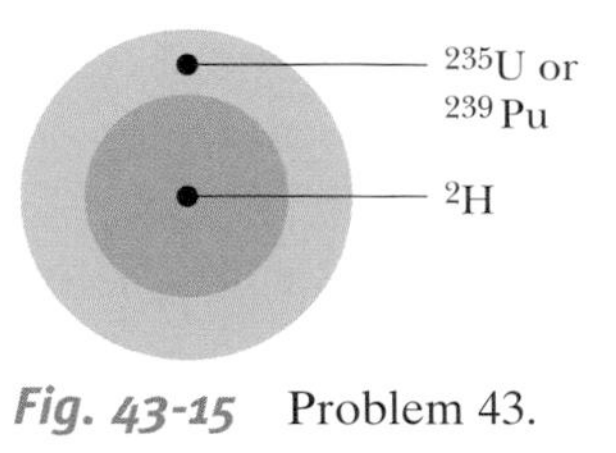

Fig. 43-15 Problem 43.

$$5\ ^2H \rightarrow ^3He + ^4He + ^1H + 2n.$$

(a) Calculate Q for the fusion reaction. For needed atomic masses, see Problem 36. (b) Calculate the rating (see Problem 16) of the fusion part of the bomb if it contains 500 kg of deuterium, 30.0% of which undergoes fusion.

sec. 43-8 Controlled Thermonuclear Fusion

•44 Verify the Q values reported in Eqs. 43-13, 43-14, and 43-15. The needed masses are

^1H	1.007 825 u	^4He	4.002 603 u
^2H	2.014 102 u	n	1.008 664 u
^3H	3.016 049 u		

••45 Roughly 0.0150% of the mass of ordinary water is due to "heavy water," in which one of the two hydrogens in an H_2O molecule is replaced with deuterium, ^2H. How much average fusion power could be obtained if we "burned" all the ^2H in 1.00 liter of water in 1.00 day by somehow causing the deuterium to fuse via the reaction $^2H + ^2H \rightarrow ^3He + n$? SSM

Additional Problems

46 The effective Q for the proton-proton cycle of Fig. 43-10 is 26.2 MeV. (a) Express this as energy per kilogram of hydrogen consumed. (b) The power of the Sun is 3.9×10^{26} W. If its energy derives from the proton-proton cycle, at what rate is it losing hydrogen? (c) At what rate is it losing mass? (d) Account for the difference in the results for (b) and (c). (e) The mass of the Sun is 2.0×10^{30} kg. If it loses mass at the constant rate calculated in (c), how long will it take to lose 0.10% of its mass?

47 At the center of the Sun, the density is 1.5×10^5 kg/m^3 and the composition is essentially 35% hydrogen by mass and 65% helium by mass. (a) What is the density of protons there? (b) What is the ratio of this density to the density of particles in an ideal gas at standard temperature (0°C) and pressure (1.01×10^5 Pa)?

48 In the deuteron-triton fusion reaction of Eq. 43-15, what is the kinetic energy of (a) the alpha particle and (b) the neutron? Neglect the relatively small kinetic energies of the two combining particles.

49 Many fear that nuclear power reactor technology will increase the likelihood of nuclear war because reactors can be used not only to produce electrical energy but also, as a byproduct through neutron capture with inexpensive ^{238}U, to make ^{239}Pu, which is a "fuel" for nuclear bombs. What simple series of reactions involving neutron capture and beta decay would yield this plutonium isotope?

50 Expressions for the Maxwell speed distribution for molecules in a gas are given in Chapter 19. (a) Show that the *most probable energy* is given by

$$K_p = \tfrac{1}{2}kT.$$

Verify this result with the energy distribution curve of Fig. 43-9, for which $T = 1.5 \times 10^7$ K. (b) Show that the *most probable speed* is given by

$$v_p = \sqrt{\frac{2kT}{m}}.$$

Find its value for protons at $T = 1.5 \times 10^7$ K. (c) Show that the *energy corresponding to the most probable speed* (which is not the same as the most probable energy) is

$$K_{v,p} = kT.$$

Locate this quantity on the curve of Fig. 43-9.

51 Repeat Problem 13 taking into account nonfission neutron capture by the ^{235}U.

52 Verify that, as reported in Table 43-1, fissioning of the ^{235}U in 1.0 kg of UO_2 (enriched so that ^{235}U is 3.0% of the total uranium) could keep a 100 W lamp burning for 690 y.

53 Verify that, as stated in Section 43-2, neutrons in equilibrium with matter at room temperature, 300 K, have an average kinetic energy of about 0.04 eV.

54 From information given in the text, determine the approximate height of the Coulomb barrier for (a) the alpha decay of ^{238}U and (b) the fission of ^{235}U by thermal neutrons.

44 Quarks, Leptons, and the Big Bang

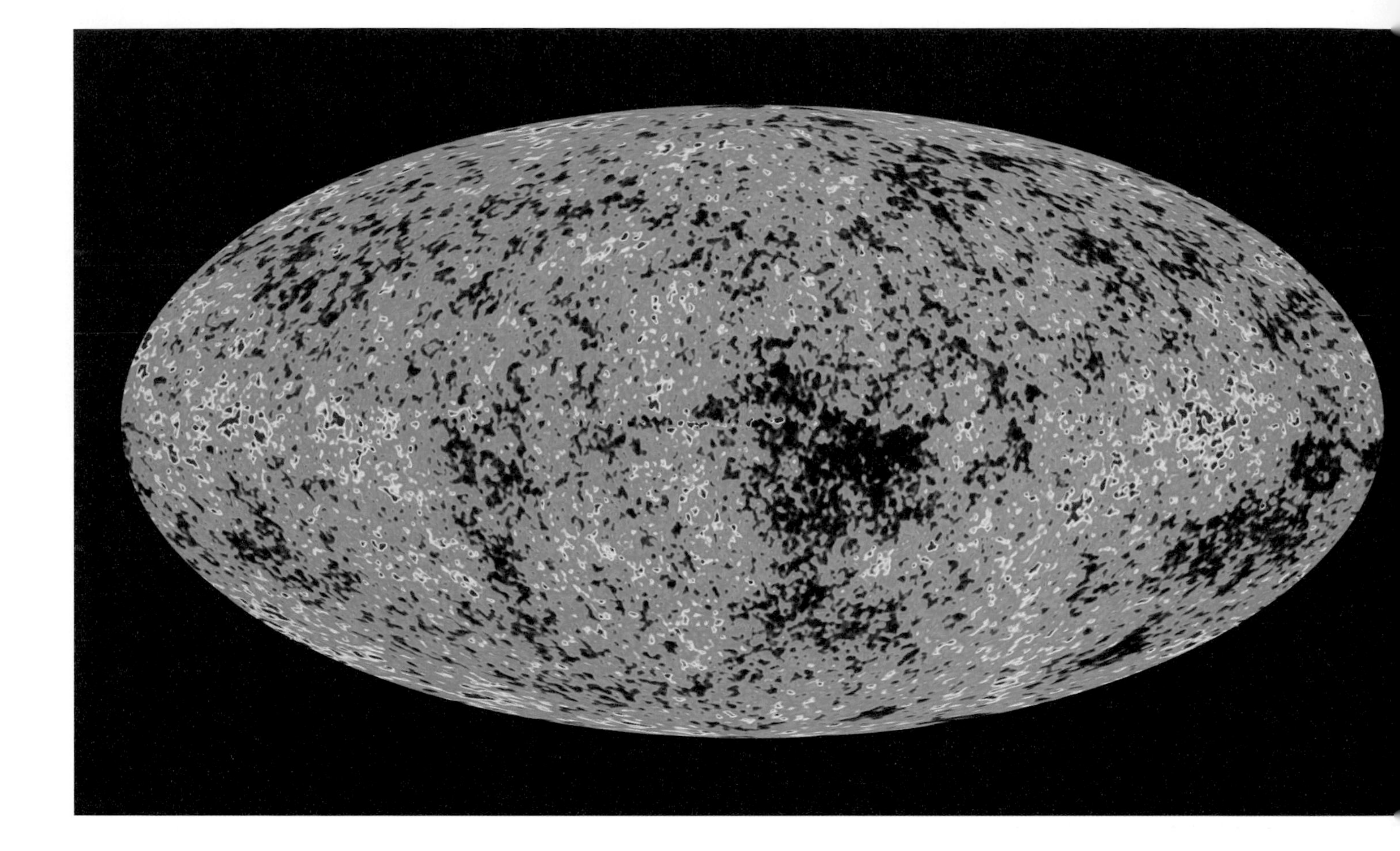

This color-coded image is effectively a photograph of the universe when it was only 379 000 y old, which was about 13.7 × 10^9 y ago. This is what you would have seen then as you looked away in all directions (the view has been condensed to this oval). Patches of light from collections of atoms stretch across the "sky," but galaxies, stars, and planets have not yet formed.

How can such a photograph of the early universe be taken?

The answer is in this chapter.

44-1 What Is Physics?

Physicists often refer to the theories of relativity and quantum physics as "modern physics," to distinguish them from the theories of Newtonian mechanics and Maxwellian electromagnetism, which are lumped together as "classical physics." As the years go by, the word "modern" seems less and less appropriate for theories whose foundations were laid down in the opening years of the 20th century. Nevertheless, the label hangs on.

In this closing chapter we consider two lines of investigation that are truly "modern" but at the same time have the most ancient of roots. They center around two deceptively simple questions:

What is the universe made of?

How did the universe come to be the way it is?

Progress in answering these questions has been rapid in the last few decades.

Many new insights are based on experiments carried out with large particle accelerators. However, as they bang particles together at higher and higher energies using larger and larger accelerators, physicists come to realize that no conceivable Earth-bound accelerator can generate particles with energies great enough to test the ultimate theories of physics. There has been only one source of particles with these energies, and that was the universe itself within the first millisecond of its existence.

In this chapter you will encounter a host of new terms and a veritable flood of particles with names that you should not try to remember. If you are temporarily bewildered, you are sharing the bewilderment of the physicists who lived through these developments and who at times saw nothing but increasing complexity with little hope of understanding. If you stick with it, however, you will come to share the excitement physicists felt as marvelous new accelerators poured out new results, as the theorists put forth ideas each more daring than the last, and as clarity finally sprang from obscurity.

The main message of this book is that, although humans have learned a lot about the physics of the world, there are still grand mysteries left.

44-2 Particles, Particles, Particles

In the 1930s, there were many scientists who thought that the problem of the ultimate structure of matter was well on the way to being solved. The atom could be understood in terms of only three particles—the electron, the proton, and the neutron. Quantum physics accounted well for the structure of the atom and for radioactive alpha decay. The neutrino had been postulated and, although not yet observed, had been incorporated by Enrico Fermi into a successful theory of beta decay. There was hope that quantum theory applied to protons and neutrons would soon account for the structure of the nucleus. What else was there?

The euphoria did not last. The end of that same decade saw the beginning of a period of discovery of new particles that continues to this day. The new particles have names and symbols such as *muon* (μ), *pion* (π), *kaon* (K), and *sigma* (Σ). All the new particles are unstable; that is, they spontaneously transform into other types of particles according to the same functions of time that apply to unstable nuclei. Thus, if N_0 particles of any one type are present in a sample at time $t = 0$, the number N of those particles present at some later time t is given by Eq. 42-15,

$$N = N_0 e^{-\lambda t}, \tag{44-1}$$

the rate of decay R, from an initial value of R_0, is given by Eq. 42-16,

$$R = R_0 e^{-\lambda t}, \tag{44-2}$$

(a) (b)

(a) The detector in Brookhaven National Laboratory's relativistic heavy ion collider (RHIC) is a gigantic and complex engineering marvel that is designed to detect particles at extreme energies. (b) Large as it is, the detector is still smaller than the collider itself, which lies under the circle shown at the top of the photograph. The collider is a circular accelerator with a circumference of 3.9 km.

and the half-life $T_{1/2}$, decay constant λ, and mean life τ are related by Eq. 42-18,

$$T_{1/2} = \frac{\ln 2}{\lambda} = \tau \ln 2. \tag{44-3}$$

The half-lives of the new particles range from about 10^{-6} s to 10^{-23} s. Indeed, some of the particles last so briefly that they cannot be detected directly but can only be inferred from indirect evidence.

These new particles are commonly produced in head-on collisions between protons or electrons accelerated to high energies in accelerators at places like Brookhaven National Laboratory (on Long Island, New York), Fermilab (near Chicago), CERN (near Geneva, Switzerland), SLAC (at Stanford University in California), and DESY (near Hamburg, Germany). They are discovered with particle detectors that have grown in sophistication until they rival the size and complexity of entire accelerators of only a few decades ago.

Today there are several hundred known particles. Naming them has strained the resources of the Greek alphabet, and most are known only by an assigned number in a periodically issued compilation. To make sense of this array of particles, we look for simple physical criteria by which we can place the particles in categories. The result is known as the **Standard Model** of particles. Although this model is continuously challenged by theorists, it remains our best scheme of understanding all the particles discovered to date.

To explore the Standard Model, we make the following three rough cuts among the known particles: fermion or boson, hadron or lepton, particle or antiparticle? Let's now look at the categories one by one.

Fermion or Boson?

All particles have an intrinsic angular momentum called **spin,** as we discussed for electrons, protons, and neutrons in Section 32-7. Generalizing the notation of that section, we can write the component of spin $\vec{S}$ in any direction (assume the component to be along a z axis) as

$$S_z = m_s\hbar \qquad \text{for } m_s = s, s-1, \ldots, -s, \tag{44-4}$$

in which $\hbar$ is $h/2\pi$, m_s is the *spin magnetic quantum number,* and s is the *spin quantum number.* This last can have either positive half-integer values ($\frac{1}{2}, \frac{3}{2}, \ldots$) or nonnegative integer values (0, 1, 2, . . .). For example, an electron has the value $s = \frac{1}{2}$. Hence the spin of an electron (measured along any direction, such as the z direction) can have the values

$$S_z = \tfrac{1}{2}\hbar \qquad \text{(spin up)}$$

or $$S_z = -\tfrac{1}{2}\hbar \quad \text{(spin down).}$$

Confusingly, the term *spin* is used in two ways: It properly means a particle's intrinsic angular momentum $\vec{S}$, but it is often used loosely to mean the particle's spin quantum number s. In the latter case, for example, an electron is said to be a spin-$\frac{1}{2}$ particle.

Particles with half-integer spin quantum numbers (like electrons) are called **fermions,** after Fermi, who (simultaneously with Paul Dirac) discovered the statistical laws that govern their behavior. Like electrons, protons and neutrons also have $s = \frac{1}{2}$ and are fermions.

Particles with zero or integer spin quantum numbers are called **bosons,** after Indian physicist Satyendra Nath Bose, who (simultaneously with Albert Einstein) discovered the governing statistical laws for *those* particles. Photons, which have $s = 1$, are bosons; you will soon meet other particles in this class.

This may seem a trivial way to classify particles, but it is very important for this reason:

Fermions obey the Pauli exclusion principle, which asserts that only a single particle can be assigned to a given quantum state. Bosons *do not* obey this principle. Any number of bosons can occupy a given quantum state.

We saw how important the Pauli exclusion principle is when we "built up" the atoms by assigning (spin-$\frac{1}{2}$) electrons to individual quantum states. Using that principle led to a full accounting of the structure and properties of atoms of different types and of solids such as metals and semiconductors.

Because bosons do *not* obey the Pauli principle, those particles tend to pile up in the quantum state of lowest energy. In 1995 a group in Boulder, Colorado, succeeded in producing a condensate of about 2000 rubidium-87 atoms—they are bosons—in a single quantum state of approximately zero energy.

For this to happen, the rubidium has to be a vapor with a temperature so low and a density so great that the de Broglie wavelengths of the individual atoms are greater than the average separation between the atoms. When this condition is met, the wave functions of the individual atoms overlap and the entire assembly becomes a single quantum system (one big atom) called a *Bose–Einstein condensate.* Figure 44-1 shows that, as the temperature of the rubidium vapor is

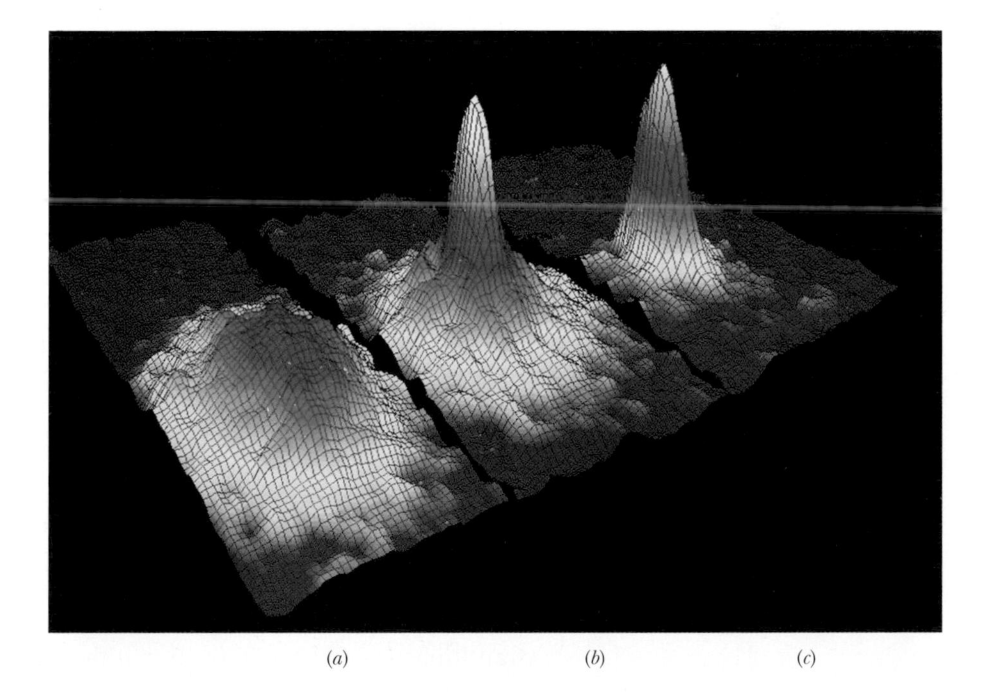

Fig. 44-1 Three plots of the particle speed distribution in a vapor of rubidium-87 atoms. The temperature of the vapor is successively reduced from plot (*a*) to plot (*c*). Plot (*c*) shows a sharp peak centered around zero speed; that is, all the atoms are in the same quantum state. The achievement of such a Bose–Einstein condensate, often called the Holy Grail of atomic physics, was finally recorded in 1995.

lowered to about 1.70×10^{-7} K, the atoms do indeed "collapse" into a single sharply defined state corresponding to approximately zero speed.

Hadron or Lepton?

We can also classify particles in terms of the four fundamental forces that act on them. The *gravitational force* acts on *all* particles, but its effects at the level of subatomic particles are so weak that we need not consider that force (at least not in today's research). The *electromagnetic force* acts on all *electrically charged* particles; its effects are well known, and we can take them into account when we need to; we largely ignore this force in this chapter.

We are left with the *strong force,* which is the force that binds nucleons together, and the *weak force,* which is involved in beta decay and similar processes. The weak force acts on all particles, the strong force only on some.

We can, then, roughly classify particles on the basis of whether the strong force acts on them. Particles on which the strong force acts are called **hadrons.** Particles on which the strong force does *not* act, leaving the weak force as the dominant force, are called **leptons.** Protons, neutrons, and pions are hadrons; electrons and neutrinos are leptons.

We can make a further distinction among the hadrons because some of them are bosons (we call them **mesons**); the pion is an example. The other hadrons are fermions (we call them **baryons**); the proton is an example.

Particle or Antiparticle?

In 1928 Dirac predicted that the electron e^- should have a positively charged counterpart of the same mass and spin. The counterpart, the *positron* e^+, was discovered in cosmic radiation in 1932 by Carl Anderson. Physicists then gradually realized that *every* particle has a corresponding **antiparticle.** The members of such pairs have the same mass and spin but opposite signs of electric charge (if they are charged) and opposite signs of quantum numbers that we have not yet discussed.

At first, *particle* was used to refer to the common particles such as electrons, protons, and neutrons, and *antiparticle* referred to their rarely detected counterparts. Later, for the less common particles, the assignment of *particle* and *antiparticle* was made so as to be consistent with certain conservation laws that we shall discuss later in this chapter. (Confusingly, both particles and antiparticles are sometimes called particles when no distinction is needed.) We often, but not always, represent an antiparticle by putting a bar over the symbol for the particle. Thus, p is the symbol for the proton, and $\bar{p}$ (pronounced "p bar") is the symbol for the antiproton.

When a particle meets its antiparticle, the two can *annihilate* each other. That is, the particle and antiparticle disappear and their combined energies reappear in other forms. For an electron annihilating with a positron, this energy reappears as two gamma-ray photons:

$$e^- + e^+ \rightarrow \gamma + \gamma. \tag{44-5}$$

If the electron and positron are stationary when they annihilate, their total energy is their total mass energy, and that energy is then shared equally by the two photons. To conserve momentum and because photons cannot be stationary, the photons fly off in opposite directions.

Large numbers of antihydrogen atoms (each with an antiproton and positron instead of a proton and electron in a hydrogen atom) are now being manufactured and studied at CERN. The Standard Model predicts that a transition in an antihydrogen atom (say, between the first excited state and the ground state) is identical to the same transition in a hydrogen atom. Thus, any difference in the transitions would clearly signal that the Standard Model is erroneous; no difference has yet been spotted.

An assembly of antiparticles, such as an antihydrogen atom, is often called *antimatter* to distinguish it from an assembly of common particles (*matter*). (The terms can easily be confusing when the word "matter" is used to describe anything that has mass.) We can speculate that future scientists and engineers may construct objects of antimatter. However, no evidence suggests that nature has already done this on an astronomical scale because all stars and galaxies appear to consist largely of matter and not antimatter. This is a perplexing observation because it means that when the universe began, some feature biased the conditions toward matter and away from antimatter. (For example, electrons are common but positrons are not.) This bias is still not well understood.

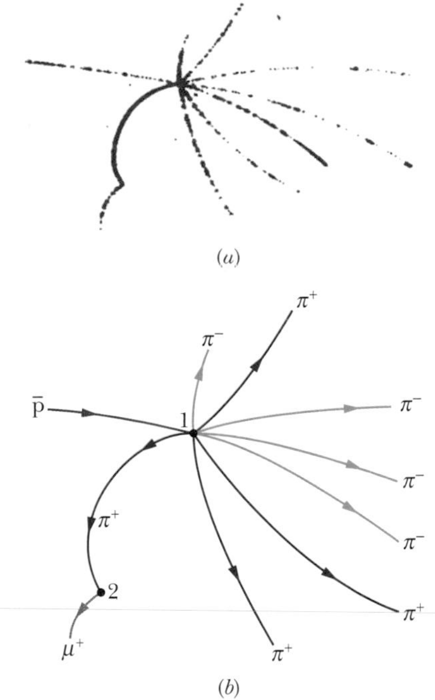

Fig. 44-2 (*a*) A bubble-chamber photograph of a series of events initiated by an antiproton that enters the chamber from the left. (*b*) The tracks redrawn and labeled for clarity. The dots at points 1 and 2 indicate the sites of specific secondary events that are described in the text. The tracks are curved because a magnetic field present in the chamber exerts a deflecting force on each moving charged particle.

44-3 An Interlude

Before pressing on with the task of classifying the particles, let us step aside for a moment and capture some of the spirit of particle research by analyzing a typical particle event—namely, that shown in the bubble-chamber photograph of Fig. 44-2*a*.

The tracks in this figure consist of bubbles formed along the paths of electrically charged particles as they move through a chamber filled with liquid hydrogen. We can identify the particle that makes a particular track by—among other means—measuring the relative spacing between the bubbles. The chamber lies in a uniform magnetic field that deflects the tracks of positively charged particles counterclockwise and the tracks of negatively charged particles clockwise. By measuring the radius of curvature of a track, we can calculate the momentum of the particle that made it. Table 44-1 shows some properties of the particles and antiparticles that participated in the event of Fig. 44-2*a*, including those that did not make tracks. Following common practice, we express the masses of the particles listed in Table 44-1—and in all other tables in this chapter—in the unit MeV/c^2. The reason for this notation is that the rest energy of a particle is needed more often than its mass. Thus, the mass of a proton is shown in Table 44-1 to be 938.3 MeV/c^2. To find the proton's rest energy, multiply this mass by c^2 to obtain 938.3 MeV.

The general tools used for the analysis of photographs like Fig. 44-2*a* are the laws of conservation of energy, linear momentum, angular momentum, and electric charge, along with other conservation laws that we have not yet discussed. Figure 44-2*a* is actually one of a stereo pair of photographs so that, in practice, these analyses are carried out in three dimensions.

The event of Fig. 44-2*a* is triggered by an energetic antiproton (p̄) that, generated in an accelerator at the Lawrence Berkeley Laboratory, enters the chamber from the left. There are three separate subevents; one occurs at point 1 in Fig. 44-2*b*, the second occurs at point 2, and the third occurs out of the frame of the figure. Let's examine each:

1. *Proton-Antiproton Annihilation.* At point 1 in Fig. 44-2*b*, the initiating antiproton (blue track) slams into a proton of the liquid hydrogen in the cham-

TABLE 44-1 The Particles or Antiparticles Involved in the Event of Fig. 44-2

Particle	Symbol	Charge q	Mass (MeV/c^2)	Spin Quantum Number s	Identity	Mean Life (s)	Antiparticle
Neutrino	ν	0	$\approx 1 \times 10^{-7}$	$\frac{1}{2}$	Lepton	Stable	$\bar{\nu}$
Electron	e^-	-1	0.511	$\frac{1}{2}$	Lepton	Stable	e^+
Muon	μ^-	-1	105.7	$\frac{1}{2}$	Lepton	2.2×10^{-6}	μ^+
Pion	π^+	$+1$	139.6	0	Meson	2.6×10^{-8}	π^-
Proton	p	$+1$	938.3	$\frac{1}{2}$	Baryon	Stable	p̄

ber, and the result is mutual annihilation. We can tell that annihilation occurred while the incoming antiproton was in flight because most of the particles generated in the encounter move in the forward direction—that is, toward the right in Fig. 44-2. From the principle of conservation of linear momentum, the incoming antiproton must have had a forward momentum when it underwent annihilation.

The total energy involved in the collision of the antiproton and the proton is the sum of the antiproton's kinetic energy and the two (identical) rest energies of those two particles (2×938.3 MeV, or 1876.6 MeV). This is enough energy to create a number of lighter particles and give them kinetic energy. In this case, the annihilation produces four positive pions (red tracks in Fig. 44-2*b*) and four negative pions (green tracks). (For simplicity, we assume that no gamma-ray photons, which would leave no tracks because they lack electric charge, are produced.) Thus we conclude that the annihilation process is

$$\mathrm{p} + \bar{\mathrm{p}} \rightarrow 4\pi^+ + 4\pi^-. \qquad (44\text{-}6)$$

We see from Table 44-1 that the positive pions (π^+) are *particles* and the negative pions (π^-) are *antiparticles.* The reaction of Eq. 44-6 is a *strong interaction* (it involves the strong force) because all the particles involved are hadrons.

Let us check whether electric charge is conserved in the reaction. To do so, we can write the electric charge of a particle as qe, in which q is a **charge quantum number.** Then determining whether electric charge is conserved in a process amounts to determining whether the initial net charge quantum number is equal to the final net charge quantum number. In the process of Eq. 44-6, the initial net charge number is $1 + (-1)$, or 0, and the final net charge number is $4(1) + 4(-1)$, or 0. Thus, charge *is* conserved.

For the energy balance, note from above that the energy available from the p-$\bar{\mathrm{p}}$ annihilation process is at least the sum of the proton and antiproton rest energies, 1876.6 MeV. The rest energy of a pion is 139.6 MeV, which means the rest energies of the eight pions amount to 8×139.6 MeV, or 1116.8 MeV. This leaves at least about 760 MeV to distribute among the eight pions as kinetic energy. Thus, the requirement of energy conservation is easily met.

2. *Pion Decay.* Pions are unstable particles and decay with a mean lifetime of 2.6×10^{-8} s. At point 2 in Fig. 44-2*b*, one of the positive pions comes to rest in the chamber and decays spontaneously into an antimuon μ^+ (purple track) and a neutrino ν:

$$\pi^+ \rightarrow \mu^+ + \nu. \qquad (44\text{-}7)$$

The neutrino, being uncharged, leaves no track. Both the antimuon and the neutrino are leptons; that is, they are particles on which the strong force does not act. Thus, the decay process of Eq. 44-7, which is governed by the weak force, is described as a *weak interaction.* From Table 44-1, the rest energy of an antimuon is 105.7 MeV and the rest energy of a neutrino is approximately 0. Thus, an energy of 139.6 MeV $-$ 105.7 MeV, or 33.9 MeV, is available to share between the antimuon and the neutrino as kinetic energy.

Let us check whether spin angular momentum is conserved in the process of Eq. 44-7. This amounts to determining whether the net component S_z of spin angular momentum along some arbitrary z axis can be conserved by the process. The spin quantum numbers s of the particles in the process are 0 for the pion π^+ and $\frac{1}{2}$ for both the antimuon μ^+ and the neutrino ν. Thus, for π^+, the component S_z must be $0\hbar$, and for μ^+ and ν, it can be either $+\frac{1}{2}\hbar$ or $-\frac{1}{2}\hbar$.

The net component S_z is conserved by the process of Eq. 44-7 if there is *any* way in which the initial S_z ($= 0\hbar$) can be equal to the final net S_z. We see that if one of the products, either μ^+ or ν, has $S_z = +\frac{1}{2}\hbar$ and the other has $S_z = -\frac{1}{2}\hbar$, then their final net value is $0\hbar$. Thus, because S_z can be conserved, the decay process of Eq. 44-7 *can* occur.

From Eq. 44-7, we also see that the net charge is conserved by the process: before the process the net charge quantum number is +1, and after the process it is +1 + 0 = +1.

3. *Muon Decay.* Muons (whether μ^- or μ^+) are also unstable, decaying with a mean life of 2.2×10^{-6} s. Although the decay products are not shown in Fig. 44-2*b*, the antimuon produced in the reaction of Eq. 44-7 comes to rest and decays spontaneously according to

$$\mu^+ \rightarrow e^+ + \nu + \bar{\nu}. \qquad (44\text{-}8)$$

The rest energy of the antimuon is 105.7 MeV, and that of the positron is only 0.511 MeV, leaving 105.2 MeV to be shared as kinetic energy among the three particles produced in the decay process of Eq. 44-8.

You may wonder: Why *two* neutrinos in Eq. 44-8? Why not just one, as in the pion decay in Eq. 44-7? One answer is that the spin quantum numbers of the antimuon, the positron, and the neutrino are each $\frac{1}{2}$; with only one neutrino, the net component S_z of spin angular momentum could not be conserved in the antimuon decay of Eq. 44-8. In Section 44-4 we shall discuss another reason.

Sample Problem 44-1

A stationary positive pion can decay according to

$$\pi^+ \rightarrow \mu^+ + \nu.$$

What is the kinetic energy of the antimuon μ^+? What is the kinetic energy of the neutrino?

Solution: The **Key Idea** here is that the pion decay process must conserve both total energy and total linear momentum. Let us first write the conservation of total energy (rest energy mc^2 plus kinetic energy K) for the decay process as

$$m_\pi c^2 + K_\pi = m_\mu c^2 + K_\mu + m_\nu c^2 + K_\nu.$$

Because the pion was stationary, its kinetic energy K_π is zero. Then, using the masses listed for m_π, m_μ, and m_ν in Table 44-1, we find

$$\begin{aligned} K_\mu + K_\nu &= m_\pi c^2 - m_\mu c^2 - m_\nu c^2 \\ &= 139.6 \text{ MeV} - 105.7 \text{ MeV} - 0 \\ &= 33.9 \text{ MeV}, \end{aligned} \qquad (44\text{-}9)$$

where we have approximated m_ν as zero.

We cannot solve Eq. 44-9 for either K_μ or K_ν separately, and so let us next apply the principle of conservation of linear momentum to the decay process. Because the pion is stationary when it decays, that principle requires that the muon and neutrino move in opposite directions after the decay. Assume that their motion is along an axis. Then, for components along that axis, we can write the conservation of linear momentum for the decay as

$$p_\pi = p_\mu + p_\nu,$$

which, with $p_\pi = 0$, gives us

$$p_\mu = -p_\nu. \qquad (44\text{-}10)$$

We want to relate these momenta p_μ and $-p_\nu$ to the kinetic energies K_μ and K_ν so that we can solve for the kinetic energies. Because we have no reason to believe that classical physics can be applied to the motion of the muon and neutrino, we use Eq. 37-54, the momentum–kinetic energy relation from special relativity:

$$(pc)^2 = K^2 + 2Kmc^2. \qquad (44\text{-}11)$$

From Eq. 44-10, we know that

$$(p_\mu c)^2 = (p_\nu c)^2. \qquad (44\text{-}12)$$

Substituting from Eq. 44-11 for each side of Eq. 44-12 yields

$$K_\mu^2 + 2K_\mu m_\mu c^2 = K_\nu^2 + 2K_\nu m_\nu c^2.$$

Approximating the neutrino mass to be $m_\nu = 0$, substituting $K_\nu = 33.9$ MeV $- K_\mu$ from Eq. 44-9, and then solving for K_μ, we find

$$\begin{aligned} K_\mu &= \frac{(33.9 \text{ MeV})^2}{(2)(33.9 \text{ MeV} + m_\mu c^2)} \\ &= \frac{(33.9 \text{ MeV})^2}{(2)(33.9 \text{ MeV} + 105.7 \text{ MeV})} \\ &= 4.12 \text{ MeV}. \end{aligned} \qquad \text{(Answer)}$$

The kinetic energy of the neutrino is then, from Eq. 44-9,

$$\begin{aligned} K_\nu &= 33.9 \text{ MeV} - K_\mu = 33.9 \text{ MeV} - 4.12 \text{ MeV} \\ &= 29.8 \text{ MeV}. \end{aligned} \qquad \text{(Answer)}$$

We see that, although the magnitudes of the momenta of the two recoiling particles are the same, the neutrino gets the larger share (88%) of the kinetic energy.

Sample Problem 44-2

Stationary protons in a bubble chamber are bombarded by energetic negative pions, and the following reaction occurs:

$$\pi^- + \mathrm{p} \rightarrow \mathrm{K}^- + \Sigma^+.$$

The rest energies of these particles are

π^-	139.6 MeV	K^-	493.7 MeV
p	938.3 MeV	Σ^+	1189.4 MeV

What is the Q of the reaction?

Solution: The **Key Idea** here is that the Q of a reaction is

$$Q = \begin{pmatrix}\text{initial total}\\ \text{mass energy}\end{pmatrix} - \begin{pmatrix}\text{final total}\\ \text{mass energy}\end{pmatrix}.$$

For the given reaction, we find

$$\begin{aligned} Q &= (m_\pi c^2 + m_\mathrm{p} c^2) - (m_\mathrm{K} c^2 + m_\Sigma c^2) \\ &= (139.6\ \mathrm{MeV} + 938.3\ \mathrm{MeV}) \\ &\quad - (493.7\ \mathrm{MeV} + 1189.4\ \mathrm{MeV}) \\ &= -605\ \mathrm{MeV}. \end{aligned} \qquad \text{(Answer)}$$

The minus sign means that the reaction is *endothermic;* that is, the incoming pion (π^-) must have a kinetic energy greater than a certain threshold value if the reaction is to occur. The threshold energy is greater than 605 MeV because linear momentum must be conserved, which means that the kaon (K^-) and the sigma (Σ^+) must not only be created but also must be given some kinetic energy. A relativistic calculation whose details are beyond our scope shows that the threshold energy for the reaction is 907 MeV.

44-4 The Leptons

In this and the next section, we discuss some of the particles of one of our classification schemes: lepton or hadron. We begin with the leptons, those particles on which the strong force does *not* act. So far, we have encountered the familiar electron and the neutrino that accompanies it in beta decay. The muon, whose decay is described in Eq. 44-8, is another member of this family. Physicists gradually learned that the neutrino that appears in Eq. 44-7, associated with the production of a muon, is *not the same particle* as the neutrino produced in beta decay, associated with the appearance of an electron. We call the former the **muon neutrino** (symbol ν_μ) and the latter the **electron neutrino** (symbol ν_e) when it is necessary to distinguish between them.

These two types of neutrino are known to be different particles because, if a beam of muon neutrinos (produced from pion decay as in Eq. 44-7) strikes a solid target, *only muons*—and never electrons—are produced. On the other hand, if electron neutrinos (produced by the beta decay of fission products in a nuclear reactor) strike a solid target, *only electrons*—and never muons—are produced.

Another lepton, the **tau,** was discovered at SLAC in 1975; its discoverer, Martin Perl, shared the 1995 Nobel Prize in physics. The tau has its own associated neutrino, different still from the other two. Table 44-2 lists all the leptons (both particles and antiparticles); all have a spin quantum number s of $\frac{1}{2}$.

TABLE 44-2 **The Leptons**[a]

Family	Particle	Symbol	Mass (MeV/c^2)	Charge q	Antiparticle
Electron	Electron	e^-	0.511	−1	e^+
	Electron neutrino[b]	ν_e	$\approx 1 \times 10^{-7}$	0	$\bar{\nu}_e$
Muon	Muon	μ^-	105.7	−1	μ^+
	Muon neutrino[b]	ν_μ	$\approx 1 \times 10^{-7}$	0	$\bar{\nu}_\mu$
Tau	Tau	τ^-	1777	−1	τ^+
	Tau neutrino[b]	ν_τ	$\approx 1 \times 10^{-7}$	0	$\bar{\nu}_\tau$

[a]All leptons have spin quantum numbers of $\frac{1}{2}$ and are thus fermions.

[b]The neutrino masses have not been well determined.

There are reasons for dividing the leptons into three families, each consisting of a particle (electron, muon, or tau), its associated neutrino, and the corresponding antiparticles. Furthermore, there are reasons to believe that there are *only* the three families of leptons shown in Table 44-2. Leptons have no internal structure and no measurable dimensions; they are believed to be truly pointlike fundamental particles when they interact with other particles or with electromagnetic waves.

The Conservation of Lepton Number

According to experiment, particle interactions involving leptons obey a conservation law for a quantum number called the **lepton number** L. Each (normal) particle in Table 44-2 is assigned $L = +1$, and each antiparticle is assigned $L = -1$. All other particles, which are not leptons, are assigned $L = 0$. Also according to experiment,

> In all particle interactions, the net lepton number *for each family* is separately conserved.

Thus, there are actually three lepton numbers L_e, L_μ, L_τ, and the net of *each* must remain unchanged during any particle interaction. This experimental fact is called the law of **conservation of lepton number.**

We can illustrate this law by reconsidering the antimuon decay process shown in Eq. 44-8, which we now write more fully as

$$\mu^+ \rightarrow e^+ + \nu_e + \bar{\nu}_\mu. \tag{44-13}$$

Consider this first in terms of the muon family of leptons. The μ^+ is an antiparticle (see Table 44-2) and thus has the muon lepton number $L_\mu = -1$. The two particles e^+ and ν_e do not belong to the muon family and thus have $L_\mu = 0$. This leaves $\bar{\nu}_\mu$ on the right which, being an antiparticle, also has the muon lepton number $L_\mu = -1$. Thus, both sides of Eq. 44-13 have the same net muon lepton number—namely, $L_\mu = -1$; if they did not, the μ^+ would not decay by this process.

No members of the electron family appear on the left in Eq. 44-13; so there the net electron lepton number must be $L_e = 0$. On the right side of Eq. 44-13, the positron, being an antiparticle (again see Table 44-2), has the electron lepton number $L_e = -1$. The electron neutrino ν_e, being a particle, has the electron number $L_e = +1$. Thus, the net electron lepton number for these two particles on the right in Eq. 44-13 is also zero; the electron lepton number is also conserved in the process.

Because no members of the tau family appear on either side of Eq. 44-13, we must have $L_\tau = 0$ on each side. Thus, each of the lepton quantum numbers L_μ, L_e, and L_τ remains unchanged during the decay process of Eq. 44-13, their constant values being -1, 0, and 0, respectively. This example is but one illustration of the conservation of lepton number; this law holds for all particle interactions.

CHECKPOINT 1 (a) The π^+ meson decays by the process $\pi^+ \rightarrow \mu^+ + \nu$. To what lepton family does the neutrino ν belong? (b) Is this neutrino a particle or an antiparticle? (c) What is its lepton number?

44-5 The Hadrons

We are now ready to consider hadrons (baryons and mesons), those particles whose interactions are governed by the strong force. We start by adding another conservation law to our list: conservation of baryon number.

To develop this conservation law, let us consider the proton decay process

$$p \rightarrow e^+ + \nu_e. \tag{44-14}$$

This process *never* happens. We should be glad that it does not because otherwise all protons in the universe would gradually change into positrons, with disastrous consequences for us. Yet this decay process does not violate the conservation laws involving energy, linear momentum, or lepton number.

We account for the apparent stability of the proton—and for the absence of many other processes that might otherwise occur—by introducing a new quantum number, the **baryon number** B, and a new conservation law, the **conservation of baryon number:**

> To every baryon we assign $B = +1$. To every antibaryon we assign $B = -1$. To all particles of other types we assign $B = 0$. A particle process cannot occur if it changes the net baryon number.

In the process of Eq. 44-14, the proton has a baryon number of $B = +1$ and the positron and neutrino both have a baryon number of $B = 0$. Thus, the process does not conserve baryon number and cannot occur.

CHECKPOINT 2 This mode of decay for a neutron is *not* observed:

$$n \rightarrow p + e^-.$$

Which of the following conservation laws does this process violate: (a) energy, (b) angular momentum, (c) linear momentum, (d) charge, (e) lepton number, (f) baryon number? The masses are $m_n = 939.6\ \text{MeV}/c^2$, $m_p = 938.3\ \text{MeV}/c^2$, and $m_e = 0.511\ \text{MeV}/c^2$.

Sample Problem 44-3

Determine whether a stationary proton can decay according to the scheme

$$p \rightarrow \pi^0 + \pi^+.$$

Properties of the proton and the π^+ pion are listed in Table 44-1. The π^0 pion has zero charge, zero spin, and a mass energy of 135.0 MeV.

Solution: The **Key Idea** here is to see whether the proposed decay violates any of the conservation laws we have discussed. For electric charge, we see that the net charge quantum number is initially +1 and finally 0 + 1, or +1. Thus, charge is conserved by the decay. Lepton number is also conserved, because none of the three particles is a lepton and thus each lepton number is zero.

Linear momentum can also be conserved: Because the proton is stationary, with zero linear momentum, the two pions must merely move in opposite directions with equal magnitudes of linear momentum (so that their total linear momentum is also zero) to conserve linear momentum. The fact that linear momentum *can* be conserved means that the process does not violate the conservation of linear momentum.

Is there energy for the decay? Because the proton is stationary, that question amounts to asking whether the proton's mass energy is sufficient to produce the mass energies and kinetic energies of the pions. To answer, we evaluate the Q of the decay:

$$\begin{aligned} Q &= \begin{pmatrix}\text{initial total}\\ \text{mass energy}\end{pmatrix} - \begin{pmatrix}\text{final total}\\ \text{mass energy}\end{pmatrix} \\ &= m_p c^2 - (m_0 c^2 + m_+ c^2) \\ &= 938.3\ \text{MeV} - (135.0\ \text{MeV} + 139.6\ \text{MeV}) \\ &= 663.7\ \text{MeV}. \end{aligned}$$

The fact that Q is positive indicates that the initial mass energy exceeds the final mass energy. Thus, the proton *does* have enough mass energy to create the pair of pions.

Is spin angular momentum conserved by the decay? This amounts to determining whether the net component S_z of spin angular momentum along some arbitrary z axis can be conserved by the decay. The spin quantum numbers s of the particles in the process are $\frac{1}{2}$ for the proton and 0 for both pions. Thus, for the proton the component S_z can be either $+\frac{1}{2}\hbar$ or $-\frac{1}{2}\hbar$ and for each pion it is $0\hbar$. We see that there is no way that S_z can be conserved. Hence, spin angular momentum is not conserved, and the proposed decay of the proton cannot occur.

The decay also violates the conservation of baryon number: The proton has a baryon number of $B = +1$, and both pions have a baryon number of $B = 0$. Thus, nonconservation of baryon number is another reason the proposed decay cannot occur.

Sample Problem 44-4

A particle called xi-minus and having the symbol Ξ^- decays as follows:

$$\Xi^- \rightarrow \Lambda^0 + \pi^-.$$

The Λ^0 particle (called lambda-zero) and the π^- particle are both unstable. The following decay processes occur in *cascade* until only relatively stable products remain:

$$\Lambda^0 \rightarrow p + \pi^- \qquad \pi^- \rightarrow \mu^- + \bar{\nu}_\mu$$
$$\mu^- \rightarrow e^- + \nu_\mu + \bar{\nu}_e.$$

(a) Is the Ξ^- particle a lepton or a hadron? If the latter, is it a baryon or a meson?

Solution: The **Key Idea** for answering the first question is that only three families of leptons exist (Table 44-2) and none include the Ξ^- particle. Thus, the Ξ^- must be a hadron.

The **Key Idea** for answering the second question is to determine the baryon number of the Ξ^- particle. If it is +1 or −1, then the Ξ^- is a baryon. If, instead, it is 0, then the Ξ^- is a meson. To see, let us the write the overall decay scheme, from the initial Ξ^- to the final relatively stable products, as

$$\Xi^- \rightarrow p + 2(e^- + \bar{\nu}_e) + 2(\nu_\mu + \bar{\nu}_\mu). \qquad (44\text{-}15)$$

On the right side, the proton has a baryon number of +1 and each electron and neutrino has a baryon number of 0. Thus, the net baryon number of the right side is +1. That must then be the baryon number of the lone Ξ^- particle on the left side. We conclude that the Ξ^- particle is a baryon.

(b) Does the decay process conserve the three lepton numbers?

Solution: The **Key Idea** here is that any process must separately conserve the net lepton number for each lepton family of Table 44-2. Let us first consider the electron lepton number L_e, which is +1 for the electron e^-, −1 for the anti-electron neutrino $\bar{\nu}_e$, and 0 for the other particles in the overall decay of Eq. 44-15. We see that the net L_e is 0 before the decay and $2[+1 + (-1)] + 2(0 + 0) = 0$ after the decay. Thus, the net electron lepton number *is* conserved. You can similarly show that the net muon lepton number and the net tau lepton number are also conserved.

(c) What can you say about the spin of the Ξ^- particle?

Solution: The **Key Idea** here is that the overall decay scheme of Eq. 44-15 must conserve the net spin component S_z. Thus, we can determine the spin component S_z of the Ξ^- particle on the left side of Eq. 44-15 by considering the S_z components of the nine particles on the right side. All nine of those particles are spin-$\frac{1}{2}$ particles and thus can have S_z of either $+\frac{1}{2}\hbar$ or $-\frac{1}{2}\hbar$. No matter how we choose between those two possible values of S_z, the net S_z for those nine particles must be a *half-integer* times $\hbar$. Thus, the Ξ^- particle must have S_z of a *half-integer* times $\hbar$, and that means that its spin quantum number s must be a half-integer. (Actually, the quantum number is $\frac{1}{2}$.)

44-6 Still Another Conservation Law

Particles have intrinsic properties in addition to the ones we have listed so far: mass, charge, spin, lepton number, and baryon number. The first of these additional properties was discovered when researchers observed that certain new particles, such as the kaon (K) and the sigma (Σ), always seemed to be produced in pairs. It seemed impossible to produce only one of them at a time. Thus, if a beam of energetic pions interacts with the protons in a bubble chamber, the reaction

$$\pi^+ + p \rightarrow K^+ + \Sigma^+ \qquad (44\text{-}16)$$

often occurs. The reaction

$$\pi^+ + p \rightarrow \pi^+ + \Sigma^+, \qquad (44\text{-}17)$$

which violates no conservation law known in the early days of particle physics, never occurs.

It was eventually proposed (by Murray Gell-Mann in the United States and independently by K. Nishijima in Japan) that certain particles possess a new property, called **strangeness,** with its own quantum number S and its own conservation law. (Be careful not to confuse the symbol S here with spin.) The name *strangeness* arises from the fact that, before the identities of these particles were pinned down, they were known as "strange particles," and the label stuck.

The proton, neutron, and pion have $S = 0$; that is, they are not "strange." It was proposed, however, that the K^+ particle has strangeness $S = +1$ and that Σ^+ has $S = -1$. In the reaction of Eq. 44-16, the net strangeness is initially zero and finally zero; thus, the reaction conserves strangeness. However, in the reac-

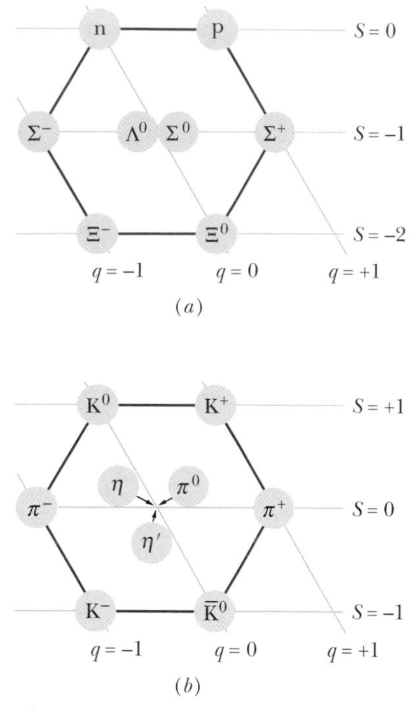

Fig. 44-3 (*a*) The eightfold way pattern for the eight spin-$\frac{1}{2}$ baryons listed in Table 44-3. The particles are represented as disks on a strangeness–charge plot, using a sloping axis for the charge quantum number. (*b*) A similar pattern for the nine spin-zero mesons listed in Table 44-4.

tion shown in Eq. 44-17, the final net strangeness is -1; thus, that reaction does not conserve strangeness and cannot occur. Apparently, then, we must add one more conservation law to our list—the **conservation of strangeness:**

> Strangeness is conserved in interactions involving the strong force.

It may seem heavy-handed to invent a new property of particles just to account for a little puzzle like that posed by Eqs. 44-16 and 44-17. However, strangeness and its quantum number soon revealed themselves in many other areas of particle physics, and strangeness is now fully accepted as a legitimate particle attribute, on a par with charge and spin.

Do not be misled by the whimsical character of the name. Strangeness is no more mysterious a property of particles than is charge. Both are properties that particles may (or may not) have; each is described by an appropriate quantum number. Each obeys a conservation law. Still other properties of particles have been discovered and given even more whimsical names, such as *charm* and *bottomness,* but all are perfectly legitimate properties. Let us see, as an example, how the new property of strangeness "earns its keep" by leading us to uncover important regularities in the properties of the particles.

44-7 The Eightfold Way

There are eight baryons—the neutron and the proton among them—that have a spin quantum number of $\frac{1}{2}$. Table 44-3 shows some of their other properties. Figure 44-3*a* shows the fascinating pattern that emerges if we plot the strangeness of these baryons against their charge quantum number, using a sloping axis for the charge quantum numbers. Six of the eight form a hexagon with the two remaining baryons at its center.

Let us turn now from the hadrons called baryons to the hadrons called mesons. Nine with a spin of zero are listed in Table 44-4. If we plot them on a sloping strangeness–charge diagram, as in Fig. 44-3*b*, the same fascinating pattern emerges! These and related plots, called the **eightfold way** patterns,* were pro-

*The name is a borrowing from Eastern mysticism. The "eight" refers to the eight quantum numbers (only a few of which we have defined here) that are involved in the symmetry-based theory that predicts the existence of the patterns.

TABLE 44-3
Eight Spin-$\frac{1}{2}$ Baryons

Particle	Symbol	Mass (MeV/c^2)	Quantum Numbers: Charge q	Quantum Numbers: Strangeness S
Proton	p	938.3	+1	0
Neutron	n	939.6	0	0
Lambda	Λ^0	1115.6	0	−1
Sigma	Σ^+	1189.4	+1	−1
Sigma	Σ^0	1192.5	0	−1
Sigma	Σ^-	1197.3	−1	−1
Xi	Ξ^0	1314.9	0	−2
Xi	Ξ^-	1321.3	−1	−2

TABLE 44-4
Nine Spin-Zero Mesons[a]

Particle	Symbol	Mass (MeV/c^2)	Quantum Numbers: Charge q	Quantum Numbers: Strangeness S
Pion	π^0	135.0	0	0
Pion	π^+	139.6	+1	0
Pion	π^-	139.6	−1	0
Kaon	K^+	493.7	+1	+1
Kaon	K^-	493.7	−1	−1
Kaon	K^0	497.7	0	+1
Kaon	$\bar{K}^0$	497.7	0	−1
Eta	η	547.5	0	0
Eta prime	η'	957.8	0	0

[a]All mesons are bosons, having spins of 0, 1, 2, ... The ones listed here all have a spin of 0.

posed independently in 1961 by Murray Gell-Mann at the California Institute of Technology and by Yuval Ne'eman at Imperial College, London. The two patterns of Fig. 44-3 are representative of a larger number of symmetrical patterns in which groups of baryons and mesons can be displayed.

The symmetry of the eightfold way pattern for the spin-$\frac{3}{2}$ baryons (not shown here) calls for ten particles arranged in a pattern like that of the tenpins in a bowling alley. However, when the pattern was first proposed, only nine such particles were known; the "headpin" was missing. In 1962, guided by theory and the symmetry of the pattern, Gell-Mann made a prediction in which he essentially said:

> *There exists a spin-$\frac{3}{2}$ baryon with a charge of −1, a strangeness of −3, and a rest energy of about 1680 MeV. If you look for this omega minus particle (as I propose to call it), I think you will find it.*

A team of physicists headed by Nicholas Samios of the Brookhaven National Laboratory took up the challenge and found the "missing" particle, confirming all its predicted properties. Nothing beats prompt experimental confirmation for building confidence in a theory!

The eightfold way patterns bear the same relationship to particle physics that the periodic table does to chemistry. In each case, there is a pattern of organization in which vacancies (missing particles or missing elements) stick out like sore thumbs, guiding experimenters in their searches. In the case of the periodic table, its very existence strongly suggests that the atoms of the elements are not fundamental particles but have an underlying structure. Similarly, the eightfold way patterns strongly suggest that the mesons and the baryons must have an underlying structure, in terms of which their properties can be understood. That structure can be explained in terms of the *quark model,* which we now discuss.

44-8 The Quark Model

In 1964 Gell-Mann and George Zweig independently pointed out that the eightfold way patterns can be understood in a simple way if the mesons and the baryons are built up out of subunits that Gell-Mann called **quarks.** We deal first with three of them, called the *up quark* (symbol u), the *down quark* (symbol d), and the *strange quark* (symbol s), and we assign to them the properties displayed in Table 44-5. (The names of the quarks, along with those assigned to three other quarks that we shall meet later, have no meaning other than as convenient labels.

TABLE 44-5
The Quarks[a]

			Quantum Numbers			
Particle	Symbol	Mass (MeV/c^2)	Charge q	Strangeness S	Baryon Number B	Antiparticle
Up	u	5	$+\frac{2}{3}$	0	$+\frac{1}{3}$	$\bar{u}$
Down	d	10	$-\frac{1}{3}$	0	$+\frac{1}{3}$	$\bar{d}$
Charm	c	1500	$+\frac{2}{3}$	0	$+\frac{1}{3}$	$\bar{c}$
Strange	s	200	$-\frac{1}{3}$	−1	$+\frac{1}{3}$	$\bar{s}$
Top	t	175 000	$+\frac{2}{3}$	0	$+\frac{1}{3}$	$\bar{t}$
Bottom	b	4300	$-\frac{1}{3}$	0	$+\frac{1}{3}$	$\bar{b}$

[a]All quarks (including antiquarks) have spin $\frac{1}{2}$ and thus are fermions. The quantum numbers q, S, and B for each antiquark are the negatives of those for the corresponding quark.

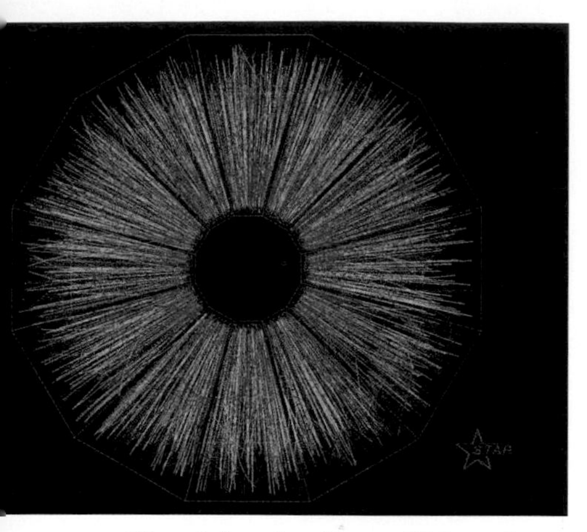

The violent head-on collision of two 30 GeV beams of gold atoms in the RHIC accelerator at the Brookhaven National Laboratory. In the moment of collision, a gas of individual quarks and gluons was created.

Collectively, these names are called the *quark flavors*. We could just as well call them vanilla, chocolate, and strawberry instead of up, down, and strange.)

The fractional charge quantum numbers of the quarks may jar you a little. However, withhold judgment until you see how neatly these fractional charges account for the observed integer charges of the mesons and the baryons. In all normal situations, whether here on Earth or in an astronomical process, quarks are always bound up together in twos or threes for reasons that are still not well understood. Let us call this our normal rule for quark combinations and look next at some recently detected exceptions.

One exception is that the collision of a high-energy gamma ray with a neutron can produce a short-lived pentaquark, which is either a tightly bound system of five quarks or a system in which a kaon (two quarks) and a neutron (three quarks) are bound together. Presumably the energy of the collision can produce two new quarks to join the three quarks that were already in the neutron target. Evidence for a bound system of four quarks has also been reported.

Another exception to the normal rule occurred in experiments at the RHIC particle collider at the Brookhaven National Laboratory. At the spot where two high-energy beams of gold nuclei collided head-on, the kinetic energy of the particles was so large that it matched the kinetic energy of particles that were present soon after the beginning of the universe (as we discuss in Section 44-14). The protons and neutrons of the gold nuclei were ripped apart to form a momentary gas of individual quarks. (The gas also contained gluons, the particles that normally hold quarks together, as we discuss in Section 44-9.) These experiments at RHIC and similar ones at CERN may be the first time that quarks have been set free of one another since the universe began.

Having noted these exceptions, let's now return to the normal rule of quark combinations to see how quarks form all known baryons and mesons.

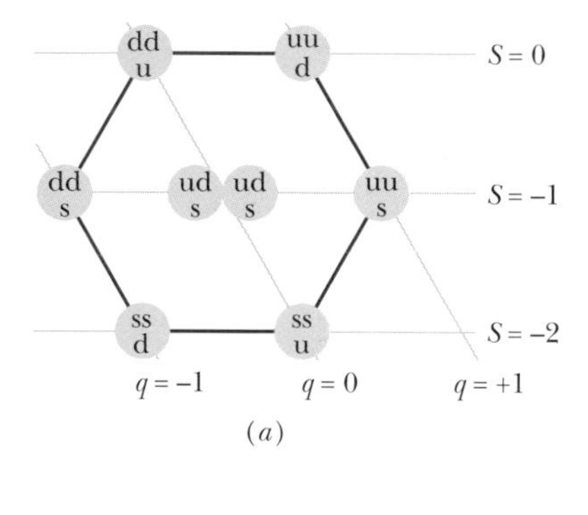

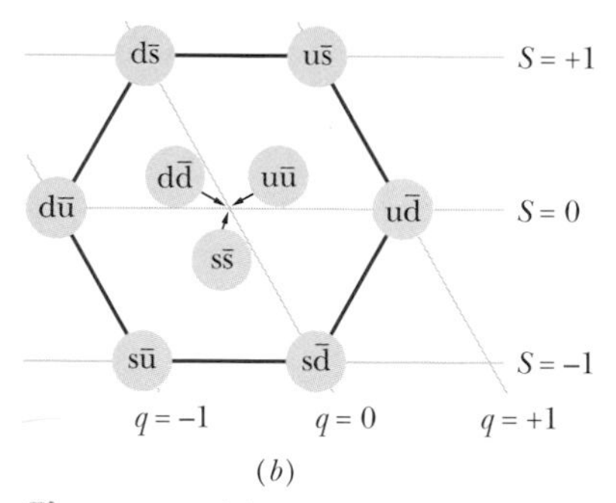

Fig. 44-4 (*a*) The quark compositions of the eight spin-$\frac{1}{2}$ baryons plotted in Fig. 44-3*a*. (Although the two central baryons share the same quark structure, they are different particles. The sigma is an excited state of the lambda, decaying into the lambda by emission of a gamma-ray photon.) (*b*) The quark compositions of the nine spin-zero mesons plotted in Fig. 44-3*b*.

Quarks and Baryons

Each baryon is a combination of three quarks; some of the combinations are given in Fig. 44-4*a*. With regard to baryon number, we see that any three quarks (each with $B = +\frac{1}{3}$) yield a proper baryon (with $B = +1$).

Charges also work out, as we can see from three examples. The proton has a quark composition of uud, and so its charge quantum number is

$$q(\text{uud}) = \tfrac{2}{3} + \tfrac{2}{3} + (-\tfrac{1}{3}) = +1.$$

The neutron has a quark composition of udd, and its charge quantum number is therefore

$$q(\text{udd}) = \tfrac{2}{3} + (-\tfrac{1}{3}) + (-\tfrac{1}{3}) = 0.$$

The Σ^- (sigma-minus) particle has a quark composition of dds, and its charge quantum number is therefore

$$q(\text{dds}) = -\tfrac{1}{3} + (-\tfrac{1}{3}) + (-\tfrac{1}{3}) = -1.$$

The strangeness quantum numbers work out as well. You can check this by using Table 44-3 for the Σ^- strangeness number and Table 44-5 for the strangeness numbers of the dds quarks.

Quarks and Mesons

Mesons are quark–antiquark pairs; some of their compositions are given in Fig. 44-4*b*. The quark–antiquark model is consistent with the fact that mesons are not baryons; that is, mesons have a baryon number $B = 0$. The baryon number for a quark is $+\frac{1}{3}$ and for an antiquark is $-\frac{1}{3}$; thus, the combination of baryon numbers in a meson is zero.

Consider the meson π^+, which consists of an up quark u and an antidown quark $\bar{d}$. We see from Table 44-5 that the charge quantum number of the up quark is $+\frac{2}{3}$ and that of the antidown quark is $+\frac{1}{3}$ (the sign is opposite that of the down quark). This adds nicely to a charge quantum number of $+1$ for the π^+ meson; that is,

$$q(u\bar{d}) = \tfrac{2}{3} + \tfrac{1}{3} = +1.$$

All the charge and strangeness quantum numbers of Fig. 44-4*b* agree with those of Table 44-4 and Fig. 44-3*b*. Convince yourself that all possible up, down, and strange quark–antiquark combinations are used and that all known spin-zero mesons are accounted for. Everything fits.

CHECKPOINT 3 Is a combination of a down quark (d) and an antiup quark ($\bar{u}$) called (a) a π^0 meson, (b) a proton, (c) a π^- meson, (d) a π^+ meson, or (e) a neutron?

A New Look at Beta Decay

Let us see how beta decay appears from the quark point of view. In Eq. 42-24, we presented a typical example of this process:

$$^{32}\text{P} \rightarrow {}^{32}\text{S} + e^- + \nu.$$

After the neutron was discovered and Fermi had worked out his theory of beta decay, physicists came to view the fundamental beta-decay process as the changing of a neutron into a proton inside the nucleus, according to the scheme

$$n \rightarrow p + e^- + \bar{\nu}_e,$$

in which the neutrino is identified more completely. Today we look deeper and see that a neutron (udd) can change into a proton (uud) by changing a down quark into an up quark. We now view the fundamental beta-decay process as

$$d \rightarrow u + e^- + \bar{\nu}_e.$$

Thus, as we come to know more and more about the fundamental nature of matter, we can examine familiar processes at deeper and deeper levels. We see too that the quark model not only helps us to understand the structure of particles but also clarifies their interactions.

Still More Quarks

There are other particles and other eightfold way patterns that we have not discussed. To account for them, it turns out that we need to postulate three more quarks, the *charm quark* c, the *top quark* t, and the *bottom quark* b. Thus, a total of six quarks exist, as listed in Table 44-5.

Note that three quarks are exceptionally massive, the most massive of them (top) being almost 170 times more massive than a proton. To generate particles that contain such quarks, with such large mass energies, we must go to higher and higher energies, which is the reason that these three quarks were not discovered earlier.

The first particle containing a charm quark to be observed was the J/Ψ meson, whose quark structure is $c\bar{c}$. It was discovered simultaneously and independently in 1974 by groups headed by Samuel Ting at the Brookhaven National Laboratory and Burton Richter at Stanford University.

The top quark defied all efforts to generate it in the laboratory until 1995, when its existence was finally demonstrated in the Tevatron, a large particle accelerator at Fermilab. In this accelerator, protons and antiprotons, each with an energy of 0.9 TeV ($= 9 \times 10^{11}$ eV), are made to collide at the centers of two large particle detectors. In a very few cases, the colliding protons generate a top–antitop ($t\bar{t}$) quark pair, which *very* quickly decays into particles that can be detected and thus can be used to infer the existence of the top–antitop pair.

Look back for a moment at Table 44-5 (the quark family) and Table 44-2 (the lepton family) and notice the neat symmetry of these two "six-packs" of particles, each dividing naturally into three corresponding two-particle families. In terms of what we know today, the quarks and the leptons seem to be truly fundamental particles having no internal structure.

Sample Problem 44-5

The Ξ^- (xi-minus) particle has a spin quantum number s of $\frac{1}{2}$, a charge quantum number q of -1, and a strangeness quantum number S of -2. Also, it is known not to contain a bottom quark. What combination of quarks makes up the Ξ^-?

Solution: From Sample Problem 44-4, we know that the Ξ^- is a baryon. One **Key Idea** then is that it must consist of three quarks (not two as for a meson).

Let us next consider the strangeness $S = -2$ of the Ξ^-. A **Key Idea** here is that only the strange quark s and the antistrange quark $\bar{s}$ have nonzero values of strangeness (see Table 44-5). Further, because only the strange quark s has a *negative* value of strangeness, Ξ^- must contain that quark. In fact, for Ξ^- to have a strangeness of -2, it must contain two strange quarks.

To determine the third quark, call it x, we can consider the other known properties of Ξ^-. Its charge quantum number q is -1, and the charge quantum number q of each strange quark is $-\frac{1}{3}$. Thus, the third quark x must have a charge quantum number of $-\frac{1}{3}$, so that we can have

$$q(\Xi^-) = q(\text{ssx}) = -\tfrac{1}{3} + (-\tfrac{1}{3}) + (-\tfrac{1}{3}) = -1.$$

Besides the strange quark, the only quarks with $q = -\frac{1}{3}$ are the down quark d and bottom quark b. Because the problem statement ruled out a bottom quark, the third quark must be a down quark. This conclusion is also consistent with the baryon quantum numbers:

$$B(\Xi^-) = B(\text{ssd}) = \tfrac{1}{3} + \tfrac{1}{3} + \tfrac{1}{3} = +1.$$

Thus, the quark composition of the Ξ^- particle is ssd.

44-9 The Basic Forces and Messenger Particles

We turn now from cataloging the particles to considering the forces between them.

The Electromagnetic Force

At the atomic level, we say that two electrons exert electromagnetic forces on each other according to Coulomb's law. At a deeper level, this interaction is described by a highly successful theory called **quantum electrodynamics** (QED). From this point of view, we say that each electron senses the presence of the other by exchanging photons with it.

We cannot detect these photons because they are emitted by one electron and absorbed by the other a very short time later. Because of their undetectable existence, we call them **virtual photons.** Because they communicate between the two interacting charged particles, we sometimes call these photons *messenger particles.*

If a stationary electron emits a photon and remains itself unchanged, energy is not conserved. The principle of conservation of energy is saved, however, by the uncertainty principle, written in the form

$$\Delta E \cdot \Delta t \approx \hbar, \tag{44-18}$$

as discussed in Sample Problem 42-10. Here we interpret this relation to mean that you can "overdraw" an amount of energy ΔE, violating conservation of energy, *provided* you "return" it within an interval Δt given by $\hbar/\Delta E$ so that the violation cannot be detected. The virtual photons do just that. When, say, electron A emits a virtual photon, the overdraw in energy is quickly set right when that electron receives a virtual photon from electron B, and the violation of the principle of conservation of energy for the electron pair is hidden by the inherent uncertainty.

The Weak Force

A theory of the weak force, which acts on all particles, was developed by analogy with the theory of the electromagnetic force. The messenger particles that transmit the weak force between particles, however, are not (massless) photons but massive particles, identified by the symbols W and Z. The theory was so successful that it revealed the electromagnetic force and the weak force as being different aspects of a single **electroweak force.** This accomplishment is a logical extension of the work of Maxwell, who revealed the electric and magnetic forces as being different aspects of a single *electromagnetic* force.

The electroweak theory was specific in predicting the properties of the messenger particles. Their charges and masses, for example, were predicted to be

Particle	Charge	Mass
W	$\pm e$	80.6 GeV/c^2
Z	0	91.2 GeV/c^2

Recall that the proton mass is only 0.938 GeV/c^2; these are massive particles! The 1979 Nobel Prize in physics was awarded to Sheldon Glashow, Steven Weinberg, and Abdus Salam for their development of the electroweak theory.

The theory was confirmed in 1983 by Carlo Rubbia and his group at CERN, who experimentally verified both messenger particles and found that their masses agreed with the predicted values. The 1984 Nobel Prize in physics went to Rubbia and Simon van der Meer for this brilliant experimental work.

Some notion of the complexity of particle physics in this day and age can be found by looking at an earlier particle physics experiment that led to the Nobel Prize in physics—the discovery of the neutron. This vitally important discovery was a "tabletop" experiment, employing particles emitted by naturally occurring radioactive materials as projectiles; it was reported in 1932 under the title "Possible Existence of a Neutron," the single author being James Chadwick.

The discovery of the W and Z messenger particles in 1983, by contrast, was carried out at a large particle accelerator, about 7 km in circumference and operating in the range of several hundred billion electron-volts. The principal particle detector alone weighed 20 MN. The experiment employed more than 130 physicists from 12 institutions in 8 countries, along with a large support staff.

The Strong Force

A theory of the strong force—that is, the force that acts between quarks to bind hadrons together—has also been developed. The messenger particles in this case are called **gluons** and, like the photon, they are predicted to be massless. The theory assumes that each "flavor" of quark comes in three varieties that, for convenience, have been labeled *red, yellow,* and *blue.* Thus, there are three up quarks, one of each color, and so on. The antiquarks also come in three colors, which we call *antired, antiyellow,* and *antiblue.* You must not think that quarks are actually colored, like tiny jelly beans. The names are labels of convenience, but (for once) they do have a certain formal justification, as you will see.

The force acting between quarks is called a **color force** and the underlying theory, by analogy with quantum electrodynamics (QED), is called **quantum chromodynamics** (QCD). Apparently, quarks can be assembled only in combinations that are *color-neutral.*

There are two ways to bring about color neutrality. In the theory of actual colors, red + yellow + blue yields white, which is color-neutral, and we use the same scheme in dealing with quarks. Thus we can assemble three quarks to form a baryon, provided one is a yellow quark, one is a red quark, and one is a blue quark. Antired + antiyellow + antiblue is also white, so that we can assemble three antiquarks (of the proper anticolors) to form an antibaryon. Finally, red + antired, or yellow + antiyellow, or blue + antiblue also yields white. Thus, we can assemble a quark–antiquark combination to form a meson. The color-neutral rule does not permit any other combination of quarks, and none are observed.

The color force not only acts to bind together quarks as baryons and mesons, but it also acts between such particles, in which case it has traditionally been called the strong force. Hence, not only does the color force bind together quarks to form protons and neutrons, but it also binds together the protons and neutrons to form nuclei.

Einstein's Dream

The unification of the fundamental forces of nature into a single force—which occupied Einstein's attention for much of his later life—is very much a current focus of research. We have seen that the weak force has been successfully combined with electromagnetism so that they may be jointly viewed as aspects of a single *electroweak force.* Theories that attempt to add the strong force to this combination—called *grand unification theories* (GUTs)—are being pursued actively. Theories that seek to complete the job by adding gravity—sometimes called *theories of everything* (TOE)—are at an encouraging but speculative stage at this time.

44-10 A Pause for Reflection

Let us put what you have just learned in perspective. If all we are interested in is the structure of the world around us, we can get along nicely with the electron, the neutrino, the neutron, and the proton. As someone has said, we can operate "Spaceship Earth" quite well with just these particles. We can see a few of the more exotic particles by looking for them in the cosmic rays; however, to see most of them, we must build massive accelerators and look for them at great effort and expense.

The reason we must go to such effort is that—measured in energy terms—we live in a world of very low temperatures. Even at the center of the Sun, the value of kT is only about 1 keV. To produce the exotic particles, we must be able to accelerate protons or electrons to energies in the GeV and TeV range and higher.

Once upon a time the temperature everywhere *was* high enough to provide such energies. That time of extremely high temperatures occurred in the **big bang** beginning of the universe, when the universe (and both space and time) came into existence. Thus, one reason scientists study particles at high energies is to understand what the universe was like just after it began.

As we shall discuss shortly, *all* of space within the universe was initially tiny in extent, and the temperature of the particles within that space was incredibly high. With time, however, the universe expanded and cooled to lower temperatures, eventually to the size and temperature we see today.

Actually, the phrase "we see today" is complicated: When we look out into space, we are actually looking back in time because the light from the stars and galaxies has taken a long time to reach us. The most distant objects that we can detect are **quasars** (*qua*si*s*tell*ar* objects), which are the extremely bright cores of galaxies that are as much as 13×10^9 ly from us. Each such core contains a gigantic black hole; as material (gas and even stars) is pulled into one of those black holes, the material heats up and radiates a tremendous amount of light, enough for us to detect in spite of the huge distance. We therefore "see" a quasar not as it looks today but rather as it once was, when that light began its journey to us billions of years ago.

44-11 The Universe Is Expanding

As we saw in Section 37-10, it is possible to measure the relative speeds at which galaxies are approaching us or receding from us by measuring the shifts in the wavelength of the light they emit. If we look only at distant galaxies, beyond our immediate galactic neighbors, we find an astonishing fact: They are *all* moving away (receding) from us!

In 1929 Edwin P. Hubble established a connection between the apparent speed of recession v of a galaxy and its distance r from us—namely, that they are directly proportional:

$$v = Hr \qquad \text{(Hubble's law)}, \tag{44-19}$$

in which H is called the **Hubble constant.** The value of H is usually measured in the unit kilometers per second-megaparsec (km/s · Mpc), where the megaparsec is a length unit commonly used in astrophysics and astronomy:

$$1 \text{ Mpc} = 3.084 \times 10^{19} \text{ km} = 3.260 \times 10^6 \text{ ly}. \tag{44-20}$$

The Hubble constant H has not had the same value since the universe began. Determining its current value is extremely difficult because doing so involves measurements of very distant galaxies. However, evidence in the image that opens this chapter has allowed researchers to nail down the value of H to a value of

$$H = 71.0 \text{ km/s} \cdot \text{Mpc} = 21.8 \text{ mm/s} \cdot \text{ly}. \tag{44-21}$$

We interpret the recession of the galaxies to mean that the universe is expanding, much as the raisins in what is to be a loaf of raisin bread grow farther apart as the dough expands. Observers on all other galaxies would find that distant galaxies were rushing away from them also, in accordance with Hubble's law. In keeping with our analogy, we can say that no raisin (galaxy) has a unique or preferred view.

Hubble's law is consistent with the hypothesis that the universe began with the big bang and has been expanding ever since. If we assume that the rate of expansion has been constant (that is, the value of H has been constant), then we can estimate the age T of the universe by using Eq. 44-19. Let us also assume that since the big bang, any given part of the universe (say, a galaxy) has been receding from our location at a speed v given by Eq. 44-19. Then the time required for the given part to recede a distance r is

$$T = \frac{r}{v} = \frac{r}{Hr} = \frac{1}{H} \qquad \text{(estimated age of universe)}. \tag{44-22}$$

For the value of H in Eq. 44-21, T works out to be 13.8×10^9 y. Much more sophisticated studies of the expansion of the universe put T at 13.7×10^9 y.

Sample Problem 44-6

The wavelength shift in the light from a particular quasar indicates that the quasar has a recessional speed of 2.8×10^8 m/s (which is 93% of the speed of light). Approximately how far from us is the quasar?

Solution: The **Key Idea** here is to apply Hubble's law to the given speed v. From Eqs. 44-19 and 44-21, we find

$$r = \frac{v}{H} = \frac{2.8 \times 10^8 \text{ m/s}}{21.8 \text{ mm/s} \cdot \text{ly}}(1000 \text{ mm/m})$$
$$= 12.8 \times 10^9 \text{ ly}. \qquad \text{(Answer)}$$

This is only an approximation because the quasar has not always been receding from our location at the same speed v; that is, H has not had its current value throughout the time during which the universe has been expanding.

Sample Problem 44-7

A particular emission line detected in the light from a galaxy has a detected wavelength $\lambda_{det} = 1.1\lambda$, where λ is the proper wavelength of the line. What is the galaxy's distance from us?

Solution: One **Key Idea** here is to assume that Hubble's law ($v = Hr$) applies to the recession of the galaxy. A second **Key Idea** is to assume that the astronomical Doppler shift of Eq. 37-36 ($v = c\,\Delta\lambda/\lambda$, for $v \ll c$) applies to the shift in wavelength due to the recession. We can then set the right side of these two equations equal to each other to write

$$Hr = \frac{c\,\Delta\lambda}{\lambda}, \qquad (44\text{-}23)$$

which leads us to

$$r = \frac{c\,\Delta\lambda}{H\lambda}. \qquad (44\text{-}24)$$

In this equation,

$$\Delta\lambda = \lambda_{det} - \lambda = 1.1\lambda - \lambda = 0.1\lambda.$$

Substituting this into Eq. 44-24 then gives us

$$r = \frac{c(0.1\lambda)}{H\lambda} = \frac{0.1c}{H}$$
$$= \frac{(0.1)(3.0 \times 10^8 \text{ m/s})}{21.8 \text{ mm/s} \cdot \text{ly}}(1000 \text{ mm/m})$$
$$= 1.4 \times 10^9 \text{ ly}. \qquad \text{(Answer)}$$

44-12 The Cosmic Background Radiation

In 1965 Arno Penzias and Robert Wilson, of what was then the Bell Telephone Laboratories, were testing a sensitive microwave receiver used for communications research. They discovered a faint background "hiss" that remained unchanged in intensity no matter where their antenna was pointed. It soon became clear that Penzias and Wilson were observing a **cosmic background radiation,** generated in the early universe and filling all space almost uniformly. Currently this radiation has a maximum intensity at a wavelength of 1.1 mm, which lies in the microwave region of electromagnetic radiation (or light, for short). The wavelength distribution of this radiation matches the wavelength distribution of light that would be emitted by a laboratory enclosure with walls at a temperature of 2.7 K. Thus, for the cosmic background radiation, we say that the enclosure is the entire universe and that the universe is at an (average) temperature of 2.7 K. For their discovery of the cosmic background radiation, Penzias and Wilson were awarded the 1978 Nobel Prize in physics.

As we discuss in Section 44-14, the cosmic background radiation is now known to be light that has been in flight across the universe since shortly after the universe began billions of years ago. When the universe was even younger, light could scarcely go any significant distance without being scattered by all the individual, high-speed particles along its path. If a light ray started from, say, point A, it would be scattered in so many directions that if you could have intercepted part of it, you would have not been able to tell that it originated at point A. However, after the particles began to form atoms, the scattering of light greatly decreased. A light ray from point A might then be able to travel for billions of years without being scattered. This light is the cosmic background radiation.

As soon as the nature of the radiation was recognized, researchers wondered, "Can we use this incoming radiation to distinguish the points at which it originated, so that we then can produce an image of the early universe, back when atoms first formed and light scattering largely ceased?" The answer is yes. The image that opens this chapter is such an image.

44-13 Dark Matter

At the Kitt Peak National Observatory in Arizona, Vera Rubin and her coworker Kent Ford measured the rotational rates of a number of distant galaxies. They did so by measuring the Doppler shifts of bright clusters of stars located within each galaxy at various distances from the galactic center. As Fig. 44-5 shows, their results were surprising: The orbital speed of stars at the outer visible edge of the galaxy is about the same as that of stars close to the galactic center.

As the solid curve in Fig. 44-5 attests, that is not what we would expect to find if all the mass of the galaxy were represented by visible light. Nor is the pattern found by Rubin and Ford what we find in the solar system. For example, the orbital speed of Pluto (the planet most distant from the Sun) is only about one-tenth that of Mercury (the planet closest to the Sun).

The only explanation for the findings of Rubin and Ford that is consistent with Newtonian mechanics is that a typical galaxy contains much more matter than what we can actually see. In fact, the visible portion of a galaxy represents only about 5 to 10% of the total mass of the galaxy. In addition to these studies of galactic rotation, many other observations lead to the conclusion that the universe abounds in matter that we cannot see. This unseen matter is called **dark matter** because either it does not emit light or its light emission is too dim for us to detect.

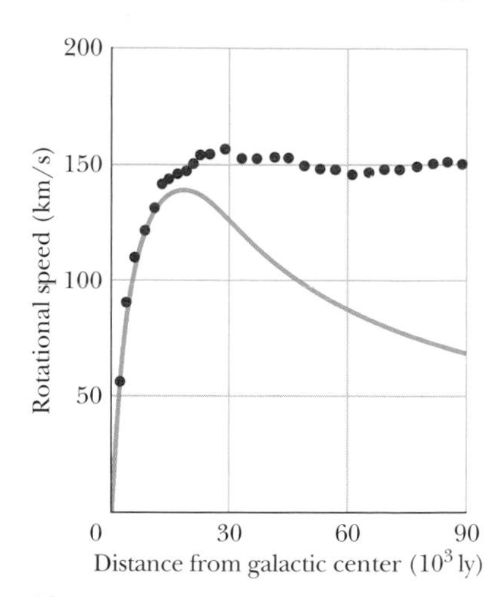

Fig. 44-5 The rotational speed of stars in a typical galaxy as a function of their distance from the galactic center. The theoretical solid curve shows that if a galaxy contained only the mass that is visible, the observed rotational speed would drop off with distance at large distances. The dots are the experimental data, which show that the rotational speed is approximately constant at large distances.

Normal matter (such as stars, planets, dust, and molecules) is often called **baryonic matter** because its mass is primarily due to the combined mass of the protons and neutrons (baryons) it contains. (The mass of the electrons is neglected because the mass of an electron is so small relative to the mass of a proton or a neutron.) Some of the normal matter, such as burned-out stars and dim interstellar gas, is part of the dark matter in a galaxy.

However, according to various calculations, this dark normal matter is only a small part of the total dark matter. The rest is called **nonbaryonic dark matter** because it does not contain protons and neutrons. We know of only one member of this type of dark matter—the neutrinos. Although the mass of a neutrino is very small relative to the mass of a proton or neutron, the number of neutrinos in a galaxy is huge and thus the total mass of the neutrinos is large. Nevertheless, calculations indicate that not even the total mass of the neutrinos is enough to account for the total mass of the nonbaryonic dark matter. In spite of over a hundred years in which elementary particles have been detected and studied, the particles that make up the rest of this type of dark matter are undetected and their nature is unknown.

44-14 The Big Bang

In 1985, a physicist remarked at a scientific meeting:

> *It is as certain that the universe started with a big bang about 15 billion years ago as it is that the Earth goes around the Sun.*

This strong statement suggests the level of confidence in which the big bang theory, first advanced by Belgian physicist Georges Lemaître, is held by those who study these matters.

You must not imagine that the big bang was like the explosion of some gigantic firecracker and that, in principle at least, you could have stood to one side and watched. There was no "one side" because the big bang represents the beginning of spacetime itself. From the point of view of our present universe, there is no position in space to which you can point and say, "The big bang happened there." It happened everywhere.

Moreover, there was no "before the big bang," because time *began* with that creation event. In this context, the word "before" loses its meaning. We can, however, conjecture about what went on during succeeding intervals of time after the big bang.

$t \approx$ **10^{-43} s.** This is the earliest time at which we can say anything meaningful about the development of the universe. It is at this moment that the concepts of space and time come to have their present meanings and the laws of physics as we know them become applicable. At this instant, the entire universe (that is, the *entire* spatial extent of the universe) is much smaller than a proton and its temperature is about 10^{32} K.

$t \approx$ **10^{-34} s.** By this moment the universe has undergone a tremendously rapid inflation, increasing in size by a factor of about 10^{30}. It has become a hot soup of photons, quarks, and leptons at a temperature of about 10^{27} K, which is too hot for protons and neutrons to form.

$t \approx$ **10^{-4} s.** Quarks can now combine to form protons and neutrons and their antiparticles. The universe has now cooled to such an extent by continued (but much slower) expansion that photons lack the energy needed to break up these new particles. Particles of matter and antimatter collide and annihilate each other. There is a slight excess of matter, which, failing to find annihilation partners, survives to form the world of matter that we know today.

$t \approx$ **1 min.** The universe has now cooled enough so that protons and neutrons, in colliding, can stick together to form the low-mass nuclei ^{2}H, ^{3}He, ^{4}He, and ^{7}Li. The predicted relative abundances of these nuclides are just what we observe in the universe today. Also, there is plenty of radiation present at $t \approx 1$ min, but this light cannot travel far before it interacts with a nucleus. Thus the universe is opaque.

$t \approx$ **379 000 y.** The temperature has now fallen to 2970 K, and electrons can stick to bare nuclei when the two collide, forming atoms. Because light does not interact appreciably with (uncharged) particles, such as neutral atoms, the light is now free to travel great distances. This radiation forms the cosmic background radiation discussed in Section 44-12. Atoms of hydrogen and helium, under the influence of gravity, begin to clump together, starting the formation of galaxies and stars.

Early measurements suggested that the cosmic background radiation is uniform in all directions, implying that 379 000 y after the big bang all matter in the universe was uniformly distributed. This finding was most puzzling because matter in the present universe is not uniformly distributed, but instead is collected in galaxies, clusters of galaxies, and superclusters of galactic clusters. There are also vast *voids* in which there is relatively little matter, and there are regions so crowded with matter that they are called *walls.* If the big bang theory of the beginning of the universe is even approximately correct, the seeds for this nonuniform distribution of matter must have been in place before the universe was 379 000 y old and now should show up as a nonuniform distribution of the microwave background radiation.

In 1992, measurements made by NASA's Cosmic Background Explorer (COBE) satellite revealed that the background radiation is, in fact, not perfectly uniform. In 2003, measurements by NASA's Wilkinson Microwave Anisotropy Probe (WMAP) greatly increased our resolution of this nonuniformity. The image resulting from the WMAP measurements opens this chapter and is effectively a color-coded photograph of the universe when the universe was only 379 000 y old. As you can see from the variations in the colors, large-scale collecting of matter had already begun. Thus, the big bang theory and the theory of inflation at $t \approx 10^{-34}$ s are, in principle, on the right track.

The Accelerated Expansion of the Universe

Recall from Section 13-9 the statement that mass causes curvature of space. Now that we have seen that mass is a form of energy, as given by Einstein's equation $E = mc^2$, we can generalize the statement: energy can cause curvature of space. This certainly happens to the space around the energy packed into a black hole and, more weakly, to the space around any other astronomical body, but is the space of the universe as a whole curved by the energy the universe contains?

The question was answered first by the 1992 COBE measurements of the cosmic background radiation. It was then answered more definitively by the 2003 WMAP measurements that produced the image opening this chapter. The spots we see in that image are the original sources of the cosmic background radiation, and the angular distribution of the spots reveals the curvature of the universe through which the light has to travel to reach us. If adjacent spots subtend either more than 1° (Fig. 44-6*a*) or less than 1° (Fig. 44-6*b*) in the detector's view (or our view) into the universe, then the universe is curved. Analysis of the spot distribution in the WMAP image shows that the spots subtend about 1° (Fig. 44-6*c*), which means that the universe is *flat* (having no curvature). Thus, the initial curvature the universe presumably had when it began must have been flattened out by the rapid inflation the universe underwent at $t \approx 10^{-34}$ s.

This flatness poses a very difficult problem for physicists because it requires that the universe contain a certain amount of energy (as mass or otherwise). The trouble is that all estimations of the amount of energy in the universe (both in known forms and in the form of the unknown type of dark matter) fall dramatically short of the required amount. Indeed, about two-thirds of the required energy is missing from the estimations.

One theory proposed about this missing energy gave it the gothic name of *dark energy* and predicted that it has the strange property of causing the expansion of the universe to accelerate. Until 1998, determining whether the expansion is, in fact, accelerating was very difficult because it requires measuring distances to very distant astronomical bodies where the acceleration might show up.

In 1998, however, advances in astronomical technology allowed astronomers to detect a certain type of supernovae at very great distances. More important, the astronomers could measure the duration of the burst of light from such a supernova. The duration reveals the brightness of the supernova that would be seen by an observer near the supernova. By measuring the brightness of the supernova as seen from Earth, astronomers could then determine the distance to the supernova. From the redshift of the light from the galaxy containing the supernova, astronomers could also determine how fast the galaxy is receding from us. Combining all this information, they could then calculate the expansion rate of the universe. The conclusion is that the expansion is indeed accelerating as predicted by the theory of dark energy. However, we still scarcely have a clue as to what this dark energy is.

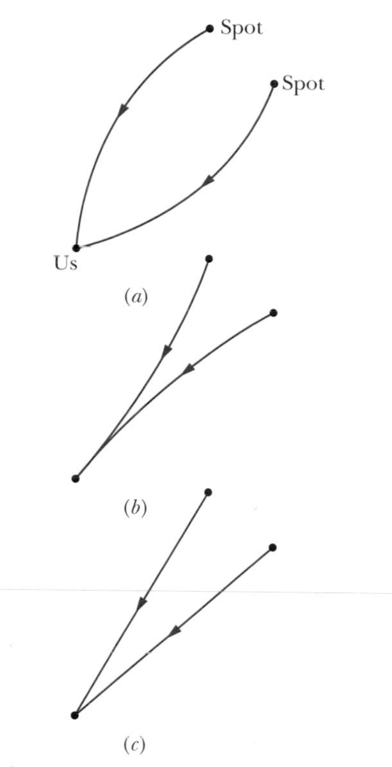

Fig. 44-6 Light rays from two adjacent spots in our view of the cosmic background radiation would reach us at an angle (*a*) greater than 1° or (*b*) less than 1° if the space along the light-ray paths through the universe were curved. (*c*) An angle of 1° means that the space is not curved.

44-15 A Summing Up

Let us, in these closing paragraphs, consider where our rapidly accumulating store of knowledge about the universe is leading us. That it provides satisfaction to a host of curiosity-motivated physicists and astronomers is beyond dispute. However, some view it as a humbling experience in that each increase in knowledge seems to reveal more clearly our own relative insignificance in the grand scheme of things. Thus, in roughly chronological order, we humans have come to realize that

Our Earth is not the center of the solar system.

Our Sun is but one star among many in our galaxy.

Our galaxy is but one of many, and our Sun is an insignificant star in it.

Our Earth has existed for perhaps only a third of the age of the universe and will surely disappear when our Sun burns up its fuel and becomes a red giant.

Our species has inhabited Earth for less than a million years—a blink in cosmological time.

Although our position in the universe may be insignificant, the laws of physics that we have discovered (uncovered?) seem to hold throughout the universe and—as far as we know—have held since the universe began and will continue to hold for all future time. At least, there is no evidence that other laws hold in other parts of the universe. Thus, until someone complains, we are entitled to stamp the laws of physics "Discovered on Earth." Much remains to be discovered: *"The universe is full of magical things, patiently waiting for our wits to grow sharper."* That declaration allows us to answer one last time the question "What is physics?" that we have explored repeatedly in this book. Physics is the gateway to those magical things.

Review & Summary

Leptons and Quarks Current research supports the view that all matter is made of six kinds of **leptons** (Table 44-2), six kinds of **quarks** (Table 44-5), and 12 **antiparticles,** one corresponding to each lepton and each quark. All these particles have spin quantum numbers equal to $\frac{1}{2}$ and are thus **fermions** (particles with half-integer spin quantum numbers).

The Interactions Particles with electric charge interact through the electromagnetic force by exchanging **virtual photons.** Leptons can interact with each other and with quarks through the **weak force,** via massive W and Z particles as messengers. In addition, quarks interact with each other through the **color force.** The electromagnetic and weak forces are different manifestations of the same force, called the **electroweak force.**

Leptons Three of the leptons (the **electron, muon,** and **tau**) have electric charge equal to $-1e$. There are also three uncharged **neutrinos** (also leptons), one corresponding to each of the charged leptons. The antiparticles for the charged leptons have positive charge.

Quarks The six quarks (up, down, strange, charm, bottom, and top, in order of increasing mass) each have baryon number $+\frac{1}{3}$ and charge equal to either $+\frac{2}{3}e$ or $-\frac{1}{3}e$. The strange quark has strangeness -1, whereas the others all have strangeness 0. These four algebraic signs are reversed for the antiquarks.

Hadrons: Baryons and Mesons Quarks combine into strongly interacting particles called **hadrons. Baryons** are hadrons with half-integer spin quantum numbers ($\frac{1}{2}$ or $\frac{3}{2}$). **Mesons** are hadrons with integer spin quantum numbers (0 or 1) and thus are **bosons.** Baryons are fermions. Mesons have baryon number equal to zero; baryons have baryon number equal to $+1$ or -1. **Quantum chromodynamics** predicts that the possible combinations of quarks are either a quark with an antiquark, three quarks, or three antiquarks (this prediction is consistent with experiment).

Expansion of the Universe Current evidence strongly suggests that the universe is expanding, with the distant galaxies moving away from us at a rate v given by **Hubble's law:**

$$v = Hr \quad \text{(Hubble's law).} \tag{44-19}$$

Here we take H, the **Hubble constant,** to have the value

$$H = 71.0 \text{ km/s} \cdot \text{Mpc} = 21.8 \text{ mm/s} \cdot \text{ly}. \tag{44-21}$$

The expansion described by Hubble's law and the presence of ubiquitous background microwave radiation reveal that the universe began in a "big bang" 13.7 billion years ago.

Questions

1 Is the direction of the magnetic field in Fig. 44-2 out of the page or into the page?

2 Not only particles such as electrons and protons but also entire atoms can be classified as fermions or bosons, depending on whether their overall spin quantum numbers are, respectively, half-integral or integral. Consider the helium isotopes ^{3}He and ^{4}He. Which of the following statements is correct? (a) Both are fermions. (b) Both are bosons. (c) ^{4}He is a fermion, and ^{3}He is a boson. (d) ^{3}He is a fermion, and ^{4}He is a boson. (The two helium electrons form a closed shell and play no role in this determination.)

3 Which of the eight pions in Fig. 44-2*b* has the least kinetic energy?

4 An electron cannot decay into two neutrinos. Which of the following conservation laws would be violated if it did: (a) energy, (b) angular momentum, (c) charge, (d) lepton number, (e) linear momentum, (f) baryon number?

5 As we have seen, the π^- meson has the quark structure $d\bar{u}$. Which of the following conservation laws would be violated if a π^- were formed, instead, from a d quark and a u quark: (a) energy, (b) angular momentum, (c) charge, (d) lepton number, (e) linear momentum, (f) baryon number?

6 A proton cannot decay into a neutron and a neutrino. Which of the following conservation laws would be violated if it did: (a) energy (assume the proton is stationary), (b) angular momentum, (c) charge, (d) lepton number, (e) linear momentum, (f) baryon number?

7 A proton has enough mass energy to decay into a shower made up of electrons, neutrinos, and their antiparticles. Which of the following conservation laws would necessarily be violated if it did: electron lepton number or baryon number?

8 Consider the neutrino whose symbol is $\bar{\nu}_\tau$. (a) Is it a quark, a lepton, a meson, or a baryon? (b) Is it a particle or an antiparticle? (c) Is it a boson or a fermion? (d) Is it stable against spontaneous decay?

9 A Σ^+ particle has these quantum numbers: strangeness $S = -1$, charge $q = +1$, and spin $s = \frac{1}{2}$. Which of the following quark combinations produces it: (a) dds, (b) $s\bar{s}$, (c) uus, (d) ssu, or (e) $uu\bar{s}$?

10 The left column that follows lists ideas from atomic physics, and the right column lists ideas from particle physics. Match the items in the two columns.

1. chemistry	a. eightfold way patterns
2. electrons	b. missing hadrons
3. periodic table	c. quantum chromodynamics
4. missing elements	d. particle physics
5. quantum mechanics	e. quarks

11 What is the lepton number of (a) π^-, (b) e^-, (c) μ^+, (d) τ^-, (e) $\bar{\nu}_\mu$?

12 List these particles in order of mass, least first: (a) proton, (b) neutrino, (c) π^+ meson, (d) strange quark, (e) tau, (f) electron, and (g) Σ^-.

13 Match the items in these two columns:

1. tau	a. quark
2. pion	b. lepton
3. proton	c. meson
4. positron	d. baryon
5. charm	e. antiparticle

Problems

SSM Solution is in the Student Solutions Manual.
WWW Solution is at http://www.wiley.com/college/halliday
• – ••• Number of dots indicates level of problem difficulty.

sec. 44-3 An Interlude

•1 A neutral pion initially at rest decays into two gamma rays: $\pi^0 \rightarrow \gamma + \gamma$. Calculate the wavelength of the gamma rays.

•2 An electron and a positron are separated by distance r. Find the ratio of the gravitational force to the electric force between them. From the result, what can you conclude concerning the forces acting between particles detected in a bubble chamber?

•3 Certain theories predict that the proton is unstable, with a half-life of about 10^{32} years. Assuming that this is true, calculate the number of proton decays you would expect to occur in one year in the water of an Olympic-sized swimming pool holding 4.32×10^5 L of water.

•4 A positively charged pion decays by Eq. 44-7: $\pi^+ \rightarrow \mu^+ + \nu$. What must be the decay scheme of the negatively charged pion? (*Hint:* The π^- is the antiparticle of the π^+.)

••5 Observations of neutrinos emitted by the supernova SN1987a (Fig. 43-11*b*) place an upper limit of 20 eV on the rest energy of the electron neutrino. If the rest energy of the electron neutrino were, in fact, 20 eV, what would be the speed difference between light and a 1.5 MeV electron neutrino?

••6 A positive tau (τ^+, rest energy = 1777 MeV) is moving with 2200 MeV of kinetic energy in a circular path perpendicular to a uniform 1.20 T magnetic field. (a) Calculate the momentum of the tau in kilogram-meters per second. Relativistic effects must be considered. (b) Find the radius of the circular path.

••7 The rest energy of many short-lived particles cannot be measured directly but must be inferred from the measured momenta and known rest energies of the decay products. Consider the ρ^0 meson, which decays by the reaction $\rho^0 \rightarrow \pi^+ + \pi^-$. Calculate the rest energy of the ρ^0 meson given that the oppositely directed momenta of the created pions each have magnitude 358.3 MeV/c. See Table 44-4 for the rest energies of the pions. SSM

••8 A neutral pion has a rest energy of 135 MeV and a mean life of 8.3×10^{-17} s. If it is produced with an initial kinetic energy of 80 MeV and decays after one mean lifetime, what is the longest possible track this particle could leave in a bubble chamber? Use relativistic time dilation.

••9 (a) A stationary particle 1 decays into particles 2 and 3, which move off with equal but oppositely directed momenta. Show that the kinetic energy K_2 of particle 2 is given by

$$K_2 = \frac{1}{2E_1}[(E_1 - E_2)^2 - E_3^2],$$

where E_1, E_2, and E_3 are the rest energies of the particles. (b) Show that the result in (a) yields the kinetic energy of the muon as calculated in Sample Problem 44-1.

sec. 44-6 Still Another Conservation Law

•10 The A_2^+ particle and its products decay according to the scheme

$$A_2^+ \to \rho^0 + \pi^+, \qquad \mu^+ \to e^+ + \nu + \bar{\nu},$$
$$\rho^0 \to \pi^+ + \pi^-, \qquad \pi^- \to \mu^- + \bar{\nu},$$
$$\pi^+ \to \mu^+ + \nu, \qquad \mu^- \to e^- + \nu + \bar{\nu}.$$

(a) What are the final stable decay products? From the evidence, (b) is the A_2^+ particle a fermion or a boson and (c) is it a meson or a baryon? (d) What is its baryon number? (*Hint:* See Sample Problem 44-4.)

•11 Which conservation law is violated in each of these proposed decays? Assume that the initial particle is stationary and the decay products have zero orbital angular momentum. (a) $\mu^- \to e^- + \nu_\mu$; (b) $\mu^- \to e^+ + \nu_e + \bar{\nu}_\mu$; (c) $\mu^+ \to \pi^+ + \nu_\mu$. SSM WWW

sec. 44-7 The Eightfold Way

•12 By examining strangeness, determine which of the following decays or reactions proceed via the strong interaction: (a) $K^0 \to \pi^+ + \pi^-$; (b) $\Lambda^0 + p \to \Sigma^+ + n$; (c) $\Lambda^0 \to p + \pi^-$; (d) $K^- + p \to \Lambda^0 + \pi^0$.

•13 The reaction $\pi^+ + p \to p + p + \bar{n}$ proceeds via the strong interaction. By applying the conservation laws, deduce the (a) charge quantum number, (b) baryon number, and (c) strangeness of the antineutron.

•14 Calculate the disintegration energy of the reactions (a) $\pi^+ + p \to \Sigma^+ + K^+$ and (b) $K^- + p \to \Lambda^0 + \pi^0$.

•15 Which conservation law is violated in each of these proposed reactions and decays? (Assume that the products have zero orbital angular momentum.) (a) $\Lambda^0 \to p + K^-$; (b) $\Omega^- \to \Sigma^- + \pi^0$ ($S = -3$, $q = -1$, $m = 1672$ MeV/c^2, and $m_s = \frac{3}{2}$ for Ω^-); (c) $K^- + p \to \Lambda^0 + \pi^+$. SSM

•16 Show that if, instead of plotting strangeness S versus charge q for the spin-$\frac{1}{2}$ baryons in Fig. 44-3*a* and for the spin-zero mesons in Fig. 44-3*b*, we plot the quantity $Y = B + S$ versus the quantity $T_z = q - \frac{1}{2}(B + S)$, we get the hexagonal patterns without using sloping axes. (The quantity Y is called *hypercharge*, and T_z is related to a quantity called *isospin*.)

••17 Use the conservation laws and Tables 44-3 and 44-4 to identify particle x in each of the following reactions, which proceed by means of the strong interaction: (a) $p + p \to p + \Lambda^0 + x$; (b) $p + \bar{p} \to n + x$; (c) $\pi^- + p \to \Xi^0 + K^0 + x$.

••18 A Σ^- particle moving with 220 MeV of kinetic energy decays according to $\Sigma^- \to \pi^- + n$. Calculate the total kinetic energy of the decay products.

••19 Consider the decay $\Lambda^0 \to p + \pi^-$ with the Λ^0 at rest. (a) Calculate the disintegration energy. What is the kinetic energy of (b) the proton and (c) the pion? (*Hint:* See Problem 9.)

sec. 44-8 The Quark Model

•20 From Tables 44-3 and 44-5, determine the identity of the baryon formed from quarks (a) ddu, (b) uus, and (c) ssd. Check your answers against the baryon octet shown in Fig. 44-3*a*.

•21 The quark makeups of the proton and neutron are uud and udd, respectively. What are the quark makeups of (a) the antiproton and (b) the antineutron?

•22 What quark combination is needed to form (a) Λ^0 and (b) Ξ^0?

•23 Using the up, down, and strange quarks only, construct, if possible, a baryon (a) with $q = +1$ and strangeness $S = -2$ and (b) with $q = +2$ and strangeness $S = 0$. SSM WWW

•24 There are 10 baryons with spin $\frac{3}{2}$. Their symbols and quantum numbers for charge q and strangeness S are as follows:

	q	S		q	S
Δ^-	−1	0	Σ^{*0}	0	−1
Δ^0	0	0	Σ^{*+}	+1	−1
Δ^+	+1	0	Ξ^{*-}	−1	−2
Δ^{++}	+2	0	Ξ^{*0}	0	−2
Σ^{*-}	−1	−1	Ω^-	−1	−3

Make a charge–strangeness plot for these baryons, using the sloping coordinate system of Fig. 44-3. Compare your plot with this figure.

••25 The spin-$\frac{3}{2}$ Σ^{*0} baryon (see Problem 24) has a rest energy of 1385 MeV (with an intrinsic uncertainty ignored here); the spin-$\frac{1}{2}$ Σ^0 baryon has a rest energy of 1192.5 MeV. If each of these particles has a kinetic energy of 1000 MeV, (a) which is moving faster and (b) by how much?

sec. 44-11 The Universe Is Expanding

•26 If Hubble's law can be extrapolated to very large distances, at what distance would the apparent recessional speed become equal to the speed of light?

•27 What is the observed wavelength of the 656.3 nm (first Balmer) line of hydrogen emitted by a galaxy at a distance of 2.40×10^8 ly? Assume that the Doppler shift of Eq. 37-36 and Hubble's law apply.

•28 In the laboratory, one of the lines of sodium is emitted at a wavelength of 590.0 nm. In the light from a particular galaxy, however, this line is seen at a wavelength of 602.0 nm. Calculate the distance to the galaxy, assuming that Hubble's law holds and that the Doppler shift of Eq. 37-36 applies.

••29 Because the apparent recessional speeds of galaxies and quasars at great distances are close to the speed of light, the relativistic Doppler shift formula (Eq. 37-31) must be used. The shift is reported as fractional red shift $z = \Delta\lambda/\lambda_0$. (a) Show that, in terms of z, the recessional speed parameter $\beta = v/c$ is given by

$$\beta = \frac{z^2 + 2z}{z^2 + 2z + 2}.$$

(b) A quasar detected in 1987 has $z = 4.43$. Calculate its speed parameter. (c) Find the distance to the quasar, assuming that Hubble's law is valid to these distances.

••30 Will the universe continue to expand forever? To attack this question, make the (reasonable?) assumption that the recessional speed v of a galaxy a distance r from us is determined only by the matter that lies inside a sphere of radius r centered on us. If the total mass inside this sphere is M, the escape speed v_e from the sphere is $v_e = \sqrt{2GM/r}$ (Eq. 13-28). (a) Show that to prevent unlimited expansion, the average density ρ inside the sphere must be at least equal to

$$\rho = \frac{3H^2}{8\pi G}.$$

(b) Evaluate this "critical density" numerically; express your answer in terms of hydrogen atoms per cubic meter. Measurements of the actual density are difficult and are complicated by the presence of dark matter.

sec. 44-12 The Cosmic Background Radiation

••31 Due to the presence everywhere of the cosmic background radiation, the minimum possible temperature of a gas in interstellar or intergalactic space is not 0 K but 2.7 K. This implies that a significant fraction of the molecules in space that can be in a low-level excited state may, in fact, be so. Subsequent de-excitation would lead to the emission of radiation that could be detected. Consider a (hypothetical) molecule with just one possible excited state. (a) What would the excitation energy have to be for 25% of the molecules to be in the excited state? (*Hint:* See Eq. 40-29.) (b) What would be the wavelength of the photon emitted in a transition back to the ground state?

sec. 44-13 Dark Matter

•32 What would the mass of the Sun have to be if Pluto (the outermost planet most of the time) were to have the same orbital speed that Mercury (the innermost planet) has now? Use data from Appendix C, express your answer in terms of the Sun's current mass M_S, and assume circular orbits.

••33 Suppose that the radius of the Sun were increased to 5.90×10^{12} m (the average radius of the orbit of Pluto), that the density of this expanded Sun were uniform, and that the planets revolved within this tenuous object. (a) Calculate Earth's orbital speed in this new configuration. (b) What is the ratio of the orbital speed calculated in (a) to Earth's present orbital speed of 29.8 km/s? Assume that the radius of Earth's orbit remains unchanged. (c) What would be Earth's new period of revolution? (The Sun's mass remains unchanged.) SSM

••34 Suppose that the matter (stars, gas, dust) of a particular galaxy, of total mass M, is distributed uniformly throughout a sphere of radius R. A star of mass m is revolving about the center of the galaxy in a circular orbit of radius $r < R$. (a) Show that the orbital speed v of the star is given by

$$v = r\sqrt{GM/R^3},$$

and therefore that the star's period T of revolution is

$$T = 2\pi\sqrt{R^3/GM},$$

independent of r. Ignore any resistive forces. (b) Next suppose that the galaxy's mass is concentrated near the galactic center, within a sphere of radius less than r. What expression then gives the star's orbital period?

sec. 44-14 The Big Bang

•35 The wavelength at which a thermal radiator at temperature T radiates electromagnetic waves most intensely is given by Wien's law: $\lambda_{max} = (2898\ \mu\text{m}\cdot\text{K})/T$. (a) Show that the energy E of a photon corresponding to that wavelength can be computed from

$$E = (4.28 \times 10^{-10}\ \text{MeV/K})T.$$

(b) At what minimum temperature can this photon create an electron–positron pair (as discussed in Section 21-6)?

•36 Use Wien's law (see Problem 35) to answer the following questions: (a) The cosmic background radiation peaks in intensity at a wavelength of 1.1 mm. To what temperature does this correspond? (b) About 379 000 y after the big bang, the universe became transparent to electromagnetic radiation. Its temperature then was 2970 K. What was the wavelength at which the background radiation was then most intense?

Additional Problems

37 How much energy would be released if Earth were annihilated by collision with an anti-Earth?

38 *A particle game.* Figure 44-7 is a sketch of the tracks made by particles in a *fictional* cloud chamber experiment (with a uniform magnetic field directed perpendicular to the page), and Table 44-6 gives *fictional* quantum numbers associated

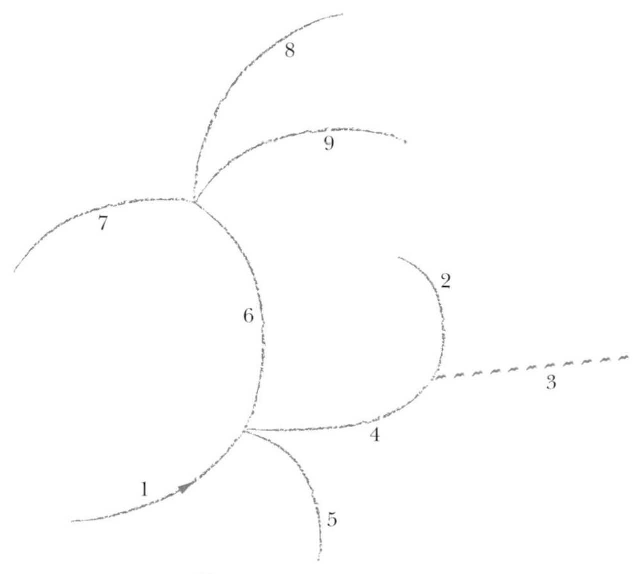

Fig. 44-7 Problem 38.

TABLE 44-6 Problem 44-38

Particle	Charge	Whimsy	Seriousness	Cuteness
A	1	1	−2	−2
B	0	4	3	0
C	1	2	−3	−1
D	−1	−1	0	1
E	−1	0	−4	−2
F	1	0	0	0
G	−1	−1	1	−1
H	3	3	1	0
I	0	6	4	6
J	1	−6	−4	−6

with the particles making the tracks. Particle A entered the chamber, leaving track 1 and decaying into three particles. Then the particle creating track 6 decayed into three other particles, and the particle creating track 4 decayed into two other particles, one of which was electrically uncharged—the path of that uncharged particle is represented by the dashed straight line. The particle that created track 8 is known to have a seriousness quantum number of zero.

By conserving the fictional quantum numbers at each decay point and by noting the directions of curvature of the tracks, identify which particle goes with track (a) 1, (b) 2, (c) 3, (d) 4, (e) 5, (f) 6, (g) 7, (h) 8, and (i) 9. One of the listed particles is not formed; the others appear only once each.

39 There is no known meson with charge quantum number $q = +1$ and strangeness $S = -1$ or with $q = -1$ and $S = +1$. Explain why in terms of the quark model.

40 Figure 44-8 is a hypothetical plot of the recessional speeds v of galaxies against their distance r from us; the best-fit straight line through the data points is shown. From this plot determine the age of the universe, assuming that Hubble's law holds and that Hubble's constant has had the same value throughout the expansion of the universe.

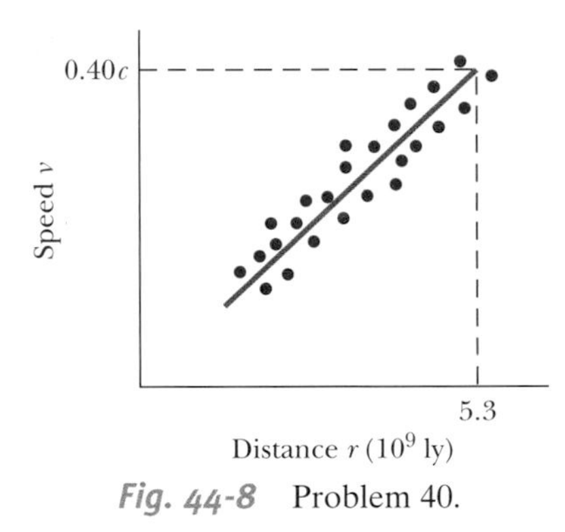

Fig. 44-8 Problem 40.

41 Figure 44-9 shows part of the experimental arrangement in which antiprotons were discovered in the 1950s. A beam of 6.2 GeV protons emerged from a particle accelerator and collided with nuclei in a copper target. According to theoretical predictions at the time, collisions between protons in the beam and the protons and neutrons in those nuclei should produce antiprotons via the reactions

$$p + p \rightarrow p + p + p + \bar{p}$$

and

$$p + n \rightarrow p + n + p + \bar{p}.$$

However, even if these reactions did occur, they would be rare compared to the reactions

$$p + p \rightarrow p + p + \pi^+ + \pi^-$$

and

$$p + n \rightarrow p + n + \pi^+ + \pi^-.$$

Thus, most of the particles produced by the collisions between the 6.2 GeV protons and the copper target were pions.

To prove that antiprotons exist and were produced by some limited number of the collisions, particles leaving the target were sent into a series of magnetic fields and detectors as shown in Fig. 44-8. The first magnetic field (M1) curved the path of any charged particle passing through it; moreover, the field was arranged so that the only particles that emerged from

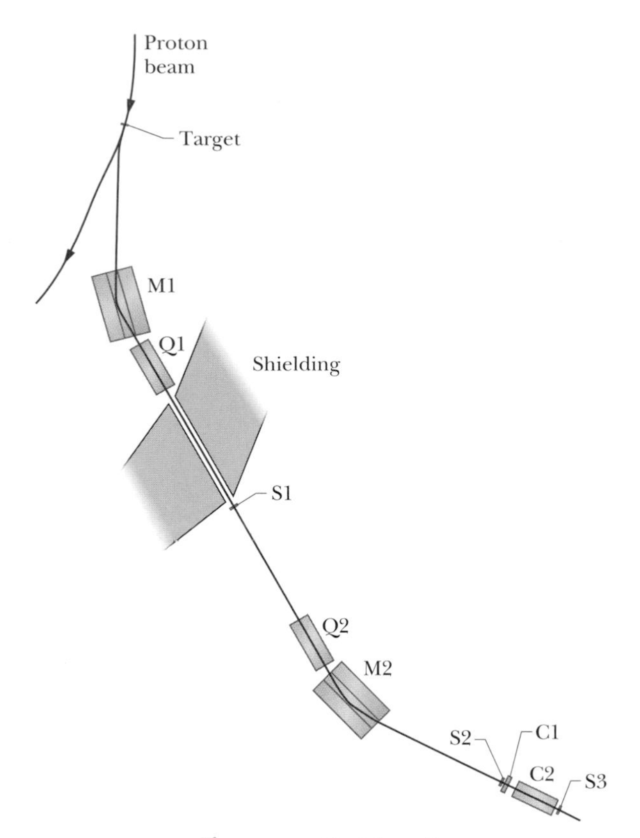

Fig. 44-9 Problem 41.

it to reach the second magnetic field (Q1) had to be negatively charged (either a $\bar{p}$ or a π^-) and have a momentum of 1.19 GeV/c. Field Q1 was a special type of magnetic field (a *quadrapole field*) that focused the particles reaching it into a beam, allowing them to pass through a hole in thick shielding to a *scintillation counter* S1. The passage of a charged particle through the counter triggered a signal, with each signal indicating the passage of either a 1.19 GeV/c π^- or (presumably) a 1.19 GeV/c $\bar{p}$.

After being refocused by magnetic field Q2, the particles were directed by magnetic field M2 through a second scintillation counter S2 and then through two *Cerenkov counters* C1 and C2. These latter detectors can be manufactured so that they send a signal only when the particle passing through them is moving with a speed that falls within a certain range. In the experiment, a particle with a speed greater than $0.79c$ would trigger C1 and a particle with a speed between $0.75c$ and $0.78c$ would trigger C2.

There were then two ways to distinguish the predicted rare antiprotons from the abundant negative pions. Both ways involved the fact that the speed of a 1.19 GeV/c $\bar{p}$ differs from that of a 1.19 GeV/c π^-: (1) According to calculations, a $\bar{p}$ would trigger one of the Cerenkov counters and a π^- would trigger the other. (2) The time interval Δt between signals from S1 and S2, which were separated by 12 m, would have one value for a $\bar{p}$ and another value for a π^-. Thus, if the correct Cerenkov counter were triggered and the time interval Δt had the correct value, the experiment would prove the existence of antiprotons.

What is the speed of (a) an antiproton with a momentum of 1.19 GeV/c and (b) a negative pion with that same momentum? (The speed of an antiproton through the Cerenkov detectors would actually be slightly less than calculated here because the antiproton would lose a little energy within the detectors.) Which Cerenkov detector was triggered by (c) an antiproton and (d) a negative pion? What time interval Δt indicated the passage of (e) an antiproton and (f) a negative pion? [Problem adapted from O. Chamberlain, E. Segrè, C. Wiegand, and T. Ypsilantis, "Observation of Antiprotons," *Physical Review,* Vol. 100, pp. 947–950 (1955).]

42 Verify that the hypothetical proton decay scheme in Eq. 44-14 does not violate the conservation law of (a) charge, (b) energy, and (c) linear momentum. (d) How about angular momentum?

43 *Cosmological redshift.* The expansion of the universe is often represented with a drawing like Fig. 44-10*a*. In that figure, we are located at the symbol labeled MW (for the Milky Way galaxy), at the origin of an r axis that extends radially away from us in any direction. Other, very distant galaxies are also represented. Superimposed on their symbols are their velocity vectors as inferred from the redshift of the light reaching us from the galaxies. In accord with Hubble's law, the speed of each galaxy is proportional to its distance from us. Such drawings can be misleading because they imply (1) that the redshifts are due to the motions of galaxies relative to us, as they rush away from us through static (stationary) space, and (2) that we are at the center of all this motion.

Actually, the expansion of the universe and the increased separation of the galaxies are due not to an outward rush of the galaxies into pre-existing space but to an expansion of space itself throughout the universe. *Space is dynamic, not static.*

Figures 44-10*b*, *c*, and *d* show a different way of representing the universe and its expansion. Each part of the figure gives part of a one-dimensional section of the universe (along an r axis); the other two spatial dimensions of the universe are not shown. Each of the three parts of the figure shows the Milky Way and six other galaxies (represented by dots); the parts are positioned along a time axis, with time increasing upward. In part *b*, at the earliest time of the three parts, the Milky Way and the six other galaxies are represented as being relatively close to one another. As time progresses upward in the figures, space expands, causing the galaxies to move apart. Note that the figure parts are drawn relative to the Milky Way, and from that observation point all the other galaxies move away because of the expansion. However, there is nothing special about the Milky Way—the galaxies also move away from any other observation point we might have chosen.

Figures 44-10*e* and *f* focus on just the Milky Way galaxy and one of the other galaxies, galaxy A, at two particular times during the expansion. In part *e*, galaxy A is a distance r from the Milky Way and is emitting a light wave of wavelength λ. In part *f*, after a time interval Δt, that light wave is being detected at Earth. Let us represent the universe's expansion rate per unit length of space with α, which we assume to be constant during time interval Δt. Then during Δt, every unit length of space (say, every meter) expands by an amount $\alpha\,\Delta t$; hence, a distance r expands by $r\alpha\,\Delta t$. The light wave of Figs. 44-10*e* and *f* travels at speed c from galaxy A to Earth. (a) Show that

$$\Delta t = \frac{r}{c - r\alpha}.$$

The detected wavelength λ' of the light is greater than the emitted wavelength λ because space expanded during time interval Δt. This increase in wavelength is called the **cosmological redshift;** it is not a Doppler effect. (b) Show that the change in wavelength $\Delta\lambda$ $(= \lambda' - \lambda)$ is given by

$$\frac{\Delta\lambda}{\lambda} = \frac{r\alpha}{c - r\alpha}.$$

(c) Expand the right side of this equation using the binomial expansion (given in Appendix E). (d) If you retain only the first term of the expansion, what is the resulting equation for $\Delta\lambda/\lambda$?

If, instead, we assume that Fig. 44-10*a* applies and that $\Delta\lambda$ is due to a Doppler effect, then from Eq. 37-36 we have

$$\frac{\Delta\lambda}{\lambda} = \frac{v}{c},$$

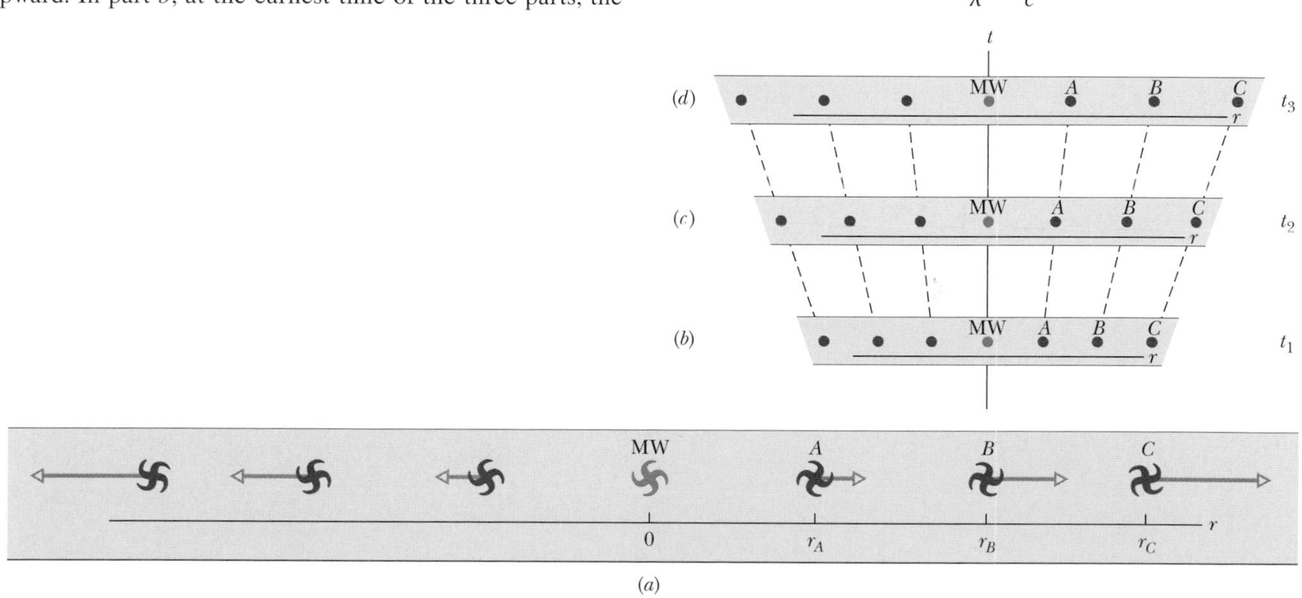

Fig. 44-10a–d Problem 43.

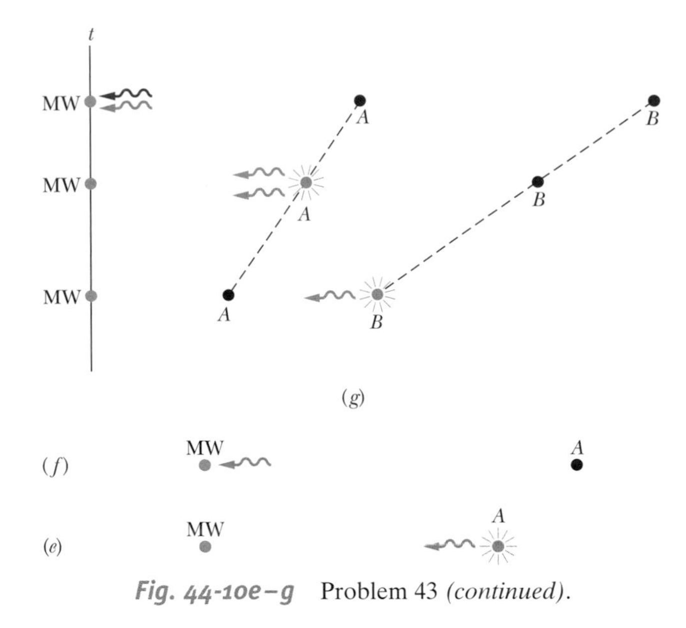

Fig. 44-10e–g Problem 43 *(continued)*.

where v is the radial velocity of galaxy A relative to Earth. (e) Using Hubble's law, compare this Doppler-effect result with the cosmological-expansion result of (d) and find a value for α. From this analysis you can see that the two results, derived with very different models about the redshift of the light we detect from distant galaxies, are compatible.

Suppose that the light we detect from galaxy A has a redshift of $\Delta\lambda/\lambda = 0.050$ and that the expansion rate of the universe has been constant at the current value given in the chapter. (f) Using the result of (b), find the distance between the galaxy and Earth when the light was emitted. Next, determine how long ago the light was emitted by the galaxy (g) by using the result of (a) and (h) by assuming that the redshift is a Doppler effect. (*Hint:* For (h), the time is just the distance at the time of emission divided by the speed of light, because if the redshift is just a Doppler effect, the distance does not change during the light's travel to us. Here the two models about the redshift of the light differ in their results.) (i) At the time of detection, what is the distance between Earth and galaxy A? (We make the assumption that galaxy A still exists; if it ceased to exist, humans would not know about its death until the last light emitted by the galaxy reached Earth.)

Now suppose that the light we detect from galaxy B (Fig. 44-10g) has a redshift of $\Delta\lambda/\lambda = 0.080$. (j) Using the result of (b), find the distance between galaxy B and Earth when the light was emitted. (k) Using the result of (a), find how long ago the light was emitted by galaxy B. (l) When the light that we detect from galaxy A was emitted, what was the distance between galaxy A and galaxy B?

44 Calculate the difference in mass, in kilograms, between the muon and pion of Sample Problem 44-1.

The International System of Units (SI)*

A

TABLE 1
The SI Base Units

Quantity	Name	Symbol	Definition
length	meter	m	". . . the length of the path traveled by light in vacuum in 1/299,792,458 of a second." (1983)
mass	kilogram	kg	". . . this prototype [a certain platinum–iridium cylinder] shall henceforth be considered to be the unit of mass." (1889)
time	second	s	". . . the duration of 9,192,631,770 periods of the radiation corresponding to the transition between the two hyperfine levels of the ground state of the cesium-133 atom." (1967)
electric current	ampere	A	". . . that constant current which, if maintained in two straight parallel conductors of infinite length, of negligible circular cross section, and placed 1 meter apart in vacuum, would produce between these conductors a force equal to 2×10^{-7} newton per meter of length." (1946)
thermodynamic temperature	kelvin	K	". . . the fraction 1/273.16 of the thermodynamic temperature of the triple point of water." (1967)
amount of substance	mole	mol	". . . the amount of substance of a system which contains as many elementary entities as there are atoms in 0.012 kilogram of carbon-12." (1971)
luminous intensity	candela	cd	". . . the luminous intensity, in a given direction, of a source that emits monochromatic radiation of frequency 540×10^{12} hertz and that has a radiant intensity in that direction of 1/683 watt per steradian." (1979)

*Adapted from "The International System of Units (SI)," National Bureau of Standards Special Publication 330, 1972 edition. The definitions above were adopted by the General Conference of Weights and Measures, an international body, on the dates shown. In this book we do not use the candela.

TABLE 2
Some SI Derived Units

Quantity	Name of Unit	Symbol	
area	square meter	m^2	
volume	cubic meter	m^3	
frequency	hertz	Hz	s^{-1}
mass density (density)	kilogram per cubic meter	kg/m^3	
speed, velocity	meter per second	m/s	
angular velocity	radian per second	rad/s	
acceleration	meter per second per second	m/s^2	
angular acceleration	radian per second per second	rad/s^2	
force	newton	N	$kg \cdot m/s^2$
pressure	pascal	Pa	N/m^2
work, energy, quantity of heat	joule	J	$N \cdot m$
power	watt	W	J/s
quantity of electric charge	coulomb	C	$A \cdot s$
potential difference, electromotive force	volt	V	W/A
electric field strength	volt per meter (or newton per coulomb)	V/m	N/C
electric resistance	ohm	Ω	V/A
capacitance	farad	F	$A \cdot s/V$
magnetic flux	weber	Wb	$V \cdot s$
inductance	henry	H	$V \cdot s/A$
magnetic flux density	tesla	T	Wb/m^2
magnetic field strength	ampere per meter	A/m	
entropy	joule per kelvin	J/K	
specific heat	joule per kilogram kelvin	$J/(kg \cdot K)$	
thermal conductivity	watt per meter kelvin	$W/(m \cdot K)$	
radiant intensity	watt per steradian	W/sr	

TABLE 3
The SI Supplementary Units

Quantity	Name of Unit	Symbol
plane angle	radian	rad
solid angle	steradian	sr

Some Fundamental Constants of Physics* B

Constant	Symbol	Computational Value	Best (1998) Value	
			Value[a]	Uncertainty[b]
Speed of light in a vacuum	c	3.00×10^{8} m/s	2.997 924 58	exact
Elementary charge	e	1.60×10^{-19} C	1.602 176 462	0.039
Gravitational constant	G	6.67×10^{-11} $m^3/s^2 \cdot kg$	6.673	1500
Universal gas constant	R	8.31 J/mol · K	8.314 472	1.7
Avogadro constant	N_A	6.02×10^{23} mol^{-1}	6.022 141 99	0.079
Boltzmann constant	k	1.38×10^{-23} J/K	1.380 650 3	1.7
Stefan–Boltzmann constant	σ	5.67×10^{-8} $W/m^2 \cdot K^4$	5.670 400	7.0
Molar volume of ideal gas at STP[d]	V_m	2.27×10^{-2} m^3/mol	2.271 098 1	1.7
Permittivity constant	ϵ_0	8.85×10^{-12} F/m	8.854 187 817 62	exact
Permeability constant	μ_0	1.26×10^{-6} H/m	1.256 637 061 43	exact
Planck constant	h	6.63×10^{-34} J · s	6.626 068 76	0.078
Electron mass[c]	m_e	9.11×10^{-31} kg	9.109 381 88	0.079
		5.49×10^{-4} u	5.485 799 110	0.0021
Proton mass[c]	m_p	1.67×10^{-27} kg	1.672 621 58	0.079
		1.0073 u	1.007 276 466 88	1.3×10^{-4}
Ratio of proton mass to electron mass	m_p/m_e	1840	1836.152 667 5	0.0021
Electron charge-to-mass ratio	e/m_e	1.76×10^{11} C/kg	1.758 820 174	0.040
Neutron mass[c]	m_n	1.68×10^{-27} kg	1.674 927 16	0.079
		1.0087 u	1.008 664 915 78	5.4×10^{-4}
Hydrogen atom mass[c]	m_{1H}	1.0078 u	1.007 825 031 6	0.0005
Deuterium atom mass[c]	m_{2H}	2.0141 u	2.014 101 777 9	0.0005
Helium atom mass[c]	m_{4He}	4.0026 u	4.002 603 2	0.067
Muon mass	m_μ	1.88×10^{-28} kg	1.883 531 09	0.084
Electron magnetic moment	μ_e	9.28×10^{-24} J/T	9.284 763 62	0.040
Proton magnetic moment	μ_p	1.41×10^{-26} J/T	1.410 606 663	0.041
Bohr magneton	μ_B	9.27×10^{-24} J/T	9.274 008 99	0.040
Nuclear magneton	μ_N	5.05×10^{-27} J/T	5.050 783 17	0.040
Bohr radius	a	5.29×10^{-11} m	5.291 772 083	0.0037
Rydberg constant	R	1.10×10^{7} m^{-1}	1.097 373 156 854 8	7.6×10^{-6}
Electron Compton wavelength	λ_C	2.43×10^{-12} m	2.426 310 215	0.0073

[a]Values given in this column should be given the same unit and power of 10 as the computational value.

[b]Parts per million.

[c]Masses given in u are in unified atomic mass units, where 1 u = $1.660\ 538\ 73 \times 10^{-27}$ kg.

[d]STP means standard temperature and pressure: 0°C and 1.0 atm (0.1 MPa).

*The values in this table were selected from the 1998 CODATA recommended values (www.physics.nist.gov).

C Some Astronomical Data

Some Distances from Earth

To the Moon*	3.82×10^{8} m	To the center of our galaxy	2.2×10^{20} m
To the Sun*	1.50×10^{11} m	To the Andromeda Galaxy	2.1×10^{22} m
To the nearest star (Proxima Centauri)	4.04×10^{16} m	To the edge of the observable universe	$\sim 10^{26}$ m

*Mean distance.

The Sun, Earth, and the Moon

Property	Unit	Sun	Earth	Moon
Mass	kg	1.99×10^{30}	5.98×10^{24}	7.36×10^{22}
Mean radius	m	6.96×10^{8}	6.37×10^{6}	1.74×10^{6}
Mean density	kg/m^3	1410	5520	3340
Free-fall acceleration at the surface	m/s^2	274	9.81	1.67
Escape velocity	km/s	618	11.2	2.38
Period of rotation[a]	—	37 d at poles[b] 26 d at equator[b]	23 h 56 min	27.3 d
Radiation power[c]	W	3.90×10^{26}		

[a]Measured with respect to the distant stars.

[b]The Sun, a ball of gas, does not rotate as a rigid body.

[c]Just outside Earth's atmosphere solar energy is received, assuming normal incidence, at the rate of 1340 W/m^2.

Some Properties of the Planets

	Mercury	Venus	Earth	Mars	Jupiter	Saturn	Uranus	Neptune	Pluto
Mean distance from Sun, 10^6 km	57.9	108	150	228	778	1430	2870	4500	5900
Period of revolution, y	0.241	0.615	1.00	1.88	11.9	29.5	84.0	165	248
Period of rotation,[a] d	58.7	-243[b]	0.997	1.03	0.409	0.426	-0.451[b]	0.658	6.39
Orbital speed, km/s	47.9	35.0	29.8	24.1	13.1	9.64	6.81	5.43	4.74
Inclination of axis to orbit	$<28°$	$\approx 3°$	23.4°	25.0°	3.08°	26.7°	97.9°	29.6°	57.5°
Inclination of orbit to Earth's orbit	7.00°	3.39°		1.85°	1.30°	2.49°	0.77°	1.77°	17.2°
Eccentricity of orbit	0.206	0.0068	0.0167	0.0934	0.0485	0.0556	0.0472	0.0086	0.250
Equatorial diameter, km	4880	12 100	12 800	6790	143 000	120 000	51 800	49 500	2300
Mass (Earth = 1)	0.0558	0.815	1.000	0.107	318	95.1	14.5	17.2	0.002
Density (water = 1)	5.60	5.20	5.52	3.95	1.31	0.704	1.21	1.67	2.03
Surface value of g,[c] m/s^2	3.78	8.60	9.78	3.72	22.9	9.05	7.77	11.0	0.5
Escape velocity,[c] km/s	4.3	10.3	11.2	5.0	59.5	35.6	21.2	23.6	1.1
Known satellites	0	0	1	2	60 + ring	31 + rings	21 + rings	11 + rings	1

[a]Measured with respect to the distant stars.

[b]Venus and Uranus rotate opposite their orbital motion.

[c]Gravitational acceleration measured at the planet's equator.

Conversion Factors

D

Conversion factors may be read directly from these tables. For example, 1 degree = 2.778×10^{-3} revolutions, so $16.7° = 16.7 \times 2.778 \times 10^{-3}$ rev. The SI units are fully capitalized. Adapted in part from G. Shortley and D. Williams, *Elements of Physics,* 1971, Prentice-Hall, Englewood Cliffs, NJ.

Plane Angle

	°	′	″	RADIAN	rev
1 degree =	1	60	3600	1.745×10^{-2}	2.778×10^{-3}
1 minute =	1.667×10^{-2}	1	60	2.909×10^{-4}	4.630×10^{-5}
1 second =	2.778×10^{-4}	1.667×10^{-2}	1	4.848×10^{-6}	7.716×10^{-7}
1 RADIAN =	57.30	3438	2.063×10^{5}	1	0.1592
1 revolution =	360	2.16×10^{4}	1.296×10^{6}	6.283	1

Solid Angle

1 sphere = 4π steradians = 12.57 steradians

Length

	cm	METER	km	in.	ft	mi
1 centimeter =	1	10^{-2}	10^{-5}	0.3937	3.281×10^{-2}	6.214×10^{-6}
1 METER =	100	1	10^{-3}	39.37	3.281	6.214×10^{-4}
1 kilometer =	10^{5}	1000	1	3.937×10^{4}	3281	0.6214
1 inch =	2.540	2.540×10^{-2}	2.540×10^{-5}	1	8.333×10^{-2}	1.578×10^{-5}
1 foot =	30.48	0.3048	3.048×10^{-4}	12	1	1.894×10^{-4}
1 mile =	1.609×10^{5}	1609	1.609	6.336×10^{4}	5280	1

1 angström = 10^{-10} m

1 nautical mile = 1852 m = 1.151 miles = 6076 ft

1 fermi = 10^{-15} m

1 light-year = 9.460×10^{12} km

1 parsec = 3.084×10^{13} km

1 fathom = 6 ft

1 Bohr radius = 5.292×10^{-11} m

1 yard = 3 ft

1 rod = 16.5 ft

1 mil = 10^{-3} in.

1 nm = 10^{-9} m

Area

	METER^2	cm^2	ft^2	in.^{2}
1 SQUARE METER =	1	10^{4}	10.76	1550
1 square centimeter =	10^{-4}	1	1.076×10^{-3}	0.1550
1 square foot =	9.290×10^{-2}	929.0	1	144
1 square inch =	6.452×10^{-4}	6.452	6.944×10^{-3}	1

1 square mile = 2.788×10^{7} ft^2 = 640 acres

1 barn = 10^{-28} m^2

1 acre = 43 560 ft^2

1 hectare = 10^{4} m^2 = 2.471 acres

Volume

	METER^3	cm^3	L	ft^3	in.^{3}
1 CUBIC METER =	1	10^6	1000	35.31	6.102×10^4
1 cubic centimeter =	10^{-6}	1	1.000×10^{-3}	3.531×10^{-5}	6.102×10^{-2}
1 liter =	1.000×10^{-3}	1000	1	3.531×10^{-2}	61.02
1 cubic foot =	2.832×10^{-2}	2.832×10^4	28.32	1	1728
1 cubic inch =	1.639×10^{-5}	16.39	1.639×10^{-2}	5.787×10^{-4}	1

1 U.S. fluid gallon = 4 U.S. fluid quarts = 8 U.S. pints = 128 U.S. fluid ounces = 231 in.^{3}

1 British imperial gallon = 277.4 in.^{3} = 1.201 U.S. fluid gallons

Mass

Quantities in the colored areas are not mass units but are often used as such. For example, when we write 1 kg "=" 2.205 lb, this means that a kilogram is a *mass* that *weighs* 2.205 pounds at a location where g has the standard value of 9.80665 m/s^2.

	g	KILOGRAM	slug	u	oz	lb	ton
1 gram =	1	0.001	6.852×10^{-5}	6.022×10^{23}	3.527×10^{-2}	2.205×10^{-3}	1.102×10^{-6}
1 KILOGRAM =	1000	1	6.852×10^{-2}	6.022×10^{26}	35.27	2.205	1.102×10^{-3}
1 slug =	1.459×10^4	14.59	1	8.786×10^{27}	514.8	32.17	1.609×10^{-2}
1 atomic mass unit =	1.661×10^{-24}	1.661×10^{-27}	1.138×10^{-28}	1	5.857×10^{-26}	3.662×10^{-27}	1.830×10^{-30}
1 ounce =	28.35	2.835×10^{-2}	1.943×10^{-3}	1.718×10^{25}	1	6.250×10^{-2}	3.125×10^{-5}
1 pound =	453.6	0.4536	3.108×10^{-2}	2.732×10^{26}	16	1	0.0005
1 ton =	9.072×10^5	907.2	62.16	5.463×10^{29}	3.2×10^4	2000	1

1 metric ton = 1000 kg

Density

Quantities in the colored areas are weight densities and, as such, are dimensionally different from mass densities. See the note for the mass table.

	slug/ft^3	KILOGRAM/ METER^3	g/cm^3	lb/ft^3	lb/in.^{3}
1 slug per foot^3 =	1	515.4	0.5154	32.17	1.862×10^{-2}
1 KILOGRAM per METER^3 =	1.940×10^{-3}	1	0.001	6.243×10^{-2}	3.613×10^{-5}
1 gram per centimeter^3 =	1.940	1000	1	62.43	3.613×10^{-2}
1 pound per foot^3 =	3.108×10^{-2}	16.02	16.02×10^{-2}	1	5.787×10^{-4}
1 pound per inch^3 =	53.71	2.768×10^4	27.68	1728	1

Time

	y	d	h	min	SECOND
1 year =	1	365.25	8.766×10^3	5.259×10^5	3.156×10^7
1 day =	2.738×10^{-3}	1	24	1440	8.640×10^4
1 hour =	1.141×10^{-4}	4.167×10^{-2}	1	60	3600
1 minute =	1.901×10^{-6}	6.944×10^{-4}	1.667×10^{-2}	1	60
1 SECOND =	3.169×10^{-8}	1.157×10^{-5}	2.778×10^{-4}	1.667×10^{-2}	1

Speed

	ft/s	km/h	METER/SECOND	mi/h	cm/s
1 foot per second =	1	1.097	0.3048	0.6818	30.48
1 kilometer per hour =	0.9113	1	0.2778	0.6214	27.78
1 METER per SECOND =	3.281	3.6	1	2.237	100
1 mile per hour =	1.467	1.609	0.4470	1	44.70
1 centimeter per second =	3.281×10^{-2}	3.6×10^{-2}	0.01	2.237×10^{-2}	1

1 knot = 1 nautical mi/h = 1.688 ft/s 1 mi/min = 88.00 ft/s = 60.00 mi/h

Force

Force units in the colored areas are now little used. To clarify: 1 gram-force (= 1 gf) is the force of gravity that would act on an object whose mass is 1 gram at a location where g has the standard value of 9.80665 m/s^2.

	dyne	NEWTON	lb	pdl	gf	kgf
1 dyne =	1	10^{-5}	2.248×10^{-6}	7.233×10^{-5}	1.020×10^{-3}	1.020×10^{-6}
1 NEWTON =	10^5	1	0.2248	7.233	102.0	0.1020
1 pound =	4.448×10^5	4.448	1	32.17	453.6	0.4536
1 poundal =	1.383×10^4	0.1383	3.108×10^{-2}	1	14.10	1.410×10^2
1 gram-force =	980.7	9.807×10^{-3}	2.205×10^{-3}	7.093×10^{-2}	1	0.001
1 kilogram-force =	9.807×10^5	9.807	2.205	70.93	1000	1

1 ton = 2000 lb

Pressure

	atm	dyne/cm^2	inch of water	cm Hg	PASCAL	lb/in.^{2}	lb/ft^2
1 atmosphere =	1	1.013×10^6	406.8	76	1.013×10^5	14.70	2116
1 dyne per centimeter^2 =	9.869×10^{-7}	1	4.015×10^{-4}	7.501×10^{-5}	0.1	1.405×10^{-5}	2.089×10^{-3}
1 inch of water[a] at 4°C =	2.458×10^{-3}	2491	1	0.1868	249.1	3.613×10^{-2}	5.202
1 centimeter of mercury[a] at 0°C =	1.316×10^{-2}	1.333×10^4	5.353	1	1333	0.1934	27.85
1 PASCAL =	9.869×10^{-6}	10	4.015×10^{-3}	7.501×10^{-4}	1	1.450×10^{-4}	2.089×10^{-2}
1 pound per inch^2 =	6.805×10^{-2}	6.895×10^4	27.68	5.171	6.895×10^3	1	144
1 pound per foot^2 =	4.725×10^{-4}	478.8	0.1922	3.591×10^{-2}	47.88	6.944×10^{-3}	1

[a]Where the acceleration of gravity has the standard value of 9.80665 m/s^2.

1 bar = 10^6 dyne/cm^2 = 0.1 MPa 1 millibar = 10^3 dyne/cm^2 = 10^2 Pa 1 torr = 1 mm Hg

Energy, Work, Heat

Quantities in the colored areas are not energy units but are included for convenience. They arise from the relativistic mass–energy equivalence formula $E = mc^2$ and represent the energy released if a kilogram or unified atomic mass unit (u) is completely converted to energy (bottom two rows) or the mass that would be completely converted to one unit of energy (rightmost two columns).

	Btu	erg	ft · lb	hp · h	JOULE	cal	kW · h	eV	MeV	kg	u
1 British thermal unit =	1	1.055×10^{10}	777.9	3.929×10^{-4}	1055	252.0	2.930×10^{-4}	6.585×10^{21}	6.585×10^{15}	1.174×10^{-14}	7.070×10^{12}
1 erg =	9.481×10^{-11}	1	7.376×10^{-8}	3.725×10^{-14}	10^{-7}	2.389×10^{-8}	2.778×10^{-14}	6.242×10^{11}	6.242×10^{5}	1.113×10^{-24}	670.2
1 foot-pound =	1.285×10^{-3}	1.356×10^{7}	1	5.051×10^{-7}	1.356	0.3238	3.766×10^{-7}	8.464×10^{18}	8.464×10^{12}	1.509×10^{-17}	9.037×10^{9}
1 horsepower-hour =	2545	2.685×10^{13}	1.980×10^{6}	1	2.685×10^{6}	6.413×10^{5}	0.7457	1.676×10^{25}	1.676×10^{19}	2.988×10^{-11}	1.799×10^{16}
1 JOULE =	9.481×10^{-4}	10^{7}	0.7376	3.725×10^{-7}	1	0.2389	2.778×10^{-7}	6.242×10^{18}	6.242×10^{12}	1.113×10^{-17}	6.702×10^{9}
1 calorie =	3.968×10^{-3}	4.1868×10^{7}	3.088	1.560×10^{-6}	4.1868	1	1.163×10^{-6}	2.613×10^{19}	2.613×10^{13}	4.660×10^{-17}	2.806×10^{10}
1 kilowatt-hour =	3413	3.600×10^{13}	2.655×10^{6}	1.341	3.600×10^{6}	8.600×10^{5}	1	2.247×10^{25}	2.247×10^{19}	4.007×10^{-11}	2.413×10^{16}
1 electron-volt =	1.519×10^{-22}	1.602×10^{-12}	1.182×10^{-19}	5.967×10^{-26}	1.602×10^{-19}	3.827×10^{-20}	4.450×10^{-26}	1	10^{-6}	1.783×10^{-36}	1.074×10^{-9}
1 million electron-volts =	1.519×10^{-16}	1.602×10^{-6}	1.182×10^{-13}	5.967×10^{-20}	1.602×10^{-13}	3.827×10^{-14}	4.450×10^{-20}	10^{-6}	1	1.783×10^{-30}	1.074×10^{-3}
1 kilogram =	8.521×10^{13}	8.987×10^{23}	6.629×10^{16}	3.348×10^{10}	8.987×10^{16}	2.146×10^{16}	2.497×10^{10}	5.610×10^{35}	5.610×10^{29}	1	6.022×10^{26}
1 unified atomic mass unit =	1.415×10^{-13}	1.492×10^{-3}	1.101×10^{-10}	5.559×10^{-17}	1.492×10^{-10}	3.564×10^{-11}	4.146×10^{-17}	9.320×10^{8}	932.0	1.661×10^{-27}	1

Power

	Btu/h	ft · lb/s	hp	cal/s	kW	WATT
1 British thermal unit per hour =	1	0.2161	3.929×10^{-4}	6.998×10^{-2}	2.930×10^{-4}	0.2930
1 foot-pound per second =	4.628	1	1.818×10^{-3}	0.3239	1.356×10^{-3}	1.356
1 horsepower =	2545	550	1	178.1	0.7457	745.7
1 calorie per second =	14.29	3.088	5.615×10^{-3}	1	4.186×10^{-3}	4.186
1 kilowatt =	3413	737.6	1.341	238.9	1	1000
1 WATT =	3.413	0.7376	1.341×10^{-3}	0.2389	0.001	1

Magnetic Field

	gauss	TESLA	milligauss
1 gauss =	1	10^{-4}	1000
1 TESLA =	10^{4}	1	10^{7}
1 milligauss =	0.001	10^{-7}	1

1 tesla = 1 weber/meter^2

Magnetic Flux

	maxwell	WEBER
1 maxwell =	1	10^{-8}
1 WEBER =	10^{8}	1

Mathematical Formulas

E

Geometry

Circle of radius r: circumference $= 2\pi r$; area $= \pi r^2$.
Sphere of radius r: area $= 4\pi r^2$; volume $= \frac{4}{3}\pi r^3$.
Right circular cylinder of radius r and height h:
area $= 2\pi r^2 + 2\pi rh$; volume $= \pi r^2 h$.
Triangle of base a and altitude h: area $= \frac{1}{2}ah$.

Quadratic Formula

If $ax^2 + bx + c = 0$, then $x = \dfrac{-b \pm \sqrt{b^2 - 4ac}}{2a}$.

Trigonometric Functions of Angle θ

$\sin\theta = \dfrac{y}{r}$ $\quad \cos\theta = \dfrac{x}{r}$

$\tan\theta = \dfrac{y}{x}$ $\quad \cot\theta = \dfrac{x}{y}$

$\sec\theta = \dfrac{r}{x}$ $\quad \csc\theta = \dfrac{r}{y}$

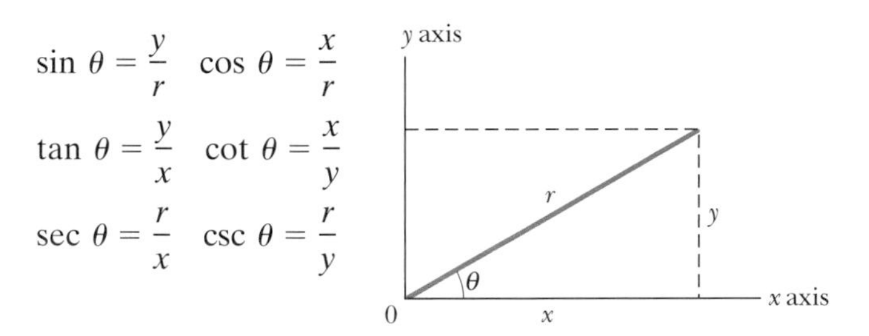

Pythagorean Theorem

In this right triangle,
$a^2 + b^2 = c^2$

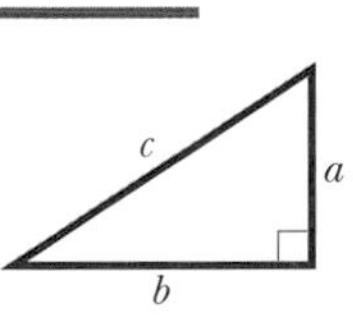

Triangles

Angles are A, B, C
Opposite sides are a, b, c
Angles $A + B + C = 180°$

$\dfrac{\sin A}{a} = \dfrac{\sin B}{b} = \dfrac{\sin C}{c}$

$c^2 = a^2 + b^2 - 2ab\cos C$

Exterior angle $D = A + C$

Mathematical Signs and Symbols

$=$ equals
$\approx$ equals approximately
$\sim$ is the order of magnitude of
$\neq$ is not equal to
$\equiv$ is identical to, is defined as
$>$ is greater than ($\gg$ is much greater than)
$<$ is less than ($\ll$ is much less than)
$\geq$ is greater than or equal to (or, is no less than)
$\leq$ is less than or equal to (or, is no more than)
$\pm$ plus or minus
$\propto$ is proportional to
Σ the sum of
x_{avg} the average value of x

Trigonometric Identities

$\sin(90° - \theta) = \cos\theta$
$\cos(90° - \theta) = \sin\theta$
$\sin\theta/\cos\theta = \tan\theta$
$\sin^2\theta + \cos^2\theta = 1$
$\sec^2\theta - \tan^2\theta = 1$
$\csc^2\theta - \cot^2\theta = 1$
$\sin 2\theta = 2\sin\theta\cos\theta$
$\cos 2\theta = \cos^2\theta - \sin^2\theta = 2\cos^2\theta - 1 = 1 - 2\sin^2\theta$
$\sin(\alpha \pm \beta) = \sin\alpha\cos\beta \pm \cos\alpha\sin\beta$
$\cos(\alpha \pm \beta) = \cos\alpha\cos\beta \mp \sin\alpha\sin\beta$
$\tan(\alpha \pm \beta) = \dfrac{\tan\alpha \pm \tan\beta}{1 \mp \tan\alpha\tan\beta}$
$\sin\alpha \pm \sin\beta = 2\sin\frac{1}{2}(\alpha \pm \beta)\cos\frac{1}{2}(\alpha \mp \beta)$
$\cos\alpha + \cos\beta = 2\cos\frac{1}{2}(\alpha + \beta)\cos\frac{1}{2}(\alpha - \beta)$
$\cos\alpha - \cos\beta = -2\sin\frac{1}{2}(\alpha + \beta)\sin\frac{1}{2}(\alpha - \beta)$

Binomial Theorem

$$(1+x)^n = 1 + \frac{nx}{1!} + \frac{n(n-1)x^2}{2!} + \cdots \qquad (x^2 < 1)$$

Exponential Expansion

$$e^x = 1 + x + \frac{x^2}{2!} + \frac{x^3}{3!} + \cdots$$

Logarithmic Expansion

$$\ln(1+x) = x - \tfrac{1}{2}x^2 + \tfrac{1}{3}x^3 - \cdots \qquad (|x| < 1)$$

Trigonometric Expansions (θ in radians)

$$\sin\theta = \theta - \frac{\theta^3}{3!} + \frac{\theta^5}{5!} - \cdots$$

$$\cos\theta = 1 - \frac{\theta^2}{2!} + \frac{\theta^4}{4!} - \cdots$$

$$\tan\theta = \theta + \frac{\theta^3}{3} + \frac{2\theta^5}{15} + \cdots$$

Cramer's Rule

Two simultaneous equations in unknowns x and y,

$$a_1x + b_1y = c_1 \quad \text{and} \quad a_2x + b_2y = c_2,$$

have the solutions

$$x = \frac{\begin{vmatrix} c_1 & b_1 \\ c_2 & b_2 \end{vmatrix}}{\begin{vmatrix} a_1 & b_1 \\ a_2 & b_2 \end{vmatrix}} = \frac{c_1b_2 - c_2b_1}{a_1b_2 - a_2b_1}$$

and

$$y = \frac{\begin{vmatrix} a_1 & c_1 \\ a_2 & c_2 \end{vmatrix}}{\begin{vmatrix} a_1 & b_1 \\ a_2 & b_2 \end{vmatrix}} = \frac{a_1c_2 - a_2c_1}{a_1b_2 - a_2b_1}.$$

Products of Vectors

Let $\hat{\imath}$, $\hat{\jmath}$, and $\hat{k}$ be unit vectors in the x, y, and z directions. Then

$$\hat{\imath}\cdot\hat{\imath} = \hat{\jmath}\cdot\hat{\jmath} = \hat{k}\cdot\hat{k} = 1, \qquad \hat{\imath}\cdot\hat{\jmath} = \hat{\jmath}\cdot\hat{k} = \hat{k}\cdot\hat{\imath} = 0,$$

$$\hat{\imath}\times\hat{\imath} = \hat{\jmath}\times\hat{\jmath} = \hat{k}\times\hat{k} = 0,$$

$$\hat{\imath}\times\hat{\jmath} = \hat{k}, \qquad \hat{\jmath}\times\hat{k} = \hat{\imath}, \qquad \hat{k}\times\hat{\imath} = \hat{\jmath}.$$

Any vector $\vec{a}$ with components a_x, a_y, and a_z along the x, y, and z axes can be written as

$$\vec{a} = a_x\hat{\imath} + a_y\hat{\jmath} + a_z\hat{k}.$$

Let $\vec{a}$, $\vec{b}$, and $\vec{c}$ be arbitrary vectors with magnitudes a, b, and c. Then

$$\vec{a}\times(\vec{b}+\vec{c}) = (\vec{a}\times\vec{b}) + (\vec{a}\times\vec{c})$$

$$(s\vec{a})\times\vec{b} = \vec{a}\times(s\vec{b}) = s(\vec{a}\times\vec{b}) \qquad (s = \text{a scalar}).$$

Let θ be the smaller of the two angles between $\vec{a}$ and $\vec{b}$. Then

$$\vec{a}\cdot\vec{b} = \vec{b}\cdot\vec{a} = a_xb_x + a_yb_y + a_zb_z = ab\cos\theta$$

$$\vec{a}\times\vec{b} = -\vec{b}\times\vec{a} = \begin{vmatrix} \hat{\imath} & \hat{\jmath} & \hat{k} \\ a_x & a_y & a_z \\ b_x & b_y & b_z \end{vmatrix}$$

$$= \hat{\imath}\begin{vmatrix} a_y & a_z \\ b_y & b_z \end{vmatrix} - \hat{\jmath}\begin{vmatrix} a_x & a_z \\ b_x & b_z \end{vmatrix} + \hat{k}\begin{vmatrix} a_x & a_y \\ b_x & b_y \end{vmatrix}$$

$$= (a_yb_z - b_ya_z)\hat{\imath} + (a_zb_x - b_za_x)\hat{\jmath} + (a_xb_y - b_xa_y)\hat{k}$$

$$|\vec{a}\times\vec{b}| = ab\sin\theta$$

$$\vec{a}\cdot(\vec{b}\times\vec{c}) = \vec{b}\cdot(\vec{c}\times\vec{a}) = \vec{c}\cdot(\vec{a}\times\vec{b})$$

$$\vec{a}\times(\vec{b}\times\vec{c}) = (\vec{a}\cdot\vec{c})\vec{b} - (\vec{a}\cdot\vec{b})\vec{c}$$

Derivatives and Integrals

In what follows, the letters u and v stand for any functions of x, and a and m are constants. To each of the indefinite integrals should be added an arbitrary constant of integration. The *Handbook of Chemistry and Physics* (CRC Press Inc.) gives a more extensive tabulation.

1. $\frac{dx}{dx} = 1$
2. $\frac{d}{dx}(au) = a\frac{du}{dx}$
3. $\frac{d}{dx}(u + v) = \frac{du}{dx} + \frac{dv}{dx}$
4. $\frac{d}{dx}x^m = mx^{m-1}$
5. $\frac{d}{dx}\ln x = \frac{1}{x}$
6. $\frac{d}{dx}(uv) = u\frac{dv}{dx} + v\frac{du}{dx}$
7. $\frac{d}{dx}e^x = e^x$
8. $\frac{d}{dx}\sin x = \cos x$
9. $\frac{d}{dx}\cos x = -\sin x$
10. $\frac{d}{dx}\tan x = \sec^2 x$
11. $\frac{d}{dx}\cot x = -\csc^2 x$
12. $\frac{d}{dx}\sec x = \tan x \sec x$
13. $\frac{d}{dx}\csc x = -\cot x \csc x$
14. $\frac{d}{dx}e^u = e^u\frac{du}{dx}$
15. $\frac{d}{dx}\sin u = \cos u\frac{du}{dx}$
16. $\frac{d}{dx}\cos u = -\sin u\frac{du}{dx}$

1. $\int dx = x$
2. $\int au\,dx = a\int u\,dx$
3. $\int (u + v)\,dx = \int u\,dx + \int v\,dx$
4. $\int x^m\,dx = \frac{x^{m+1}}{m+1} \quad (m \neq -1)$
5. $\int \frac{dx}{x} = \ln|x|$
6. $\int u\frac{dv}{dx}\,dx = uv - \int v\frac{du}{dx}\,dx$
7. $\int e^x\,dx = e^x$
8. $\int \sin x\,dx = -\cos x$
9. $\int \cos x\,dx = \sin x$
10. $\int \tan x\,dx = \ln|\sec x|$
11. $\int \sin^2 x\,dx = \frac{1}{2}x - \frac{1}{4}\sin 2x$
12. $\int e^{-ax}\,dx = -\frac{1}{a}e^{-ax}$
13. $\int xe^{-ax}\,dx = -\frac{1}{a^2}(ax + 1)e^{-ax}$
14. $\int x^2e^{-ax}\,dx = -\frac{1}{a^3}(a^2x^2 + 2ax + 2)e^{-ax}$
15. $\int_0^\infty x^ne^{-ax}\,dx = \frac{n!}{a^{n+1}}$
16. $\int_0^\infty x^{2n}e^{-ax^2}\,dx = \frac{1\cdot 3\cdot 5\cdots(2n-1)}{2^{n+1}a^n}\sqrt{\frac{\pi}{a}}$
17. $\int \frac{dx}{\sqrt{x^2 + a^2}} = \ln(x + \sqrt{x^2 + a^2})$
18. $\int \frac{x\,dx}{(x^2 + a^2)^{3/2}} = -\frac{1}{(x^2 + a^2)^{1/2}}$
19. $\int \frac{dx}{(x^2 + a^2)^{3/2}} = \frac{x}{a^2(x^2 + a^2)^{1/2}}$
20. $\int_0^\infty x^{2n+1}e^{-ax^2}\,dx = \frac{n!}{2a^{n+1}} \quad (a > 0)$
21. $\int \frac{x\,dx}{x + d} = x - d\ln(x + d)$

F Properties of the Elements

All physical properties are for a pressure of 1 atm unless otherwise specified.

Element	Symbol	Atomic Number Z	Molar Mass, g/mol	Density, g/cm^3 at 20°C	Melting Point, °C	Boiling Point, °C	Specific Heat, J/(g · °C) at 25°C
Actinium	Ac	89	(227)	10.06	1323	(3473)	0.092
Aluminum	Al	13	26.9815	2.699	660	2450	0.900
Americium	Am	95	(243)	13.67	1541	—	—
Antimony	Sb	51	121.75	6.691	630.5	1380	0.205
Argon	Ar	18	39.948	1.6626×10^{-3}	−189.4	−185.8	0.523
Arsenic	As	33	74.9216	5.78	817 (28 atm)	613	0.331
Astatine	At	85	(210)	—	(302)	—	—
Barium	Ba	56	137.34	3.594	729	1640	0.205
Berkelium	Bk	97	(247)	14.79	—	—	—
Beryllium	Be	4	9.0122	1.848	1287	2770	1.83
Bismuth	Bi	83	208.980	9.747	271.37	1560	0.122
Bohrium	Bh	107	262.12	—	—	—	—
Boron	B	5	10.811	2.34	2030	—	1.11
Bromine	Br	35	79.909	3.12 (liquid)	−7.2	58	0.293
Cadmium	Cd	48	112.40	8.65	321.03	765	0.226
Calcium	Ca	20	40.08	1.55	838	1440	0.624
Californium	Cf	98	(251)	—	—	—	—
Carbon	C	6	12.01115	2.26	3727	4830	0.691
Cerium	Ce	58	140.12	6.768	804	3470	0.188
Cesium	Cs	55	132.905	1.873	28.40	690	0.243
Chlorine	Cl	17	35.453	3.214×10^{-3} (0°C)	−101	−34.7	0.486
Chromium	Cr	24	51.996	7.19	1857	2665	0.448
Cobalt	Co	27	58.9332	8.85	1495	2900	0.423
Copper	Cu	29	63.54	8.96	1083.40	2595	0.385
Curium	Cm	96	(247)	13.3	—	—	—
Darmstadtium	Ds	110	(271)	—	—	—	—
Dubnium	Db	105	262.114	—	—	—	—
Dysprosium	Dy	66	162.50	8.55	1409	2330	0.172
Einsteinium	Es	99	(254)	—	—	—	—
Erbium	Er	68	167.26	9.15	1522	2630	0.167
Europium	Eu	63	151.96	5.243	817	1490	0.163
Fermium	Fm	100	(237)	—	—	—	—
Fluorine	F	9	18.9984	1.696×10^{-3} (0°C)	−219.6	−188.2	0.753
Francium	Fr	87	(223)	—	(27)	—	—
Gadolinium	Gd	64	157.25	7.90	1312	2730	0.234
Gallium	Ga	31	69.72	5.907	29.75	2237	0.377
Germanium	Ge	32	72.59	5.323	937.25	2830	0.322
Gold	Au	79	196.967	19.32	1064.43	2970	0.131

Element	Symbol	Atomic Number Z	Molar Mass, g/mol	Density, g/cm^3 at 20°C	Melting Point, °C	Boiling Point, °C	Specific Heat, J/(g · °C) at 25°C
Hafnium	Hf	72	178.49	13.31	2227	5400	0.144
Hassium	Hs	108	(265)	—	—	—	—
Helium	He	2	4.0026	0.1664×10^{-3}	−269.7	−268.9	5.23
Holmium	Ho	67	164.930	8.79	1470	2330	0.165
Hydrogen	H	1	1.00797	0.08375×10^{-3}	−259.19	−252.7	14.4
Indium	In	49	114.82	7.31	156.634	2000	0.233
Iodine	I	53	126.9044	4.93	113.7	183	0.218
Iridium	Ir	77	192.2	22.5	2447	(5300)	0.130
Iron	Fe	26	55.847	7.874	1536.5	3000	0.447
Krypton	Kr	36	83.80	3.488×10^{-3}	−157.37	−152	0.247
Lanthanum	La	57	138.91	6.189	920	3470	0.195
Lawrencium	Lr	103	(257)	—	—	—	—
Lead	Pb	82	207.19	11.35	327.45	1725	0.129
Lithium	Li	3	6.939	0.534	180.55	1300	3.58
Lutetium	Lu	71	174.97	9.849	1663	1930	0.155
Magnesium	Mg	12	24.312	1.738	650	1107	1.03
Manganese	Mn	25	54.9380	7.44	1244	2150	0.481
Meitnerium	Mt	109	(266)	—	—	—	—
Mendelevium	Md	101	(256)	—	—	—	—
Mercury	Hg	80	200.59	13.55	−38.87	357	0.138
Molybdenum	Mo	42	95.94	10.22	2617	5560	0.251
Neodymium	Nd	60	144.24	7.007	1016	3180	0.188
Neon	Ne	10	20.183	0.8387×10^{-3}	−248.597	−246.0	1.03
Neptunium	Np	93	(237)	20.25	637	—	1.26
Nickel	Ni	28	58.71	8.902	1453	2730	0.444
Niobium	Nb	41	92.906	8.57	2468	4927	0.264
Nitrogen	N	7	14.0067	1.1649×10^{-3}	−210	−195.8	1.03
Nobelium	No	102	(255)	—	—	—	—
Osmium	Os	76	190.2	22.59	3027	5500	0.130
Oxygen	O	8	15.9994	1.3318×10^{-3}	−218.80	−183.0	0.913
Palladium	Pd	46	106.4	12.02	1552	3980	0.243
Phosphorus	P	15	30.9738	1.83	44.25	280	0.741
Platinum	Pt	78	195.09	21.45	1769	4530	0.134
Plutonium	Pu	94	(244)	19.8	640	3235	0.130
Polonium	Po	84	(210)	9.32	254	—	—
Potassium	K	19	39.102	0.862	63.20	760	0.758
Praseodymium	Pr	59	140.907	6.773	931	3020	0.197
Promethium	Pm	61	(145)	7.22	(1027)	—	—
Protactinium	Pa	91	(231)	15.37 (estimated)	(1230)	—	—
Radium	Ra	88	(226)	5.0	700	—	—
Radon	Rn	86	(222)	9.96×10^{-3} (0°C)	(−71)	−61.8	0.092
Rhenium	Re	75	186.2	21.02	3180	5900	0.134
Rhodium	Rh	45	102.905	12.41	1963	4500	0.243
Rubidium	Rb	37	85.47	1.532	39.49	688	0.364
Ruthenium	Ru	44	101.107	12.37	2250	4900	0.239
Rutherfordium	Rf	104	261.11	—	—	—	—

Element	Symbol	Atomic Number Z	Molar Mass, g/mol	Density, g/cm^3 at 20°C	Melting Point, °C	Boiling Point, °C	Specific Heat, J/(g · °C) at 25°C
Samarium	Sm	62	150.35	7.52	1072	1630	0.197
Scandium	Sc	21	44.956	2.99	1539	2730	0.569
Seaborgium	Sg	106	263.118	—	—	—	—
Selenium	Se	34	78.96	4.79	221	685	0.318
Silicon	Si	14	28.086	2.33	1412	2680	0.712
Silver	Ag	47	107.870	10.49	960.8	2210	0.234
Sodium	Na	11	22.9898	0.9712	97.85	892	1.23
Strontium	Sr	38	87.62	2.54	768	1380	0.737
Sulfur	S	16	32.064	2.07	119.0	444.6	0.707
Tantalum	Ta	73	180.948	16.6	3014	5425	0.138
Technetium	Tc	43	(99)	11.46	2200	—	0.209
Tellurium	Te	52	127.60	6.24	449.5	990	0.201
Terbium	Tb	65	158.924	8.229	1357	2530	0.180
Thallium	Tl	81	204.37	11.85	304	1457	0.130
Thorium	Th	90	(232)	11.72	1755	(3850)	0.117
Thulium	Tm	69	168.934	9.32	1545	1720	0.159
Tin	Sn	50	118.69	7.2984	231.868	2270	0.226
Titanium	Ti	22	47.90	4.54	1670	3260	0.523
Tungsten	W	74	183.85	19.3	3380	5930	0.134
Unnamed	Uuu	111	(272)	—	—	—	—
Unnamed	Uub	112	(285)	—	—	—	—
Unnamed	Uut	113	—	—	—	—	—
Unnamed	Unq	114	(289)	—	—	—	—
Unnamed	Uup	115	—	—	—	—	—
Unnamed	Uuh	116	—	—	—	—	—
Unnamed	Uus	117	—	—	—	—	—
Unnamed	Uuo	118	(293)	—	—	—	—
Uranium	U	92	(238)	18.95	1132	3818	0.117
Vanadium	V	23	50.942	6.11	1902	3400	0.490
Xenon	Xe	54	131.30	5.495×10^{-3}	−111.79	−108	0.159
Ytterbium	Yb	70	173.04	6.965	824	1530	0.155
Yttrium	Y	39	88.905	4.469	1526	3030	0.297
Zinc	Zn	30	65.37	7.133	419.58	906	0.389
Zirconium	Zr	40	91.22	6.506	1852	3580	0.276

The values in parentheses in the column of molar masses are the mass numbers of the longest-lived isotopes of those elements that are radioactive. Melting points and boiling points in parentheses are uncertain.

The data for gases are valid only when these are in their usual molecular state, such as H_2, He, O_2, Ne, etc. The specific heats of the gases are the values at constant pressure.

Source: Adapted from J. Emsley, *The Elements,* 3rd ed., 1998, Clarendon Press, Oxford. See also www.webelements.com for the latest values and newest elements.

Periodic Table of the Elements

G

Metals

Metalloids

Nonmetals

Transition metals

THE HORIZONTAL PERIODS

	Alkali metals IA	IIA	IIIB	IVB	VB	VIB	VIIB	VIIIB			IB	IIB	IIIA	IVA	VA	VIA	VIIA	Noble gases 0
1	1 H																	2 He
2	3 Li	4 Be											5 B	6 C	7 N	8 O	9 F	10 Ne
3	11 Na	12 Mg											13 Al	14 Si	15 P	16 S	17 Cl	18 Ar
4	19 K	20 Ca	21 Sc	22 Ti	23 V	24 Cr	25 Mn	26 Fe	27 Co	28 Ni	29 Cu	30 Zn	31 Ga	32 Ge	33 As	34 Se	35 Br	36 Kr
5	37 Rb	38 Sr	39 Y	40 Zr	41 Nb	42 Mo	43 Tc	44 Ru	45 Rh	46 Pd	47 Ag	48 Cd	49 In	50 Sn	51 Sb	52 Te	53 I	54 Xe
6	55 Cs	56 Ba	57-71 *	72 Hf	73 Ta	74 W	75 Re	76 Os	77 Ir	78 Pt	79 Au	80 Hg	81 Tl	82 Pb	83 Bi	84 Po	85 At	86 Rn
7	87 Fr	88 Ra	89-103 †	104 Rf	105 Db	106 Sg	107 Bh	108 Hs	109 Mt	110 Ds	111	112	113	114	115	116	117	118

Inner transition metals

Lanthanide series *	57 La	58 Ce	59 Pr	60 Nd	61 Pm	62 Sm	63 Eu	64 Gd	65 Tb	66 Dy	67 Ho	68 Er	69 Tm	70 Yb	71 Lu
Actinide series †	89 Ac	90 Th	91 Pa	92 U	93 Np	94 Pu	95 Am	96 Cm	97 Bk	98 Cf	99 Es	100 Fm	101 Md	102 No	103 Lr

Elements 111, 112, 114, and 116 have been discovered but, as of 2003, have not yet been named. Evidence for the discovery of elements 113 and 115 has been reported. See www.webelements.com for the latest information and newest elements.

Answers

to Checkpoints and Odd-Numbered Questions and Problems

chapter 38

CP **1.** b, a, d, c **2.** (a) lithium, sodium, potassium, cesium; (b) all tie **3.** (a) same; (b)–(d) x rays **4.** (a) proton; (b) same; (c) proton **5.** same **Q** **1.** potassium **3.** (a) microwave; (b) x ray; (c) x ray **5.** positive charge builds up on the plate, inhibiting further electron emission **7.** none **9.** (a) B; (b) A; (c) A; (d) A **11.** no essential change **13.** electron **15.** (a) decreasing; (b) increasing; (c) same; (d) same **17.** a **19.** all tie **P** **1.** (a) 2.1 μm; (b) infrared **3.** 1.0×10^{45} photons/s **5.** 2.047 eV **7.** 4.7×10^{26} photons **9.** (a) infrared; (b) 1.4×10^{21} photons/s **11.** (a) 2.96×10^{20} photons/s; (b) 4.86×10^7 m; (c) 5.89×10^{18} photons/m$^2 \cdot$ s **13.** 170 nm **15.** 676 km/s **17.** (a) 1.3 V; (b) 6.8×10^2 km/s **19.** (a) 2.00 eV; (b) 0; (c) 2.00 V; (d) 295 nm **21.** (a) 382 nm; (b) 1.82 eV **23.** (a) 3.1 keV; (b) 14 keV **25.** (a) 2.73 pm; (b) 6.05 pm **27.** (a) 8.57×10^{18} Hz; (b) 3.55×10^4 eV; (c) 35.4 keV/c **29.** (a) 2.43 pm; (b) 1.32 fm; (c) 0.511 MeV; (d) 939 MeV **31.** (a) -8.1×10^{-9}%; (b) -4.9×10^{-4}%; (c) -8.9%; (d) -66% **33.** 300% **35.** (a) 2.43 pm; (b) 4.11×10^{-6}; (c) -8.67×10^{-6} eV; (d) 2.43 pm; (e) 9.78×10^{-2}; (f) -4.45 keV **37.** (a) 41.8 keV; (b) 8.2 keV **39.** 7.75 pm **41.** 4.3 μeV **43.** (a) 1.24 μm; (b) 1.22 nm; (c) 1.24 fm; (d) 1.24 fm **45.** (a) 1.9×10^{-21} kg $\cdot$ m/s; (b) 346 fm **47.** (a) 0.025 fm; (b) 2.0×10^2 **49.** neutron **51.** 9.76 kV **59.** 2.1×10^{-24} kg $\cdot$ m/s **63.** (a) 9.02×10^{-6}; (b) 3.0 MeV; (c) 3.0 MeV; (d) 7.33×10^{-8}; (e) 3.0 MeV; (f) 3.0 MeV; **65.** (a) -20%; (b) -10%; (c) $+15$% **67.** 5.9 μeV **69.** (a) 73 pm; (b) 3.4 nm; (c) yes, their average de Broglie wavelength is smaller than their average separation **73.** 0.19 m **75.** 1.7×10^{-35} m **81.** (a) cesium; (b) both

chapter 39

CP **1.** b, a, c **2.** (a) all tie; (b) a, b, c **3.** a, b, c, d **4.** $E_{1,1}$ (neither n_x nor n_y can be zero) **5.** (a) 5; (b) 7 **Q** **1.** a, c, b **3.** c **5.** (a) 18; (b) 17 **7.** equal **9.** b, c, and d **11.** (a) decrease; (b) increase **13.** $n = 1, n = 2, n = 3$ **15.** same **17.** (a) $n = 3$; (b) $n = 1$; (c) $n = 5$ **P** **1.** 1.41 **3.** 1.9 GeV **5.** 0.85 nm **7.** 0.65 eV **9.** (a) 68.7 nm; (b) 25.8 nm; (c) 13.7 nm **11.** (a) 72.2 eV; (b) 13.7 nm; (c) 17.2 nm; (d) 68.7 nm; (e) 41.2 nm; (g) 68.7 nm; (h) 25.8 nm **13.** (a) 0.050; (b) 0.10; (c) 0.0095 **15.** 59 eV **19.** 3.08 eV **21.** (a) 8; (b) 0.75; (c) 1.00; (d) 1.25; (e) 3.75; (f) 3.00; (g) 2.25 **23.** (a) 7; (b) 1.00; (c) 2.00; (d) 3.00; (e) 9.00; (f) 8.00; (g) 6.00 **25.** 4.0 **27.** (a) 12.1 eV; (b) 6.45×10^{-27} kg $\cdot$ m/s; (c) 102 nm **31.** (a) 291 nm^{-3}; (b) 10.2 nm^{-1} **33.** (a) 13.6 eV; (b) 3.40 eV **35.** (a) 13.6 eV; (b) -27.2 eV **39.** 0.68 **43.** (a) $(r^4/8a^5)[\exp(-r/a)]\cos^2\theta$; (b) $(r^4/16a^5)[\exp(-r/a)]\sin^2\theta$ **45.** (a) 0.0037; (b) 0.0054 **51.** (a) n; (b) $2\ell + 1$; (c) n^2 **53.** (b) $(2\pi/h)[2m(U_0 - E)]^{0.5}$ **55.** (b) meter$^{-2.5}$ **57.** (a) 4; (b) 2; (c) Balmer

chapter 40

CP **1.** 7 **2.** (a) decrease; (b)–(c) remain the same **3.** less **4.** A, C, B **Q** **1.** same number (10) **3.** (a) 2; (b) 8; (c) 5; (d) 50 **5.** 2, -1, 0, and 1 **7.** (a) n; (b) n and ℓ **9.** all true **11.** (a) 2; (b) 3 **13.** in addition to the quantized energy, a helium atom has kinetic energy; its total energy can equal 20.66 eV **P** **1.** (a) 3; (b) 3 **3.** (a) 3.65×10^{-34} J $\cdot$ s; (b) 3.16×10^{-34} J $\cdot$ s **5.** 24.1° **7.** (a) 4; (b) 5; (c) 2 **9.** (a) 3.46; (b) 3.46; (c) 3; (d) 3; (e) -3; (f) 30.0°; (g) 54.7°; (h) 150° **13.** (a) 54.7°; (b) 125° **15.** 72 km/s^2 **17.** 5.35 cm **19.** 44 **21.** (a) 51; (b) 53; (c) 56 **23.** 42 **25.** (a) $4p$; (b) 4; (c) $4p$; (d) 5; (e) $4p$; (f) 6 **27.** (a) (2, 0, 0, +½), (2, 0, 0, −½); (b) (2, 1, 1, +½), (2, 1, 1, −½), (2, 1, 0, +½), (2, 1, 0, −½), (2, 1, −1, +½), (2, 1, −1, −½) **31.** (a) 49.6 pm; (b) 99.2 pm **33.** (a) 35.4 pm; (b) 56.5 pm; (c) 49.6 pm **37.** 0.563 **39.** (a) 69.5 kV; (b) 17.8 pm; (c) 21.3 pm; (d) 18.5 pm **41.** 80.3 pm **43.** (a) -25%; (b) -15%; (c) -11%; (d) -7.9%; (e) -6.4%; (f) -4.7%; (g) -3.5%; (h) -2.6%; (i) -2.0%; (j) -1.5% **45.** (a) 3.60 mm; (b) 5.24×10^{17} **47.** 9.0×10^{-7} **49.** 7.3×10^{15} s^{-1} **51.** 2×10^7 **53.** (a) 3.03×10^5; (b) 1.43 GHz; (d) 3.31×10^{-6} **55.** (a) 0; (b) 68 J **57.** (a) 2.13 meV; (b) 18 T **59.** (a) no; (b) 140 nm **63.** (a) 6.0; (b) 3.2×10^6 y **67.** argon **69.** $n > 3$; $\ell = 3$; $m_\ell = +3, +2, +1, 0, -1, -2, -3$; $m_s = \pm 1/2$

chapter 41

CP **1.** (a) larger; (b) same **2.** Cleveland, metal; Boca Raton, none; Seattle, semiconductor **3.** a, b, and c **4.** b **Q** **1.** 4 **3.** much less than **5.** b, c, d (the latter due to thermal expansion) **7.** b and d **9.** none **11.** $+4e$ **13.** (a) right to left; (b) back bias **15.** blue **P** **1.** 8.49×10^{28} m^{-3} **3.** (b) 6.81×10^{27} m^{-3} eV$^{-3/2}$; (c) 1.52×10^{28} m^{-3} eV^{-1} **5.** (a) 0; (b) 0.0955 **9.** (a) 2.50×10^3 K; (b) 5.30×10^3 K **11.** (a) 6.81 eV; (b) 1.77×10^{28} m^{-3} eV^{-1}; (c) 1.59×10^{28} m^{-3} eV^{-1} **13.** (a) 1.36×10^{28} m^{-3} eV^{-1}; (b) 1.68×10^{28} m^{-3} eV^{-1}; (c) 9.01×10^{27} m^{-3} eV^{-1}; (d) 9.56×10^{26} m^{-3} eV^{-1}; (e) 1.71×10^{18} m^{-3} eV^{-1} **15.** (a) 1.0; (b) 0.99; (c) 0.50; (d) 0.014; (e) 2.4×10^{-17}; (f) 7.0×10^2 K **17.** 3 **19.** (a) 5.86×10^{28} m^{-3}; (b) 5.49 eV; (c) 1.39×10^3 km/s; (d) 0.522 nm **23.** 57.1 kJ **25.** 472 K **27.** (a) 226 nm; (b) ultraviolet **29.** (a) 1.5×10^{-6}; (b) 1.5×10^{-6} **31.** 0.22 μg **33.** (a) 4.79×10^{-10}; (b) 0.0140; (c) 0.824 **35.** 6.0×10^5 **37.** 4.20 eV **39.** 13 μm **41.** (b) 1.8×10^{28} m^{-3} eV^{-1} **43.** (a) 109.5°; (b) 238 pm **47.** 3.49×10^3 atm

chapter 42

CP **1.** ^{90}As and ^{158}Nd **2.** a little more than 75 Bq (elapsed time is a little less than three half-lives) **3.** ^{206}Pb **Q** **1.** less **3.** above **5.** (a) ^{196}Pt; (b) no **7.** (a) on $N = Z$ line; (b) positrons; (c) about 120 **9.** yes **11.** no **13.** no effect **15.** ^{209}Po **17.** d **P** **1.** 15.8 fm **3.** (a) 0.390 MeV;

(b) 4.61 MeV **5.** (a) $+7.825 \times 10^{-3}$ u; (b) $+7.290$ MeV/c^2; (c) $+8.664 \times 10^{-3}$ u; (d) $+8.071$ MeV/c^2; (e) -9.780×10^{-2} u; (f) -91.10 MeV/c^2 **7.** 13 km **9.** (a) 2.3×10^{17} kg/m^3; (b) 2.3×10^{17} kg/m^3; (d) 1.0×10^{25} C/m^3; (e) 8.8×10^{24} C/m^3 **11.** (a) 6.2 fm; (b) yes **13.** (a) 9.303%; (b) 11.71% **19.** (b) 7.92 MeV **21.** (a) 0.250; (b) 0.125 **23.** (a) 7.5×10^{16} s^{-1}; (b) 4.9×10^{16} s^{-1} **25.** (a) 64.2 h; (b) 0.125; (c) 0.0749 **27.** 5.3×10^{22} **29.** 9.0×10^8 Bq **31.** (a) 2.0×10^{20}; (b) 2.8×10^9 s^{-1} **33.** 265 mg **35.** 1.12×10^{11} y **39.** (a) 8.88×10^{10} s^{-1}; (b) 1.19×10^{15}; (c) 0.111 μg **41.** (a) 1.2×10^{-17}; (b) 0 **43.** 4.269 MeV **45.** 1.21 MeV **47.** 0.783 MeV **49.** (b) 0.961 MeV **51.** 78.3 eV **53.** (a) 1.06×10^{19}; (b) 0.624×10^{19}; (c) 1.68×10^{19}; (d) 2.97×10^9 y **55.** 1.7 mg **57.** 1.02 mg **59.** 13 mJ **61.** (a) 6.3×10^{18}; (b) 2.5×10^{11}; (c) 0.20 J; (d) 2.3 mGy; (e) 30 mSv **63.** (a) 6.6 MeV; (b) no **65.** (a) 25.4 MeV; (b) 12.8 MeV; (c) 25.0 MeV **67.** (a) 59.5 d; (b) 1.18 **69.** 730 cm^2 **71.** 600 keV **73.** 30 MeV **75.** 3.2×10^4 y **77.** ^{7}Li **79.** ^{225}Ac **85.** (a) 11.906 83 u; (b) 236.2025 u **87.** 27

chapter 43

CP 1. c and d **2.** (a) no; (b) yes; (c) no **3.** e **Q 1.** 10 000 **3.** ^{140}I, ^{105}Mo, ^{152}Nd, ^{123}In, ^{115}Pd **5.** ^{239}Np **7.** c, a, d, b **9.** 20 **11.** less than **P 1.** 3.1×10^{10} s^{-1} **3.** (a) 2.6×10^{24}; (b) 8.2×10^{13} J; (c) 2.6×10^4 y **5.** -23.0 MeV **7.** (a) 16 day^{-1}; (b) 4.3×10^8 **9.** (a) ^{153}Nd; (b) 110 MeV; (c) 60 MeV; (d) 1.6×10^7 m/s; (e) 8.7×10^6 m/s **11.** (a) 251 MeV; (b) typical fission energy is 200 MeV **13.** 462 kg **15.** 557 W **17.** (a) 1.2 MeV; (b) 3.2 kg **19.** (a) 84 kg; (b) 1.7×10^{25}; (c) 1.3×10^{25} **21.** (b) 1.0; (c) 0.89; (d) 0.28; (e) 0.019; (f) 8 **23.** 0.99938 **25.** 3.6×10^9 y **27.** 1.7×10^9 y **29.** 170 keV **31.** 1.41 MeV **35.** (a) 4.3×10^9 kg/s; (b) 3.1×10^{-4} **37.** (a) 1.8×10^{38} s^{-1}; (b) 8.2×10^{28} s^{-1} **39.** (a) 4.1 eV/atom; (b) 9.0 MJ/kg; (c) 1.5×10^3 y **41.** 1.6×10^8 y **43.** (a) 24.9 MeV; (b) 8.65 megatons TNT **45.** 14.4 kW **47.** (a) 3.1×10^{31} protons/m^3; (b) 1.2×10^6 **49.** $^{238}\text{U} + \text{n} \rightarrow {}^{239}\text{U} \rightarrow {}^{239}\text{Np} + \text{e} + \nu$, $^{239}\text{Np} \rightarrow {}^{239}\text{Pu} + \text{e} + \nu$ **51.** 6×10^2 kg

chapter 44

CP 1. (a) the muon family; (b) a particle; (c) $L_\mu = +1$ **2.** b and e **3.** c **Q 1.** into **3.** the π^+ pion whose track terminates at point 2 **5.** c, f **7.** baryon number **9.** c **11.** (a) 0; (b) $+1$; (c) -1; (d) $+1$; (e) -1 **13.** 1b, 2c, 3d, 4e, 5a **P 1.** 18.4 fm **3.** 1 **5.** 2.7 cm/s **7.** 769 MeV **11.** (a) angular momentum, L_e; (b) charge, L_μ; (c) energy, L_μ **13.** (a) 0; (b) -1; (c) 0 **15.** (a) energy; (b) strangeness; (c) charge **17.** (a) K$^+$; (b) $\bar{\text{n}}$; (c) anti-K^0 **19.** (a) 37.7 MeV; (b) 5.35 MeV; (c) 32.4 MeV **21.** (a) $\bar{u}\bar{u}\bar{d}$; (b) $\bar{u}\bar{d}\bar{d}$; **23.** (a) not possible; (b) uuu **25.** (a) Σ^0; (b) 7.51×10^6 m/s **27.** 668 nm **29.** (b) 0.934; (c) 1.28×10^{10} ly **31.** (a) 0.26 meV; (b) 4.8 mm **33.** (a) 121 m/s; (b) 0.00406; (c) 248 y **35.** (b) 2.39×10^9 K **37.** 1.08×10^{42} J **41.** (a) $0.785c$; (b) $0.993c$; (c) C2; (d) C1; (e) 51 ns; (f) 40 ns **43.** (c) $r\alpha/c + (r\alpha/c)^2 + (r\alpha/c)^3 + \cdots$; (d) $r\alpha/c$; (e) $\alpha = H$; (f) 6.5×10^8 ly; (g) 6.9×10^8 y; (h) 6.5×10^8 y; (i) 6.9×10^8 ly; (j) 1.0×10^9 ly; (k) 1.1×10^9 y; (l) 3.9×10^8 ly

Photo Credits

chapter 38

Opener: Courtesy International Business Machines Corporation. Page 1069: Courtesy A. Tonomura, J. Endo, T. Matsuda, and T. Kawasaki, Advanced Research Laboratory, Hitachi, Ltd., Kokubinju, Tokyo; H. Ezawa, Department of Physics, Gakushuin University, Mejiro, Tokyo. Page 1070 (top left): Courtesy Riber Division of Instruments, Inc. Page 1070 (top right): From PSSC film, *Matter Waves.* Reproduced courtesy Education Development Center, Newton, Massachusetts. Page 1070 (bottom): Lawrence Berkeley Laboratory/Science Photo Library/Photo Researchers.

chapter 39

Opener: Courtesy International Business Machines Corporation, Almaden Research Center, CA. Page 1094 (top): From *Scientific American,* January 1993, page 122. Reproduced with permission of Michael Steigerwald, Bell Labs-Lucent Technologies. Page 1094 (bottom): From *Scientific American,* September 1995, page 67. Image reproduced with permission of H. Temkin, Texas Tech University. Page 1095: Courtesy International Business Machines Corporation, Almaden Research Center. Page 1101: From W. Finkelburg, *Structure of Matter,* Springer-Verlag, 1964. Reproduced with permission.

chapter 40

Opener: Robert J. Herko/The Image Bank/Getty Images. Page 1113: Courtesy Warren Nagourney. Page 1122: Howard Sochurek/ Corbis Images. Page 1132: Charles O'Rear/Corbis Images.

chapter 41

Opener: Reuters New Media, Inc/Corbis Images. Page 1158: Courtesy AT&T. Page 1159: Courtesy Intel.

chapter 42

Opener: Bill Bachman/Photo Researchers. Page 1182: George Rockwin/Bruce Coleman, Inc.

chapter 43

Opener: Courtesy U.S. Department of Energy. Page 1201: V. Ivelva /Magnum Photos, Inc. Page 1202: ©AP/Wide World Photos. Page 1205: Gary Sheehan, *Birth of the Atomic Age,* 1957. Reproduced courtesy of The Chicago Historical Society. Page 1210: Courtesy Anglo Australian Telescope Board. Page 1212 (top): Courtesy Princeton University Physics Laboratory. Page 1212 (bottom): Courtesy Los Alamos National Laboratory, New Mexico. Page 1215: Courtesy Martin Marietta Energy Systems/U.S. Department of Energy.

chapter 44

Opener: Courtesy WMAP Science Team/NASA. Pages 1220 and 1232: Courtesy Brookhaven National Laboratory. Page 1221: Courtesy Michael Mathews. Page 1223: Courtesy Lawrence Berkeley Laboratory.

Index

Figures are noted by page numbers in *italics*, tables are indicated by t following the page number.

E